MOLECULAR SIMULATION METHODS FOR PREDICTING POLYMER PROPERTIES

MOLECULAR SIMULATION METHODS FOR PREDICTING POLYMER PROPERTIES

Edited by
VASSILIOS GALIATSATOS, Ph.D

A JOHN WILEY & SONS, INC. PUBLICATION

Copyright © 2005 by John Wiley & Sons, Inc. All rights reserved.

Published by John Wiley & Sons, Inc., Hoboken, New Jersey.
Published simultaneously in Canada.

No part of this publication may be reproduced, stored in a retrieval system, or transmitted in any form or by any means, electronic, mechanical, photocopying, recording, scanning, or otherwise, except as permitted under Section 107 or 108 of the 1976 United States Copyright Act, without either the prior written permission of the Publisher, or authorization through payment of the appropriate per-copy fee to the Copyright Clearance Center, Inc., 222 Rosewood Drive, Danvers, MA 01923, 978-750-8400, fax 978-646-8600, or on the web at www.copyright.com. Requests to the Publisher for permission should be addressed to the Permissions Department, John Wiley & Sons, Inc., 111 River Street, Hoboken, NJ 07030, (201) 748-6011, fax (201) 748-6008.

Limit of Liability/Disclaimer of Warranty: While the publisher and author have used their best efforts in preparing this book, they make no representations or warranties with respect to the accuracy or completeness of the contents of this book and specifically disclaim any implied warranties of merchantability or fitness for a particular purpose. No warranty may be created or extended by sales representatives or written sales materials. The advice and strategies contained herein may not be suitable for your situation. You should consult with a professional where appropriate. Neither the publisher nor author shall be liable for any loss of profit or any other commercial damages, including but not limited to special, incidental, consequential, or other damages.

For general information on our other products and services please contact our Customer Care Department within the U.S. at 877-762-2974, outside the U.S. at 317-572-3993 or fax 317-572-4002.

Wiley also publishes its books in a variety of electronic formats. Some content that appears in print, however, may not be available in electronic format.

Library of Congress Cataloging-in-Publication Data is available.

Galiatsatos, Vassilios
Molecular simulation methods for predicting polymer properties.

ISBN 0-471-46481-3

Printed in the United States of America

10 9 8 7 6 5 4 3 2 1

CONTENTS

Preface vii
Vassilios Galiatsatos

1. *Ab Initio* Polymer Quantum Theory 1
Benoît Champagne
Laboratoire de Chimie Théorique Appliquée,
Facultés Universitaires Notre-Dame de la Paix,
rue de Bruxelles 61, B-5000 Namur, Belgium

2. Quantum-Chemistry-Based Force Fields For Polymers 47
Grant D. Smith and Oleg Borodin
University of Utah, Salt Lake City, Utah, USA

3. Monte Carlo Simulations of Binary Polymer Liquids 95
Marcus Müller
Institut für Physik, WA331, Johannes Gutenberg-Universität,
D55099 Mainz, Germany

4. Mesoscopic Simulations of Polymer Mixtures 153
Olaf Evers
BASF Aktiengesellschaft, Department of Polymer Physics,
D-67056 Ludwigshafen am Rhein, Germany

5. Prediction of Mechanical Properties of Semicrystalline Polymers   **219**

A. Raphael and I. Alig
Deutsches Kunststoff-Institut, Schlossgartenstr. 6,
64289 Darmstadt, Germany

M. Kröhn
Siemens VDO Automotive AG, VDO-Str. 1,
64832 Babenhausen, Germany

6. Crosslinking Simulations in Polymer Design   **243**

Robert T. Johnston
Ethylene Elastomers R&D, DuPont Dow Elastomers LLC,
Freeport, Texas, USA

Index   **277**

PREFACE

Modeling and simulation methods have been around for long time. Considerable advances in algorithms and computer hardware have allowed the development of reliable models to study a variety of physical phenomena. The pharmaceutical industry has long relied on modeling to screen potential compounds with interesting pharmacophoric activity. The aerospace and automotive industries have likewise relied on simulation to study the behavior and the properties of parts and prototypes before production starts.

The promise of applying similar methods to design new polymers and polymer compounds is immense. At the same time the technical challenges that need to be met are considerable. Yet great progress has been achieved and scientists and engineers who work in the polymer field have a number of tools at their disposal.

The aim of this book is to provide a reference for those scientists and engineers who are interested in applying simulation methods in the field of synthetic polymers. The area has seen significant advances in the last 20 years. We are now up to the point where much valuable information may be obtained by the proper use of simulations. This book hopes to teach techniques and methods that allow the connection of polymer structure to physical properties. Chemical reactivity issues are outside the scope of the book.

Familiarity with basic modeling and mathematical techniques is assumed, as is familiarity with fundamental concepts in polymeric materials. The reader will notice some overlap among chapters, which is unavoidable. At the same time this overlap serves as a connection between the various subjects and aims to help the readers to better understand the underlying concepts. We believe that the techniques presented here are appropriate for synthetic polymers whether they are of the commodity type, engineering polymers or high-value engineering polymers.

The chapters start by addressing small atomistic systems first. Quantum mechanical methods (*ab initio* and semi-empirical) are appropriate at this level. Using theory and several assumptions about the periodic nature of polymers, one may predict physical properties. Increases in the size of the systems are studied, for example, long polymer chains and ensembles of polymers chains require the introduction of coarse graining, and a large part of the book is devoted to those methods. Polymer blends may be studied by relying on coarse-grained methods. Macroscopic three-dimensional crosslinked systems are also addressed and the latest available techniques are outlined. The readers should first determine what size system they are interested in studying and then go to the appropriate chapter.

Chapter 1, by Dr Champagne, deals with application of *ab initio* calculations to predict properties of polymers. This is an extremely challenging subject, but one with great potential. Currently only oligomers may be studied by quantum mechanical calculations because of the immense computational resources needed as the number of atoms in the molecule increases. However, *ab initio* calculations include a high level of detail on structural information and that level of sophistication is necessary for several properties, as discussed in this chapter.

Dr Champagne mentions two approaches that one may use to go from oligomers to polymers: the 'oligomer' approach and the 'band structure' approach. In the band structure approach, the wave-functions and associated energies reflect the assumption that the system under study is of infinite length and periodic in one dimension, but three-dimensional (3-D). The advantage of this approach is that it provides a direct calculation of the properties of the infinite system. In the oligomer approach, systems of increasing size are considered and their properties are extrapolated to the infinite chain limit. Such an approach can take advantage of the methods which have been developed to treat small molecules and can incorporate in a straightforward manner defects or aperiodic conditions. On the other hand, it requires a large number of calculations to be performed. As a result it can suffer from increasingly large computational requirements or difficulties associated with the extrapolation procedures. The development of direct polymer approaches has often been induced by results obtained by the oligomer approach. Similarly, oligomer properties are often compared with band structure data, for instance to assess the extrapolation procedure.

Moving up the length scale, and in order to study isolated chains, and ensembles of chains, one has to investigate molecular mechanics and dynamics methods. Force fields are central in the ability of these methods to predict properties reliably. This is especially true for industrial applications where reliability of results is critical for successful product development. Therefore the ability to accurately represent the potential energy of an ensemble of atoms is central to simulations of real materials. Various forms of classical potentials (i.e. force fields) for polymers can be found in the literature. The form of potential most appropriate for a given polymer depends largely upon the properties of interest to the simulator.

Drs Smith and Borodin address this critical topic in Chapter 2. Their interest lies in reproducing the static, thermodynamic and dynamic (transport and relaxation) properties of non-reactive polymer systems. They present a critical discussion and offer guidance for the parameterization of potentials that accurately reproduce the molecular geometry, non-bonded interactions, and conformational energies of polymers. As demonstrated in two detailed examples provided in this chapter, a relatively simple representation of the classical potential energy works remarkably well for these polymer properties. However, parameterization of even simple potentials is a challenging task. The chapter also deals with the complications arising in polar materials when many-body dipole polarization is taken into account explicitly.

By far the most convenient way to obtain a force field for any polymer is to utilize an extant one. As the authors show, force fields can be divided into three categories: (a) force fields parameterized based upon a broad training set of molecules such as small organic molecules, peptides, or amino acids, including AMBER, COMPASS, OPLS-AA and CHARMM; (b) generic potentials such as DREIDING and UNIVERSAL that are not parameterized to reproduce properties of any particular set of molecules; and (c) specialized force fields carefully parameterized to reproduce properties of a specific compound such as water, polymers, polymer aqueous solutions and polymer electrolytes. In choosing a potential, two major issues must be faced—the quality of the potential and the transferability of the potential. This chapter addresses those critical issues and gives guidance for proper implementation.

Chapter 3, written by Dr Müller, deals with the simulation of important materials such as those resulting from melt blending of individual polymer components. Anyone who works in the polymer industry knows that this is the preferred way to obtain new compounds. Polymeric materials in daily life are generally multicomponent systems. Chemically different polymers are blended to design a material which combines the favorable characteristics of the individual components. Clearly the miscibility behavior of the blend is crucial to understand and tailor properties relevant for practical applications.

Miscibility at a microscopic length scale is desirable, for example, for high tensile strength of the material. Unlike metallic alloys, however, chemically different polymers often do not mix at microscopic length scales. Rather a complicated morphology of droplets of one component dispersed into the other component forms at a mesoscopic length scale: the blend may be considered as an assembly of interfaces. While the detailed structure at this mesoscopic length scale depends strongly on the way the material is processed, the local properties of interfaces are certainly crucial for understanding the material properties. For instance, as Dr Müller explains, the interfacial width sets the length scale at which entanglements between polymers of the different components form. Experiments suggest that the mechanical strength increases if the interfacial width exceeds the entanglement length. Alternatively, the interfacial tension is important for the break-up of droplets under shear. The lower the interfacial tension is, the finer the dispersion of the two components in each other.

Chapter 3 considers coarse-grained models that do not capture the structure at the atomistic scale, but lump a small number of chemical repeat units into a monomer of the coarse-grained model. Dr Müller shows how these monomers interact via coarse-grained, simplified interactions. Electrostatic and torsional potentials are typically neglected in these models. The reduced number of degrees of freedom and the softer interactions lead to a significant computational speed-up. Hence, large system sizes and long time-scales can be studied.

Chapter 4, by Dr Evers, deals with coarse-grained models at the mesoscopic scale. The mesoscopic length scale is typically of the order of 100 nm up to micrometers. It typically links the macroscopic features to microscopic descriptors like the chemical structure of the molecules. Macroscopic parameters have no real meaning at this length scale. As in Chapter 3, a description of the system in terms of atoms and electrons makes no sense either because the mesoscopic length scale is simply too large for it. Dr Evers shows what type of information may be gained from simulations at this scale.

Polymer blends or melts of block copolymers show complex morphologies that range from including spheres, cylinders, gyroids, lamellae or even combinations of these. A simulation model for the mesoscale world should therefore be able to predict these morphologies at a resolution of about a few nanometers. The system of simulations described in this chapter has great potential to resolve a number of issues in the polymer industry.

Chapter 5, written by Drs Raphael, Kröhn and Alig, deals with how one would apply the techniques described in other chapters of the book to the prediction of the mechanical properties of semicrystalline polymers. The knowledge of the mechanical and thermal properties of semicrystalline polymers is of major importance for the simulation of technical processes, like injection or compression molding, including shrinkage and warpage effects.

Even for commodity semicrystalline polymers, like polyethylene (PE) or polypropylene (PP), structure–property relationships are not yet fully understood, because the morphology of semicrystalline polymeric materials is extremely dependent on the preparation of the sample (temperature, pressure, shear rate, cooling rate etc.). Furthermore, microscopic anisotropies resulting from processing (e.g. deformed spherulites, shish kebab structures) have not yet been included in macroscopic finite element calculations because of the lack of experimental data for mechanical or thermal properties, depending on thermal history.

The authors of Chapter 5 show how molecular simulations in combination with micromechanical models can assist in providing detailed data for a large number of different structures. Since the direct experimental determination of the mechanical properties of the crystalline phase is difficult, this opens up an interesting field for atomistic simulation, the subject of Chapter 2.

To model the complex structure of crystalline domains in an amorphous matrix, one would have to cover a volume of several cubic micrometers. Quantitative simulations of macroscopic properties of the semicrystalline superstructure at an atomistic level are still beyond the scope of current computer capacities. An alternative way to predict the mechanical behavior of semicrystalline polymers is to simulate the amorphous and crystalline phases separately at the length scale of several chains or chain fragments, and to employ micromechanical mixing rules to calculate the properties of the resulting complex system.

Chapter 5 shows how combination of molecular simulations and micromechanical models can extend the databases for process simulations, including heat history effects. This may be less expensive than experiments on a large number of samples prepared under different processing conditions. Although until now boundary effects and interphases have not been included, the authors show that the combination of molecular simulation, micromechanical models and process simulations can help to close the gap between modeling at a molecular level and the need for material properties data for applications development.

Chapter 6, by Dr Johnston, deals with a most challenging field in polymer simulations, that of predicting the structure and the properties of elastomers. Dr Johnston does a thorough job in walking the reader through the various techniques available today and provides guidance on implementing them. Dr Johnston takes the reader through the hierarchy of complexity and sophistication in the crosslinking simulation literature. Choosing the appropriate level of complexity for a simulation project involves weighing the trade-offs between accuracy and computing resources, while also factoring in experimental parameter requirements and knowledge, ease-of-use and other indirect factors.

Dr Johnston summarizes a general theoretical and experimental approach that has been found useful in conjunction with simulations to build an understanding of key parameters affecting polymer structure–network and structure–property relationships. The author shows that crosslinking simulations are widely applicable, with published work describing crosslinking simulations for polymers ranging from thermoset resins to rubber, chemistries ranging from free radical random crosslinking to endlinking of telechelic polymers, and end-use applications from dental resins to adhesives to tires.

There are a number of very good books and reviews that deal with the subject of molecular modeling and simulation of polymers. Some of them, covering the last dozen years or so, in chronological order, are:

1. *Computer Simulation of Polymers*, R.J. Roe, Editor, Prentice Hall, 1991.
2. *Computational Modeling of Polymers*, J. Bicerano, Editor, Marcel Dekker, 1992.
3. *Computer Simulation of Polymers*, E.A. Colbourn, Editor, Longman Scientific & Technical, 1994.
4. *Atomistic Modeling of Physical Properties of Polymers*, L. Monnerie and U.W. Suter, Editors, Advances in Polymer Science, 1994.
5. *Molecular Modeling of Polymer Structures and Properties*, Bruce R. Gelin, Hanser/Gardner Publishers, 1994.
6. *Computational Methods for Modeling Polymers: an Introduction*, Vassilios Galiatsatos, in *Reviews in Computational Chemistry*, Volume 6, K.B. Lipkowitz and D.B. Boyd, Editors, VCH Publishers, 1995, p. 149.
7. *Monte Carlo and Molecular Dynamics Simulations in Polymer Science*, Kurt Binder, Editor, Oxford University Press, 1995.

8. *Computational Chemistry: a Practical Guide for Applying Techniques to Real World Problems*, David Young, Wiley-Interscience 2001.
9. *Simulation Methods for Polymers*, M.J. Kotelyanskii and D.N. Theodorou, Marcel Dekker, 2004

It is our sincere hope that the book will motivate readers to include the tools of simulation and modeling in their R&D projects. We also hope that the content will stimulate the young (and old) bright minds that are out there to develop solutions to the many unsolved problems of polymer science and engineering by employing modeling.

I am personally grateful to the authors whose work is included in this book. They immediately understood the unique nature of the book and they tailored the material to serve the intended readership.

I would be very interested in hearing from readers with comments about improving the book and techniques that they would like to see addressed. Please keep in mind that we chose not to include QSPR/QSAR techniques because we think they belong to a separate book.

I would like to express my deepest thanks to Dr Ed Immergut for his encouragement to make this book a reality. He is both a mentor and a source of motivation and inspiration. I also want to thank Amy Romano, our Editor, for running a tight and effective business.

VASSILIOS GALIATSATOS

MOLECULAR SIMULATION METHODS FOR PREDICTING POLYMER PROPERTIES

1

AB INITIO POLYMER QUANTUM THEORY

BENOÎT CHAMPAGNE
Laboratoire de Chimie Théorique Appliquée, Facultés Universitaires Notre-Dame de la Paix, rue de Bruxelles 61, B-5000 Namur, Belgium

CONTENTS

1.1. Introduction	2
1.2. Polymer Band Structure and Wave-Functions	4
1.2.1. Bloch's Theorem	4
1.2.2. LCAO Hartree–Fock Methodology	5
1.2.3. Specific Aspects of Hartree–Fock Band Structure Calculations	8
1.2.4. Helical Symmetry	12
1.2.5. Unrestricted Hartree–Fock Approach	13
1.2.6. Møller–Plesset and Other Wave-Function-Based Electron Correlation Schemes	14
1.2.7. Density Functional Theory Approaches	19
1.3. Polymer Properties and Applications	20
1.3.1. Band Structure, Density of States, and Photoelectron Spectroscopy	20
1.3.2. Excitation Energies: Band Gaps and Excitons	23
1.3.3. Electronic Polarizabilities and Hyperpolarizabilities	28
1.3.4. Structural (Geometries), Vibrational, and Mechanical Properties	32
1.4. Further Aspects and Outlook	36

Molecular Simulation Methods for Predicting Polymer Properties, Edited by Vassilios Galiatsatos
ISBN 0-471-46481-3 Copyright © 2005 John Wiley & Sons, Inc.

1.1. INTRODUCTION

Since Hermann Staudinger and his work, which proved the existence of molecules with very large molecular mass, polymeric materials have replaced conventional materials in many areas, from informatics to automobiles, from packaging to medicine and from clothing to cooking to such an extent that they are now considered as commodity materials. No doubt this is related to the infinite variations of their synthesis that can produce materials with specific mechanical, thermal, optical, and electronic properties. In addition, biopolymers like DNA, RNA, proteins, or polysaccharides play vital roles for human beings.

In the search for new polymeric materials with targeted properties; that is, defining those structures or substitutions among the members of a polymer family which will display the optimal properties, computer simulation, including quantum theory, constitutes an essential step. Although the foundations of quantum mechanics were drawn in the 1920s and electronic structure calculations for polyatomic molecules were already being carried out in the 1950s using the linear combination of atomic orbital (LCAO) scheme, the first applications of quantum theory to polymers—of which the size is assumed, for computational purposes, to be infinite—appeared in the second part of the 1960s [1, 2]. In these calculations, another approximation is the assumption of stereo-regularity (Figure 1.1), although, later on, different schemes have been proposed to treat random co-polymers and defects. During the last 40 years, methods and algorithms have been elaborated in order to address the numerous aspects of determining the structures and properties of polymeric materials. The importance of polymers in everyday life combined with the many different research areas is reflected in several review articles [3–7] and monographs [8, 9]. The aim of this chapter is therefore to review the developments in the *ab initio* quantum theory of polymers with an emphasis on the most recent achievements. To be sure, the presentation of this topic is biased by the education the author has received in one of the pioneering groups in this field which, over the years, has successively focused on electron spectroscopy for chemical analysis (ESCA), conjugated polymers, linear, and nonlinear optics (NLO), and,

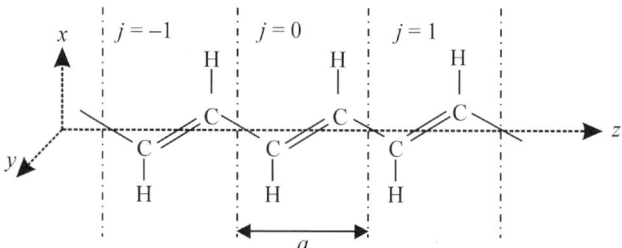

Figure 1.1. Representation of a stereoregular polymer (here, all-*trans* polyacetylene) as the repetition in one direction of the same chemical motif, the Wigner–Seitz unit cell. *a* is the unit cell length.

more recently, structural and vibrational properties, whereas the aspects related to lattice summations, integration in the first Brillouin zone and electron correlation issues have always been a subject of research.

As for small molecules, the standard (starting) theory is based on the Hartree–Fock method where any electron moves in the field of the fixed nuclei and in the mean Coulombic and exchange fields of all the other electrons. For molecular systems, any electron is associated to a one-electron wave-function called molecular orbital (MO), which can be either occupied or unoccupied. In the quantum theory of polymers, the wave-functions and associated energies reflect the fact that the system is periodic in one dimension but three-dimensional (3-D). Mimicking the solid-state physics treatment of 3-D periodic crystals—but for the unique dimension of periodicity—Bloch's theorem states the phase relation of the polymer (crystalline) analogs of the MOs at periodically related points $\phi(\mathbf{r} + ja\mathbf{e}_z) = e^{ikja}\phi(\mathbf{r})$, where a is the unit cell length and $\mathbf{e}_z$ the unit vector in the periodicity direction. j is associated with the unit cell number and is a pure number. These $\phi(\mathbf{r})$ functions, sometimes called Bloch's functions, are eigenfunctions of a translation operator. It follows that the one-electron wave-functions and energies are functions of k, the quasi-momentum of the particle. The k-dispersion curves of the one-electron energies form the band structure.

Originally, polymer band structure calculations were carried out at simplified levels of approximation, including: (i) the Hückel approach that is able to show the interplay between the opening of the gap between the occupied and unoccupied levels and the bond length or electron density alternation [9, 10]; (ii) the semi-empirical π-electron Pariser–Parr–Pople (PPP) scheme, which incorporates electron–electron interactions [2, 11]; (iii) the extended Hückel method that was adopted in the early days to treat polyethylene and polyacetylene [12]; (iv) the semiempirical techniques [13, 14]; and (v) the valence effective Hamiltonian (VEH) approach [15], which has been largely employed for studying conjugated polymers. Then, with the advent of more and more powerful computers and the elaboration of efficient algorithms, *ab initio* schemes and applications appeared. This review concerns these *ab initio* studies carried out at the reference Hartree–Fock level or by including electron correlation effects via conventional wave-function-based approaches or density functional theory schemes.

Band structure approaches—and therefore the assumption of infinite length and periodicity—are not the only techniques for characterizing the properties of polymers. The major alternative is the *oligomer approach* [16], where systems of increasing size are considered and their properties are extrapolated to the infinite chain limit. Such an approach can take advantage of the methods which have been developed for treating small molecules and can incorporate in a straightforward manner defects or aperiodicity. On the other hand, it requires a large set of calculations to be performed and therefore it can suffer from increasingly large computational needs or from difficulties associated with the extrapolation procedures, although improved extrapolation techniques have been developed [17]. This contrasts with the polymer methods that provide in a one-shot calculation the properties of the infinite system. The oligomer approach is not discussed in this review because it pertains to the class of small molecule methods. However, it is important to stress that the developments of polymer approaches have often been induced by oligomer results and, similarly, oligomer properties are often compared with band structure data, for instance to assess the extrapolation procedure.

1.2. POLYMER BAND STRUCTURE AND WAVE-FUNCTIONS

1.2.1. Bloch's Theorem

Bloch's theorem [18],

$$\phi(\mathbf{r} + ja\mathbf{e}_z) = e^{ikja}\phi(\mathbf{r}) \tag{1.1}$$

which states the phase relation between one-electron polymer or crystalline orbitals at periodically related points ensues from the 1-D periodicity of the electronic density:

$$\rho(\mathbf{r}) = \rho(\mathbf{r} + ja\mathbf{e}_z) \tag{1.2}$$

and implies that the one-electron wave-functions can be written in the form of a product between a plane-wave and a function presenting the periodicity of the direct space:

$$\phi(k,\mathbf{r}) = e^{ikz}v(k,\mathbf{r}) \tag{1.3}$$

When there is only one function per unit cell, the (nonnormalized) Bloch's function reads:

$$\phi(k,\mathbf{r}) = \theta_\nu(k,\mathbf{r}) = \sum_{j=-\infty}^{\infty} e^{ikja}\chi_\nu(\mathbf{r} - \mathbf{R}_\nu - ja\mathbf{e}_z) = \sum_{j=-\infty}^{\infty} e^{ikja}\chi_\nu^j(\mathbf{r}) \tag{1.4}$$

where $\chi_\nu(\mathbf{r} - \mathbf{R}_\nu - ja\mathbf{e}_z)$ is the atomic orbital χ_ν located in the jth unit cell and centered in $\mathbf{R}_\nu$. These functions and their associated energies are thus functions of k, a continuous index having dimensions of an inverse length. The k-dispersion curves of the one-particle energy are called 'energy bands' and form the band structure. They are generally plotted in the interval going from $k = 0$ to $k = \pi/a$, denoted half of the first Brillouin zone. Indeed, inspecting equations (1.1) and (1.4) reveals that the wave-functions and energies present a periodicity of $2\pi/a$, the periodicity of the reciprocal space. This $[-\pi/a, \pi/a]$ interval is denoted the first Brillouin zone and is the equivalent of the Wigner–Seitz unit cell in the direct space. Further symmetry enables to restrict the representation to the positive k-values.

The Born–von Kàrmàn [19] cyclic boundary conditions are generally introduced. They impose a periodicity of $(2N+1)a$ on Bloch's functions or, more simply, impose to the latter to be identical in the 0th and $2N+1$th unit cells, with N tending to infinity:

$$\begin{aligned}\phi[k,\mathbf{r}+(2N+1)a\mathbf{e}_z] &= e^{ik[z+(2N+1)a]}v[k,\mathbf{r}+(2N+1)a\mathbf{e}_z] \\ &= e^{ikz}e^{ik(2N+1)a}v(k,\mathbf{r}) = e^{ikz}v(k,\mathbf{r}) = \phi(k,\mathbf{r})\end{aligned} \quad (1.5)$$

so that the k-values are now discrete:

$$k = \frac{2\pi}{(2N+1)a}\kappa \quad (1.6)$$

with κ belonging to the $[-N, N]$ interval. The j lattice summation in equation (1.4) is therefore limited to the $[-N, N]$ interval.

1.2.2. LCAO Hartree–Fock Methodology

In general, there are ω atomic orbitals in each unit cell and the corresponding crystalline orbitals are built as linear combinations of these atomic orbitals or normalized θ functions:

$$\begin{aligned}\phi_n(k,\mathbf{r}) &= \sum_{\nu=1}^{\omega} C_{\nu,n}(k) \frac{1}{\sqrt{2N+1}} \sum_{j=-N}^{N} e^{ikja} \chi_\nu^j(\mathbf{r}) \\ &= \sum_{\nu=1}^{\omega} C_{\nu,n}(k) \theta_\nu(k,\mathbf{r})\end{aligned} \quad (1.7)$$

where n is the band index. In principle, as written above, N tends to infinity but in practice it is finite and defines the short-range region consisting of $2N+1$ unit cells. The orbital energies and LCAO coefficients, $C_{\nu,n}(k)$, depend on the quasi-momentum k. At the restricted Hartree–Fock level, that is, following the application of the Ritz variational principle to the energy of a wave-function having the form of a Slater determinant built from doubly occupied crystalline orbitals of the form displayed in equation (7), for each k-point, the $C_{\nu,n}(k)$ are obtained from an iterative solution of [8, 9]:

$$\mathbf{F}(k)\,\mathbf{C}(k) = \mathbf{S}(k)\mathbf{C}(k)\,\varepsilon(k) \quad (1.8)$$

where the elements of the k-dependent Fock matrix, $\mathbf{F}(k)$, are simply a generalization to periodic systems of the molecular orbital expressions and are given by:

$$F_{\mu,\nu}(k) = \sum_{j=-N}^{N} e^{ikja} F_{\mu,\nu}^{0,j} \quad (1.9)$$

and a similar expression can be written for $\mathbf{S}(k)$. $F_{\mu,\nu}^{0,j}$ is a direct space Fock matrix element between the atomic orbitals χ_μ of the reference cell (0) and χ_ν of the jth cell, and is given by:

$$\begin{aligned}
F_{\mu,\nu}^{0,j} &= H_{\mu,\nu}^{0,j} + \sum_{l=-N}^{N}\sum_{\rho=1}^{\omega}\sum_{\sigma=1}^{\omega} P_{\sigma,\rho}^{0,l} \sum_{h=-\infty}^{\infty} G_{\mu,\nu,\rho,\sigma}^{0,j,h,h+l} \\
&\quad - \frac{1}{2}\sum_{l=-N}^{N}\sum_{h=-N}^{N}\sum_{\rho=1}^{\omega}\sum_{\sigma=1}^{\omega} P_{\sigma,\rho}^{0,l+j-h} G_{\mu,\rho,\nu,\sigma}^{0,h,j,j+l} \\
&= \frac{-1}{2}\langle \chi_\mu^0|\nabla^2|\chi_\nu^j\rangle - \frac{1}{2}\sum_{h=-\infty}^{\infty}\sum_{A=1}^{N_C}\left\langle \chi_\mu^0 \left| \frac{Q_A}{|\mathbf{r} - \mathbf{R}_A - ha\mathbf{e}_z|} \right| \chi_\nu^j \right\rangle \\
&\quad + \sum_{l=-N}^{N}\sum_{\rho=1}^{\omega}\sum_{\sigma=1}^{\omega} P_{\sigma,\rho}^{0,l} \sum_{h=-\infty}^{\infty} G_{\mu,\nu,\rho,\sigma}^{0,j,h,h+l} \\
&\quad - \frac{1}{2}\sum_{l=-N}^{N}\sum_{h=-N}^{N}\sum_{\rho=1}^{\omega}\sum_{\sigma=1}^{\omega} P_{\sigma,\rho}^{0,l+j-h} G_{\mu,\rho,\nu,\sigma}^{0,h,j,j+l}
\end{aligned} \quad (1.10)$$

with Q_A and $\mathbf{R}_A$ being the nuclear charge and position in the reference unit cell of atom A. N_C is the number of atoms in the reference unit cell while $P_{\sigma,\rho}^{0,l}$, the density matrix element obtained by integration over the first Brillouin zone, reads:

$$P_{\sigma,\rho}^{0,l} = \frac{a}{\pi}\int_{-\pi/a}^{\pi/a} dk \left[\sum_{n=1}^{N_d} C_{\sigma,n}(k)C_{\rho,n}^*(k)\right] e^{ikla} \quad (1.11)$$

N_d is the number of doubly occupied bands. The two-electron integrals in equation (1.10) can be expressed as:

$$G_{\mu,\nu,\rho,\sigma}^{0,j,h,h+l} = \int d\mathbf{r} \int d\mathbf{r}' \chi_\mu^0(\mathbf{r})\chi_\nu^j(\mathbf{r}) \frac{1}{|\mathbf{r} - \mathbf{r}'|} \chi_\rho^h(\mathbf{r}')\chi_\sigma^{h+l}(\mathbf{r}') \quad (1.12)$$

The sum over h runs from $-\infty$ to $+\infty$ in the Coulomb terms (these are the second and third terms on the right-hand side of the second equality), whereas it is restricted to the short-range region in the exchange term (the last term on the right-hand side; see Figure 1.2). Different schemes have been proposed to evaluate efficiently the different integrals and, in particular, the two-electron integrals. In addition to the simple use of molecular algorithms where the loops over the lattice unit cells are completely outside the integral calculation, one can take advantage of the cell-independent character of many quantities. This is the case for the algorithm implemented by Mosley et al. [20] based on the McMurchie–Davidson technique [21], where the lattice loops are placed between the shell and contraction loops. A more recent and efficient algorithm, implemented by Jacquemin et al. [22] takes advantage of the structure of the Fock matrix elements as well as of modern molecular integral technologies. So, for any batch of two-electron integrals of the

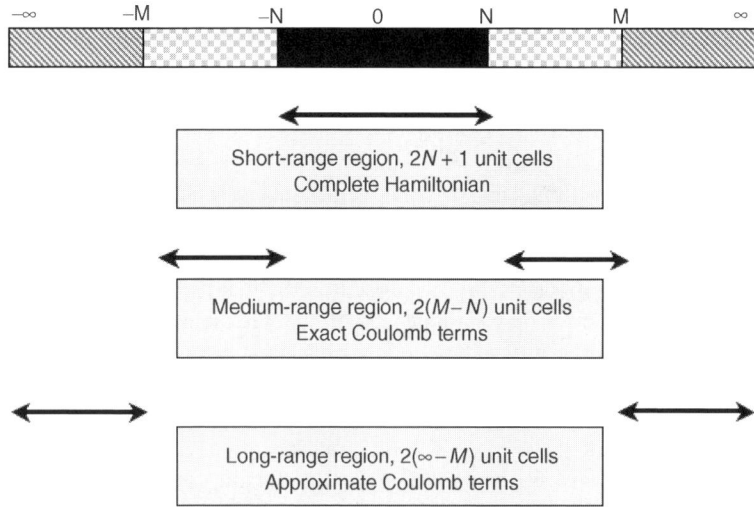

Figure 1.2. Sketch of the Namur threshold scheme for band structure calculations.

same angular momentum and contraction length, the Coulomb packing [sum over h from $-\infty$ to $+\infty$ in the third term of the second right-hand side of equation (1.10)] is performed—and the long-range corrections (see next paragraph) are added—before the costly Hermite to Cartesian integral transformation. The procedure for the exchange integrals is different: an extra loop during the shell-quartet selection is added to ensure that all integrals in the batch have exactly the same number of useful exchange integrals. In this new scheme are also employed (i) the Schwartz inequality expression due to Gill [23], employing only one-electron integrals, to discard the integrals corresponding to large inter-cell distance, (ii) the polymer 8-fold integral symmetry [20], and (iii) a semi-direct approach where the (packed) Coulomb integrals are calculated and stored on disk prior to SCF cycles while the exchange integrals are computed at each cycle, taking advantage of the values of the density matrix elements in the integral discarding techniques. Further-improved schemes using the point group symmetry have been proposed by Wilson-Zicovitch and Dovesi [24].

With the exception of the additional integration in the first Brillouin zone [equation (1.11)], the one-particle wave-functions (k-dependent LCAO coefficients) and corresponding energies are thus obtained in a self-consistent way similar to that of molecular calculations. The total Hartree–Fock energy per unit cell is given by the following expression:

$$E_{\text{UC}}^{\text{HF}} = \frac{1}{2} \sum_{j=-N}^{N} \sum_{\mu=1}^{\omega} \sum_{\nu=1}^{\omega} P_{\nu,\mu}^{0,j} [H_{\mu,\nu}^{0,j} + F_{\mu,\nu}^{0,j}] + \frac{1}{2} \sum_{h=-\infty}^{\infty} \sum_{A=1}^{N_C} \sum_{B=1}^{N_C}{}' \frac{Q_A Q_B}{|\mathbf{R}_A - \mathbf{R}_B - ha\mathbf{e}_z|}$$

(1.13)

where the second term on the right-hand side corresponds to the nuclear–nuclear repulsion. $E_{\text{UC}}^{\text{HF}}$ is the quantity that is minimized during a geometry optimization procedure.

1.2.3. Specific Aspects of Hartree–Fock Band Structure Calculations

1.2.3.1. Lattice Summations. The convergence of the lattice summations in equation (1.10) is problematic because of the long-range nature of the interactions between electron densities. Taken separately, the two Coulomb terms in equation (1.10) diverge when summed over h [25, 26] but, if the two are combined, then the divergent contributions cancel, provided the unit cell is neutral. Nevertheless, an accurate estimation of the energy requires a very large number of unit cells to be taken into account. This is why, to limit this number, the multipole expansion technique has been proposed [26]. It is based on the fact that unit cells with negligible overlap can interact via multipoles rather than via the full Fock operator. As a consequence, in the procedure elaborated by the Namur group the sums over h in the Coulomb terms are explicitly evaluated for the $2M+1$ cells ($M \geq 2N$) that define the medium-range region while the remainder, covering the $]-\infty, -M[\cup]M, \infty[$ domain, is evaluated by means of multipoles (Figure 1.2). In addition to the energy and Fock matrix elements [26], formulas for the long-range region terms have also been determined for energy gradients [27–29] and hessians [30] using Taylor series expansions which include successive multipolar interactions.

The behavior of the two-electron Coulomb term [the third term of the second equality of equation (1.10)], as a function of l, can be directly related to the corresponding overlap terms, $S_{p,\sigma}^{h,h+l} = S_{p,\sigma}^{0,l}$, and, for that reason, an exponentially converging behavior with respect to the distance between the unit cells is guaranteed. The same is true for all the other Coulomb and exchange lattice summations with the exception of the sum over j in equation (1.9) for the exchange term. So far, with the exception of model or ideal cases [31], no medium- or long-range general and efficient correction schemes have yet been implemented for the exchange. As a consequence, in the Namur thresholding scheme [32], the Coulomb and exchange integrals are not treated on an equal footing* and truncation errors are generally introduced into the Fock matrix elements. A correct balance between Coulomb and exchange requires the exchange contributions at the extremities of the short-range $[-N, N]$ region to be negligible. The errors on the Fock matrix elements depend upon the convergence behavior of the product between the two-electron exchange integrals and the density matrix elements. For non-metallic chains (nonzero gap), the exchange decreases exponentially [32, 33] as a result of the exponential decay of the one-particle density matrix [34], since the exchange energy is a simple functional of the one-particle density matrix. Although the decay is exponential, it can be slow and a large number of unit cells may be necessary to converge the exchange and therefore provide a balanced treatment of Coulomb and exchange [32]. This was shown for a polyyne chain with a very small bond length alternation (BLA). It has also been shown that the addition of diffuse functions tends to slow the convergence and that, for a given system and basis set, the square root of the band gap is the key parameter to assess the convergence. Typical values of N to reach

---

*Nearest-neighbor, next-nearest-neighbor approaches could, on the other hand, be considered as treating the Coulomb and exchange terms on an equal footing but they introduce substantial errors.

convergence range from 10 for saturated systems to 20–40 for conjugated chains. An alternative to evaluating exchange exactly consists of using the Fourier transform method developed by Delhalle et al. [35], which is the only available *ab initio* method, although it is so far limited to s-type atomic orbitals. At the Møller–Plesset second-order (MP2) perturbation theory level, a detailed investigation of the convergence of the lattice summations has been performed by Sun and Bartlett [36]. Section 1.2.6 discusses these aspects further.

1.2.3.2. Integration Procedures. In integrating the density matrix from reciprocal to real space by means of equation (1.11), one encounters another source of inaccuracy which arises from the fact that, for large l, the integrand oscillates rapidly. Several strategies have originally been described for dealing with this issue [37], while a recent paper compare their pros and cons as a function of the accuracy required and the computational resources available [22]. In practice, equation (1.11) is transformed into an integration over half the first Brillouin zone:

$$\begin{aligned} P_{\sigma,\rho}^{0,l} &= \frac{a}{\pi} \int_{-\pi/a}^{\pi/a} dk \left[\sum_{n=1}^{N_d} C_{\sigma,n}(k) C_{\rho,n}^*(k) \right] e^{ikla} \\ &= \frac{a}{\pi} \int_{-\pi/a}^{\pi/a} dk P_{\sigma,\rho}(k) e^{ikla} \\ &= \frac{2a}{\pi} \int_0^{\pi/a} dk \{\Re[P_{\sigma,\rho}(k)] \cos(kla) - \Im[P_{\sigma,\rho}(k)], \sin(kla)\} \end{aligned} \quad (1.14)$$

and the integration is performed numerically. For a given accuracy (ideally identical for all l values), the smallest number of k-points should be used because, for each of them, equation (1.8) has to be solved. The integration techniques can be classified into two categories: (i) the *nonoscillatory techniques*, which integrate the product of the density matrix elements by the exponential as a whole; and (ii) the *oscillatory techniques*, which consider separately the trigonometric components.

The Gauss–Legendre (GL) [38] and Clenshaw–Curtis (CC) [39] quadratures do not take into account explicitly the cosine and sine parts of equation (1.14), so that they are expected, when using few integration points, to lead to incorrect results for large l because the trigonometric functions oscillate rapidly. This is illustrated in Figure 1.3 in the case of a band structure calculation on all-*trans* polyethylene using the 6–31G basis set. There are, however, noticeable differences between the GL and CC schemes: (i) Within the CC scheme, the $P_{\sigma,\rho}(k)$ obtained in a quadrature using N_k points can be used in a $2N_k$ quadrature, whereas this is not the case with GL because the $2N_k$ Legendre polynomial do not include the zeros of an N_k Legendre polynomial; and (ii) GL is more accurate than CC for a given value of N_k^-.

The Filon [40–43] and Alaylioglu, Evans, and Hyslop (AEH) quadratures take explicitly into account the trigonometric functions. The Filon scheme consists of interpolating the nonoscillatory part of the integrand using a second-degree polynomial and then in integrating the result numerically. In the AEH scheme, $P_{\sigma,\rho}(k)$

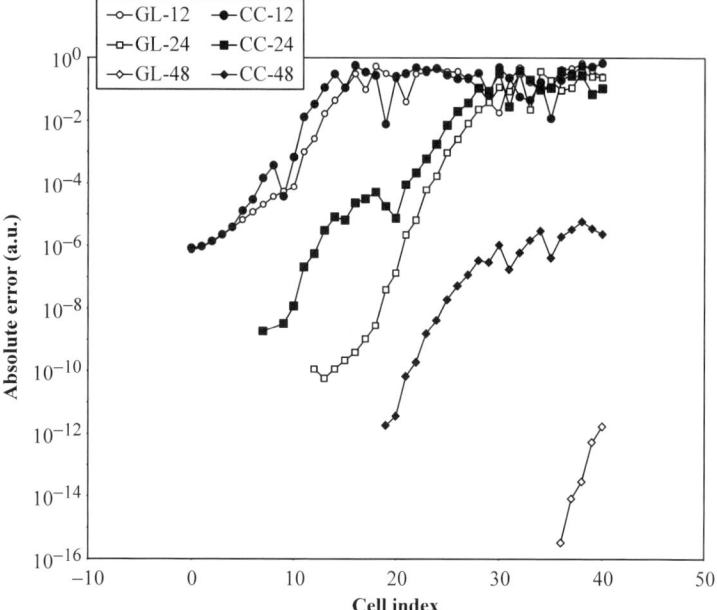

Figure 1.3. Absolute error on the $P_{9,9}^{0,l}$ element of polyethylene, calculated using the 6–31G basis set, as a function of l for nonoscillatory techniques employing different numbers of integration points. When the error is smaller than 10^{-16} a.u., no point is drawn for the clarity of the figure. Reprinted from Jacquemin et al. [22] by permission of John Wiley & Sons Inc.

is fitted with a Chebyshev polynomial [39]. The Filon quadrature is very efficient for slowly varying density matrix elements but requires the use of many integration points to reach high accuracies (Figure 1.4). AEH is a very accurate scheme, especially when employing pretabulated basic integral values [22].

1.2.3.3. Quasi-linear Dependencies. Quasi-linear dependencies, also called pseudolinear dependencies, may occur when the overlap matrices possess eigenvalues tending towards zero, and therefore create convergence difficulties of the SCF procedure. This can be understood by describing the procedure for solving equation (1.8) which consists of first transforming it into a classical eigenvalue problem, in other words, orthogonalizing the atomic orbitals:

$$\mathbf{G}(k)\mathbf{U}(k) = \mathbf{U}(k)\varepsilon(k) \tag{1.15}$$

where the $\mathbf{G}(k)$ and $\mathbf{U}(k)$ matrices are defined using a transformation matrix, $\mathbf{T}(k)$,

$$\mathbf{G}(k) = \mathbf{T}^{\dagger}(k)\,\mathbf{F}(k)\,\mathbf{T}(k) \tag{1.16}$$

$$\mathbf{U}(k) = \mathbf{T}^{-1}(k)\,\mathbf{C}(k) \tag{1.17}$$

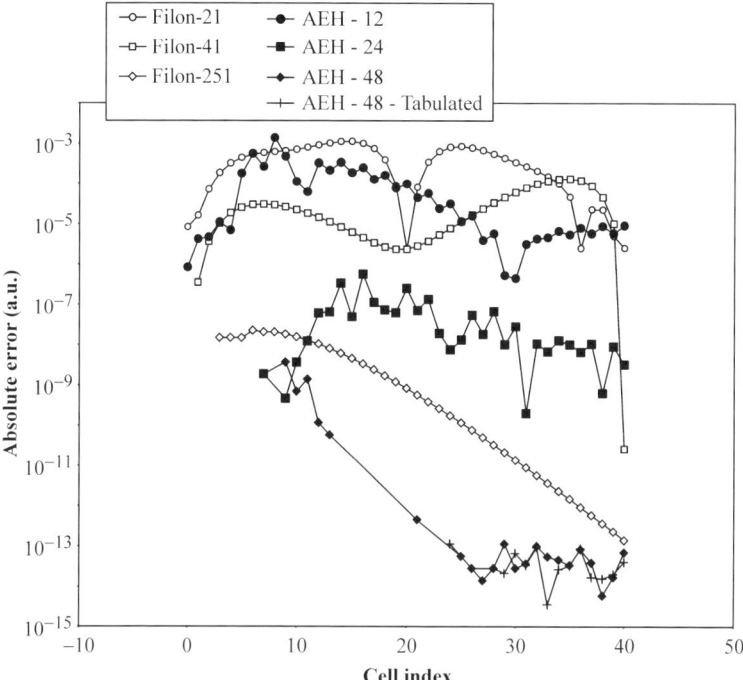

Figure 1.4. Absolute error on the largest element of the density matrix of polyethylene ($P^{0,l}_{9,9}$), calculated using the 6–31G basis set, as a function of l for oscillatory techniques employing different numbers of integration points. When the error is smaller than 10^{-16} a.u., no point is drawn for clarity of presentation. Reprinted from Jacquemin et al. [22] by permission of John Wiley & Sons Inc.

which must satisfy the following condition:

$$\mathbf{T}^\dagger(k)\, \mathbf{S}(k)\, \mathbf{T}(k) = \mathbf{1} \qquad (1.18)$$

In the canonical orthogonolization procedure due to Löwdin [44], $\mathbf{T}(k)$ is defined by:

$$\mathbf{T}(k) = \mathbf{W}(k)\Lambda^{-1/2}(k) \qquad (1.19)$$

where $\mathbf{W}(k)$ and $\Lambda(k)$ are the eigenvector and eigenvalue matrices of the overlap matrix:

$$\mathbf{S}(k)\mathbf{W}(k) = \mathbf{W}(k)\Lambda(k) \qquad (1.20)$$

Following the analysis due to Suhai et al. [45], when eigenvalues of $\mathbf{S}(k)$ are very small, columns of $\mathbf{T}(k)$ contain very large quantities [equation (1.19)], the inaccuracies in the Fock matrix elements (initially due to the truncation of the exchange term) are largely multiplied and, subsequently, cycle after cycle, the errors can increase and the SCF procedure can be oscillating, unstable, or converging slowly. For polymers, instabilities of the SCF procedure have been detected when elements of $\Lambda(k)$ are smaller than 10^{-2}. This is referred to as a quasi-linear-dependency problem, to distinguish it from the linear dependencies that are found in molecules when eigenvalues of $\mathbf{S}(k)$ are smaller than 10^{-5}–10^{-6}. The convergence difficulties in the SCF procedure can be removed in different ways: (i) by employing efficient integration schemes and lattice summation algorithms; (ii) by increasing the exponents of the most diffuse functions to get more localized functions [46]; and (iii) by removing those columns of $\mathbf{W}(k)$ associated with the smallest eigenvalues of $\mathbf{S}(k)$. The latter procedure is not theoretically correct because it may strongly deteriorate the basis set and lead to poor estimates of the polymer properties. Quasi-linear-dependency problems are not restricted to the SCF procedure of wave-function determination but have also been encountered in coupled-perturbed Hartree–Fock (CPHF) and time-dependent Hartree–Fock (TDHF) procedures, as discussed in Section 1.3.3.

1.2.4. Helical Symmetry

In addition to translational symmetry that characterizes 1-D infinite periodic systems, many polymers exist in helical conformations (e.g. DNA, polypropylene). The geometrical characteristics of a helix are contained in the L^*M/t notation. L is the number of skeletal atoms in the asymmetric unit cell and M represents the number of asymmetric units per t turns of the helix in the translational unit cell. The reduced unit cell length is therefore given by $\tau = a/M$. For these systems, the translational unit cell is generally very large and reliable *ab initio* calculations constitute a formidable task so that taking advantage of the helical symmetry would reduce considerably the computational task. Such a scheme has been proposed by Imamura and Fujita [47], Ukrainskii [48], as well as Blumen and Merkel [49]. In such a case, the LCAO expression [equation (1.7)] is rewritten in the form:

$$\phi_n(k, \mathbf{r}) = \sum_{v=1}^{\omega} C_{v,n}(k) \frac{1}{\sqrt{2N+1}} \sum_{j=-N}^{N} e^{ikj\tau} \chi_v \{\mathbf{D}(-j\alpha)[\mathbf{r} - S^j(\mathbf{R}_v)]\} \qquad (1.21)$$

where $\chi_v\{\mathbf{D}(-j\alpha)[\mathbf{r} - S^j(\mathbf{R}_v)]\}$ is the vth atomic orbital of the jth asymmetric unit cell, taking into account the screw operation defined by:

$$S^j(\mathbf{r}) = \mathbf{D}(j\alpha)(\mathbf{r}) + j\tau \mathbf{e}_z \qquad (1.22)$$

α is $2\pi t/M$ and $-2\pi t/M$ for right- and left-handed helices, respectively. $\mathbf{D}(\alpha)$ is a matrix representing a clockwise rotation around the z-axis. Using helical symmetry reduces the size of the matrices and the number of integrals but it also simplifies the interpretation of the results. Further aspects related to helical symmetry can be found in several works [50, 51] and helix-specific multipole expansion expressions for the long-range Coulomb contributions have also been worked out [52].

Further use of the symmetry (glide plane) can also be considered by adopting the general scheme developed for 3-D periodic structures [24].

1.2.5. Unrestricted Hartree–Fock Approach

In restricted Hartree–Fock (RHF) theory, the wave-function is determined by requiring the spin-orbitals to have the same spatial parts for α and β spins. Thus, the wave-function is an eigenfunction of both the total ($\hat{S}^2$) and projected ($\hat{S}_z$) spin operators and the wave-function transforms as an irreducible representation of the molecular point group. On the other hand, in the unrestricted Hartree–Fock (UHF) theory, also called the different orbitals for different spins (DODS) method, this restriction is removed and pairs of electrons of different spins are allowed to occupy different spatial orbitals. As a consequence, the wave-function is usually not an eigenfunction of the total spin operator and will not transform as an irreducible representation. Indeed, the wave-function for a given spin multiplicity contains contributions of higher spin multiplicity, which is referred to as spin contaminations. At the UHF level, the density matrix [equation (1.11)] contains two parts:

$$P^{0,l}_{\sigma,\rho}(\text{total}) = P^{0,l}_{\sigma,\rho}(\alpha) + P^{0,l}_{\sigma,\rho}(\beta) \tag{1.23}$$

with

$$P^{0,l}_{\sigma,\rho}(\alpha) = \frac{a}{2\pi} \int_{-\pi/a}^{\pi/a} dk \left[\sum_{n_\alpha=1}^{N_\alpha} C_{\sigma,n_\alpha}(k) C^*_{\rho,n_\alpha}(k) \right] e^{ikla} \tag{1.24}$$

where the sum runs over all the bands of α spin. There are also two eigenvalue problems [equation (1.8)] and two sets of Fock matrix elements. For the α-type,

$$F^{0,j}_{\mu,\nu}(\alpha) = H^{0,j}_{\mu,\nu} + \sum_{l=-N}^{N} \sum_{\rho=1}^{\omega} \sum_{\sigma=1}^{\omega} P^{0,l}_{\sigma,\rho}(\text{total}) \sum_{h=-\infty}^{\infty} G^{0,j,h,h+l}_{\mu,\nu,\rho,\sigma}$$

$$- \sum_{l=-N}^{N} \sum_{h=-N}^{N} \sum_{\rho=1}^{\omega} \sum_{\sigma=1}^{\omega} P^{0,l+j-h}_{\sigma,\rho}(\alpha) G^{0,h,j,j+l}_{\mu,\rho,\nu,\sigma} \tag{1.25}$$

Only the exchange part is different for the β-analog of equation (1.25). The computation time will therefore be increased by a factor of 2 in what concerns the diagonalization procedure, the Brillouin zone integration, and the exchange evaluation. Among the applications of an UHF CO approach, the study by Poulsen et al. [53] describes the effects of the interaction between triplet O_2 molecules and *trans*-1,4-polybutadiene chain on the band structure. High-spin polymers [54–56] constitute another field of application for UHF-based approaches.

1.2.6. Møller–Plesset and Other Wave-Function-Based Electron Correlation Schemes

In order to predict accurate band structures and properties, the next step after the Hartree–Fock theory, which suffers from the correlation defect, is to consider perturbation theory within the Møller–Plesset (MP) partitioning scheme or its infinite-order generalization, the coupled cluster (CC) theory. Configuration interaction approaches are not suitable because their practical (truncated) versions are usually not size-consistent. Kunz and co-workers [57] were the first to propose the use of MP perturbation theory (MPPT), also called many-body perturbation theory (MBPT), for infinite periodic systems. Within this scheme, the total energy per unit cell of a system is expressed as:

$$\begin{aligned} E_{UC} &= E_{UC}^{(0)} + E_{UC}^{(1)} + E_{UC}^{(2)} + E_{UC}^{(3)} + \cdots \\ &= E_{UC}^{HF} + E_{UC}^{(2)} + E_{UC}^{(3)} + \cdots \end{aligned} \quad (1.26)$$

where the first correction to the HF energy is of second order in terms of electron–electron interactions. The polymeric expression for $E^{(2)}$ is similar to the molecular case formula which contains summations over the one-particle occupied and unoccupied levels. However, in 1-D periodic systems, the one-electron wave-functions are labeled by both the band index and the quasi-momentum, and, following the work of Sun and Bartlett [58], the expression of the second-order MP energy of a closed-shell system reads:

$$E^{(2)} = \frac{1}{(2N+1)^2} \sum_{i,j}^{N_d} \sum_{a,b}^{N_{du}} \sum_{k_i,k_a,k_b}^{BZ}$$
$$\times \frac{2|Q(i,j,a,b;k_i,k_a,k_b)|^2 - \Re[Q(i,j,a,b;k_i,k_a,k_b)Q^*(i,j,b,a;k_i,k_b,k_a)]}{\varepsilon_i(k_i) + \varepsilon_j(T[k_a + k_b - k_i]) - \varepsilon_a(k_a) - \varepsilon_b(k_b)} \quad (1.27)$$

where N_{du} is the number of (doubly) unoccupied bands and $T(k)$ is a function which moves the variable k to the first Brillouin zone by a n-fold translation of $2\pi/a$. The Q elements read:

$$Q(i,j,a,b;k_i,k_a,k_b) = \sum_{\mu,\nu,\rho,\sigma=1}^{\omega} C_{\mu,i}^*(k_i) C_{\nu,a}(k_a) C_{\rho,j}^*(T[k_a + k_b - k_i]) C_{\sigma,b}(k_b)$$
$$\times \sum_{j,l=-N}^{N} \sum_{h=-\infty}^{\infty} e^{ik_a ja} e^{i[k_i - k_a]ha} e^{ik_b la} G_{\mu,\nu,\rho,\sigma}^{0,j,h,h+l} \quad (1.28)$$

They correspond to modified two-electron integrals between crystalline orbitals. Considering the $2N+1$ k-values in the first Brillouin zone, the k-summations in equation (1.27) can be replaced by integrations times the $(2N+1)a/2\pi$ factor such that the second-order correction to the unit cell energy reads:

$$
\begin{aligned}
E_{\text{UC}}^{(2)} = \left(\frac{a}{2\pi}\right)^3 \sum_{i,j}^{N_\text{d}} \sum_{a,b}^{N_\text{du}} \int_{-\pi/a}^{\pi/a} dk_i \int_{-\pi/a}^{\pi/a} dk_a \int_{-\pi/a}^{\pi/a} dk_b \\
\times \frac{2|Q(i,j,a,b;k_i,k_a,k_b)|^2 - \Re[Q(i,j,a,b;k_i,k_a,k_b)Q^*(i,j,b,a;k_i,k_b,k_a)]}{\varepsilon_i(k_i) + \varepsilon_j([k_a + k_b - k_i]) - \varepsilon_a(k_a) - \varepsilon_b(k_b)}
\end{aligned}
$$

(1.29)

These $E_{\text{UC}}^{(2)}$ expressions can easily be generalized to the unrestricted case. As for the Fock matrix elements and the HF energy per unit cell, $E_{\text{UC}}^{(2)}$ involves lattice summations and integrations over the quasi-momentum. Sun and Bartlett [36, 58, 59] performed several investigations on the convergence of $E_{\text{UC}}^{(2)}$ with N, which defines the limit of the lattice summations for the j, h, and l cell indices. First, keeping constant the parameters of the initial HF calculation, they have shown that the convergence is rather fast, whereas, varying simultaneously the HF and MP2 parameters, the convergence is much slower [60]. Indeed, $E_{\text{UC}}^{(2)}$ converges as $1/N^3$ [36, 59], whereas the convergence of the HF exchange can be slower, although it follows an exponential decrease. As pointed out and illustrated for the case of polyacetylene by Sun and Bartlett [59], the knowledge of this convergence rate provides via extrapolation a tool to guarantee fully converged (with respect to N) MP2 results.

In fact, the first applications of the MP2 scheme to stereoregular polymers are due to Suhai in 1983 [60, 61]. In these early works, in order to estimate, within the available computational resources but with sufficient accuracy, the Q terms, it was proposed to substitute the delocalized crystalline orbitals by Wannier functions (WF) [62]:

$$ w_n^l(\mathbf{r}) = \frac{1}{\sqrt{2N+1}} \sum_{\kappa=-N}^{N} e^{-ikla} \phi_n(k,\mathbf{r}) \iff \phi_n(k,\mathbf{r}) = \frac{1}{\sqrt{2N+1}} \sum_{l=-N}^{N} e^{ikla} w_n^l(\mathbf{r}) $$

(1.30)

where $w_n^l(\mathbf{r}) = w_n(\mathbf{r} - la\mathbf{e}_z)$ is the Wannier function centered around the l unit cell and κ has been defined by equation (1.6). To improve localization of the WFs, Suhai took advantage of the fact that the COs are defined within a phase factor. The phase factors were then determined in order to minimize the spatial extension of the WFs [63]. Later, Förner [64] reformulated the WF correlation energy expressions within the orbital invariant second-order Møller–Plesset theory so as to reduce the computational needs. Third- [65] and fourth- [66] order corrections to the unit cell energy have also been generalized to infinite periodic systems. On the one hand, the errors introduced on the different energy contributions by truncating the lattice summations were shown to be small and decreasing smoothly in order to become negligible for five to 10 neighbors [65, 66]. On the other hand, Sun and

Bartlett [36] specify that the total energy per unit cell converges as $1/N^3$ and that apparent divergences vanish either by cancellation of divergent terms which go by pairs, or by integration over the reciprocal space. In particular, in applications, one should not simply pursue the convergence of the integrand with lattice summations but consider the correlation correction, that is, accounting for the integration.

Another atomic-orbital formulation of second-order Møller–Plesset theory for periodic systems has been proposed by Ayala et al. [67]. It is based on the Laplace transform of the energy denominator. In that scheme, the multidimensional k-space integration is replaced by independent Fourier transforms of weighted density matrices and the computational cost becomes independent of the number of k-points used. Recently, another scheme based on Fourier transform has been proposed by Fripiat and coworkers [68]. Other works considering the estimation of high-order electron correlation corrections for infinite periodic systems include Yu et al. [69] and Hirata and Iwata [70].

The hierarchy of CC approaches was also employed for studying stereoregular polymers. Using localized orbitals [63], this was achieved by Förner et al. [63, 71] for the CC doubles (CCD) approach as well as for its linearized version (LCCD). Employing complex COs, Hirata et al. [72, 73] implemented the CC theory with single and double substitution operators (CCSD) as well as several approximations: linearized CCSD (LCCSD), quadratic configuration interaction with single and double substitution operators (QCISD), approximate CCD (ACCD), CCD, LCCD, MP2, and MP3. Although the first application was limited to (LiH)$_\infty$ [72], it was shown that the t_1-amplitudes decay rapidly (exponentially) with the distance (L) between the correlated electrons while the t_2-amplitudes decay as $1/L^3$. Their conclusion about the $1/N^3$ convergence of the correlation energy meets those of Sun and Bartlett [36]. In a very recent investigation [73], they compared the convergence of the correlation energy with respect to the extend of the lattice sums for polyethylene, polyacetylene, and polyyne and showed it reflects the polymer electronic structure, for example the degree of π-electron conjugation. Indeed, although the asymptotic decay behaviors of the t_1- and t_2-amplitudes are invariable with the systems (exponential or $1/N^3$), the decay rate is much faster for the saturated polyethylene than for the π-conjugated polyacetylene and polyyne. Similar types of behaviors were highlighted for the exchange term [32]. It turns out from Hirata et al. [73] that the unit cell correlation energy exhibits a rapid monotonic decay with the cell index and that total correlation energies can be converged to within 10^{-5} Hartree by including eight to 10 nearest neighbors. As for MPn schemes, no specific accelerating procedure to account for the full long-range contribution has been designed so far. Nevertheless, it was also stressed in Hirata et al. [73] that, due to their spatial locality, AOs are well-suited to the concept of distance-based screening of integrals and therefore to replace the CO-based integrals and t-amplitudes, which converge slowly with respect to the energy.

Møller–Plesset—as well as CC—corrections have also been worked out for the band energies [57]. At the Hartree–Fock level, the one-electron band energies correspond to the opposite of either the ionization energies (IEs) or electron-affinities (EAs) and can be expressed as:

$$\varepsilon_i(k_i) = E_N^{HF} - E_{N-1}^{HF}(i, k_i) \tag{1.31}$$

$$\varepsilon_a(k_a) = E_{N+1}^{HF}(a, k_a) - E_N^{HF} \tag{1.32}$$

where $E_{N-1}^{HF}(i, k_i)$ and $E_{N+1}^{HF}(a, k_a)$ are total energies for the system in which one electron is removed from crystalline orbital $\phi_i(k_i)$ or added to crystalline orbital $\phi_a(k_a)$. The band gap (E_g) is defined as the energy difference between the lowest unoccupied (LU) state and the highest occupied (HO) one, that is, the difference between the LUCO and HOCO. When electron relaxation is important and/or electron correlation effects substantial, these energies become poor approximations for the IEs, EAs, and band gap and electron correlation should be included. Beyond the HF approximation, following Toyozawa's electronic polaron formalism [74], the band energies can still be defined by equations (1.31) and (1.32) but the $\varepsilon_i(k_i)$ and $\varepsilon_a(k_a)$ are now replaced by quasi-particle energies and the Hartree–Fock total energies (E^{HF}) are replaced by correlated total energies. Their expressions are given by perturbation expansions of which the first term corresponds to the one-particle HF approximation:

$$\varepsilon_i^{QP}(k_i) = \varepsilon_i(k_i) + \varepsilon_i^{(2)}(k_i) + \cdots \tag{1.33}$$

A similar expression can be written for the unoccupied states. The expression for the second-order correction reads:

$$\varepsilon_p^{(2)}(k_p) = U(p, k_p) + V(p, k_p) \tag{1.34}$$

with

$$U(p, k_p) = \sum_i^{N_d} \sum_{a,b}^{N_{du}} \int_{-\pi/a}^{\pi/a} dk_a \int_{-\pi/a}^{\pi/a} dk_b$$

$$\times \frac{2|Q(p, i, a, b; k_p, k_a, k_b)|^2 - \Re[Q(p, i, a, b; k_p, k_a, k_b) Q^*(p, i, b, a; k_p, k_b, k_a)]}{\varepsilon_p(k_p) + \varepsilon_i(T[k_a + k_b - k_p]) - \varepsilon_a(k_a) - \varepsilon_b(k_b)}$$

$$\tag{1.35}$$

and

$$V(p, k_p) = \sum_{i,j}^{N_d} \sum_a^{N_{du}} \int_{-\pi/a}^{\pi/a} dk_i \int_{-\pi/a}^{\pi/a} dk_j$$

$$\times \frac{2|Q(p, a, i, j; k_p, k_i, k_j)|^2 - \Re[Q(p, a, i, j; k_p, k_i, k_j) Q^*(p, a, j, i; k_p, k_j, k_i)]}{\varepsilon_p(k_p) + \varepsilon_a(T[k_i + k_j - k_p]) - \varepsilon_i(k_i) - \varepsilon_j(k_j)}$$

$$\tag{1.36}$$

The correction is positive for the occupied levels and negative for the unoccupied ones. As a consequence, including correlation effects at the MP2 level reduces E_g. As for the total correlated energies, quasi-particle energies were determined for higher-order MP levels [65] as well as within CC schemes [71], while different numerical procedures have been employed [58–60, 64, 67, 71]. In particular, it is interesting to note that, contrary to $E_{UC}^{(2)}$, which converges rather rapidly with the extend of the lattice summations and the number of k-points for integration, the V-part of the quasi-particle energies [equations (1.34) and (1.36)] (i) is not a smooth function of k_i and k_j, which require more integration points [58], and (ii) converges as $1/N^2$ [58, 59]. As a consequence, this requires a larger computational effort than for the total energy per unit cell. On the other hand, the U-part of equations (1.34) and (1.35) evolves as $1/N^3$.

Another approach to determine quasi-particle energy bands (and improved IEs, EAs, and E_g) consists of using one-electron propagator, also called one-particle Green's function, approaches [75]. In this scheme, the ionization energies and electron affinities are given by the poles of the one-electron propagator, which is represented in a matrix form by defining super-operators, by using the inner projection technique and then, in practice, by truncating the operator manifold and by choosing a reference state [76]. Adopting Møller–Plesset partitioning, the inverse of the electron propagator is expressed as a Dyson equation:

$$\mathbf{G}^{-1}(E) = \mathbf{G}_0^{-1}(E) - \Sigma(E) \qquad (1.37)$$

where the self-energy, $\Sigma(E)$, contains the electron correlation and relaxation effects. The $G_0^{-1}(E) = (E\mathbf{1} - \varepsilon^{HF})$ matrix is diagonal. For infinite periodic systems, the most common scheme consists of adopting the MP2 level and restricting $\Sigma(E)$ to its diagonal part [58, 77, 78]. Then, the quasi-particle energy correction reads:

$$\varepsilon_p^{(2d)}(k_p) = U^{(d)}(p, k_p) + V^{(d)}(p, k_p) \qquad (1.38)$$

with

$$U^{(d)}(p, k_p) = \sum_i^{N_d} \sum_{a,b}^{N_{du}} \int_{-\pi/a}^{\pi/a} dk_a \int_{-\pi/a}^{\pi/a} dk_b$$
$$\times \frac{2|Q(p, i, a, b; k_p, k_a, k_b)|^2 - \Re[Q(p, i, a, b; k_p, k_a, k_b) \times Q^*(p, i, b, a; k_p, k_b, k_a)]}{\varepsilon_p(k_p) + \varepsilon_p^{(2d)}(k_p) + \varepsilon_i(T[k_a + k_b - k_i]) - \varepsilon_a(k_a) - \varepsilon_b(k_b)} \qquad (1.39)$$

and

$$V^{(d)}(p, k_p) = \sum_{i,j}^{N_d} \sum_a^{N_{du}} \int_{-\pi/a}^{\pi/a} dk_i$$
$$\times \int_{-\pi/a}^{\pi/a} dk_j \frac{2|Q(p, a, i, j; k_p, k_i, k_j)|^2 - \Re[Q(p, a, i, j; k_p, k_i, k_j) \times Q^*(p, a, j, i; k_p, k_j, k_i)]}{\varepsilon_p(k_p) + \varepsilon_p^{(2d)}(k_p) + \varepsilon_a(T[k_i + k_j - k_p]) - \varepsilon_i(k_i) - \varepsilon_j(k_j)} \qquad (1.40)$$

where a superscript (d) has been added to specify that it is Dyson-related. Equations (1.39)–(1.40) differ from equations (1.35)–(1.36) by the presence in the denominator of the quasi-particle energy correction. The quasi-particle energies are thus obtained by an iterative procedure of which the first iteration corresponds to the MP2 expressions. Going to higher orders requires first the inclusion of the nondiagonal part of the self-energy, which corresponds to a computationally demanding step. The convergence of the quasi-particle energy of the one-electron propagator with respect to lattice summations has been addressed by several authors [36, 58, 59, 74, 78, 79].

Electron correlation was also calculated following a projection method, elaborated by Fulde, Stollhoff, and co-workers [80], where the Liouville space is partitioned into a relevant and a neglected part. Using the former, the electron correlation is expressed and estimated by building a set of local operators containing interatomic and intraatomic components.

1.2.7. Density Functional Theory Approaches

Band structure calculations have also been carried out using density functional theory (Kohn–Sham method) and a broad range of exchange-correlation functionals. Adopting the LCAO scheme, the Kohn–Sham orbital equations are written in a matrix form like equation (1.8), but where the Fock matrix has been replaced by the Kohn–Sham matrix of which the elements read:

$$^{KS}F_{\mu,\nu}^{0,j} = \frac{-1}{2}\langle \chi_\mu^0 | \nabla^2 | \chi_\nu^j \rangle - \frac{1}{2}\sum_{h=-\infty}^{\infty}\sum_{A=1}^{N_C}\left\langle \chi_\mu^0 \left| \frac{Q_A}{|\mathbf{r} - \mathbf{R}_A - ha\mathbf{e}_z|} \right| \chi_\nu^j \right\rangle$$

$$+ \sum_{l=-N}^{N}\sum_{\rho=1}^{\omega}\sum_{\sigma=1}^{\omega} P_{\sigma,\rho}^{0,l} \sum_{h=-\infty}^{\infty} G_{\mu,\nu,\rho,\sigma}^{0,j,h,h+l}$$

$$- \frac{a}{2}\sum_{l=-N}^{N}\sum_{h=-N}^{N}\sum_{\rho=1}^{\omega}\sum_{\sigma=1}^{\omega} P_{\sigma,\rho}^{0,l+j-h} G_{\mu,\rho,\nu,\sigma}^{0,h,j,j+l} + b\langle \chi_\mu^0 | v_{xc} | \chi_\nu^j \rangle \qquad (1.41)$$

where, through the a and b multiplicative factors, equation (1.41) accounts for both pure and hybrid DFT exchange-correlation functionals (for pure DFT exchange, $a = 0$ and $b = 1$). v_{xc} is the exchange-correlation potential of which different forms have been proposed over recent decades. Since, in their conventional form, the exchange-correlation functionals at point $\mathbf{r}$ depend solely on the density and its gradient at point $\mathbf{r}$, the $\langle \chi_\mu^0 | v_{xc} | \chi_\nu^j \rangle$ term decays more rapidly with the unit cell index j than its HF analog, ensuring a faster convergence of the lattice summation.

At the beginning of the 1980s, although at that time the most common applications of DFT to periodic systems were based on plane-waves, Mintmire and White [81] worked out an approach representing the COs as linear combinations of atom-centered Gaussian-type orbitals. For the exchange-correlation potential, they used the local density ($X\alpha$) approximation. The exchange-correlation potentials as well as the charge density were expressed as linear combinations of auxiliary

Gaussian basis sets [82]. In their approach, the frozen-core approximation is not used whereas the long-range nature of the Coulomb interaction is tackled by adopting the multipole scheme of Piela and co-workers [25, 26]. At the beginning of the 1990s and in subsequent works, helical symmetry was introduced in their *ab initio* treatment [83].

Another stream in applying DFT to 1-D periodic systems is due to Springborg and co-workers [6, 84, 85], who developed a full potential linear-muffin-tin-orbital (LMTO) method for polymers. In this approach the space is divided into atom-centered nonoverlapping (muffin-tin) spheres such that the nuclei occupy the centers and into interstitial regions. In the spheres, the potential (Coulomb + exchange-correlation) is replaced by its spherically symmetric component. This scheme employs the local density approximation and was also generalized to helical polymers [51].

Several other groups have also generalized molecular or 3-D periodic crystal DFT schemes to 1-D periodic systems. Therefore, in some cases, 1-D, 2-D, and 3-D periodic systems can be treated with the same method. Jones and co-workers used plane-wave basis sets and pseudopotentials in combination with local or gradient-corrected exchange-correlation functionals [86]. Local-density approximation band structures and subsequent properties were also determined for 3-D crystalline polymers using plane-wave/pseudopotential approaches [87]. An alternative general LCAO scheme was elaborated by te Velde and Baerends [88]. It allows the use of mixed bases combining plane-waves and Slater-type decreasing-exponential functions. The numerical precision aspects of their schemes have been addressed concerning the computation of Hamiltonian matrix elements, the integration in the first Brillouin zone, and the evaluation and processing of the Coulomb potential. te Velde and Baerends further showed that plane-waves and off-site one-center functions are not required to reach high precisions.

Hirata et al. [89] combined the basics of HF CO methods with molecular DFT to produce a theoretical treatment of 1-D periodic systems which can be viewed as an extension of the method of Mintmire and co-workers [81–83]. In particular, their general scheme can use local and nonlocal (gradient-corrected) but also hybrid exchange-correlation functionals. Starting from 3-D periodic structures, the Torino group elaborated different schemes to incorporate the effects of correlation using DFT [90, 91]. Kudin and Scuseria [92, 93] worked out a general Kohn–Sham density-functional method for periodic systems. In their scheme, the fast multipole method, which is employed for evaluating the Coulomb term, achieves linear scaling of computational time with the system size.

1.3. POLYMER PROPERTIES AND APPLICATIONS

1.3.1. Band Structure, Density of States, and Photoelectron Spectroscopy

Experimental band structures are only known for a few polymers including polyethylene and polyparaphenylene. They have been probed by Ueno, Seki, and co-workers [94, 95] by using angle-resolved ultraviolet photoelectron spectroscopy

(ARUPS). Indeed, to get the intrachain energy k-dispersion, the chains have to be correctly aligned, which turns out to be the major difficulty in ARUPS measurements. In ARUPS, the band structure is determined by assuming that only direct transitions (no change in quasi-momentum) are allowed and by approximating the final state dispersion curve by a free electron parabola in a constant inner potential. In the case of polyethylene, successive works using synchrotron radiation of increasing energy (up to 120 eV) have enabled the observation of the complete valence band structure and its successful comparison with theoretical simulations [46] (Figure 1.5). However, shift and scaling of the energy bands was necessary to account for the missing electron correlation and relaxation effects, characteristics of Koopmans' theorem. In particular, the experiments located the top of the valence band at $k = 0$ and reported a highly dispersive character and a large optical transition probability for the corresponding states. This was correlated with the nature of the CO which are built from C_{2p} orbitals directed along the chain and present the largest contribution to the polyethylene polarizability [96] (see also Section 1.3.3).

In contrast to ARUPS, XPS and UPS have been extensively used to probe the composition and structure of polymeric systems. Many simulations have also been achieved to understand their properties. Indeed, the band structure data can be transformed into a form readily comparable to experiment, for example, under the form of a density of states (DOS) which can be related to XPS and UPS experiments. The DOS is defined as the number of allowed energy levels per energy unit,

$$D(E) = \frac{a}{\pi} \sum_{n=1}^{N_d} \left| \frac{dk}{d\varepsilon_n(k)} \right|_{\varepsilon_n(k)=E} \quad (1.42)$$

A DOS histogram approach has been elaborated to compute the DOS and avoid the problem linked to the vanishing of $d\varepsilon_n(k)/dk$ at several k-points, including the extremities of half the first Brillouin zone [98]. Calculated spectra are then obtained by convoluting the results of the previous equation with a Gaussian function in order to simulate the experimental resolution in the solid state as well by taking into account specific cross-section effects. Figure 1.6 displays a comparison between the experimental and simulated UPS spectra of polypyrrole, showing that the calculations provide a good qualitative agreement with the experimental photoelectron spectroscopy data and therefore enable an unambiguous interpretation of the experimental peaks [99]. It is important to emphasize on the fact that this agreement is obtained within the Koopmans approximation, which lacks electron correlation and relaxation effects. Such an approach has been adopted many times to study a large number of polymers going from polyethylene [100] to polyfluoroethylenes [101] and conjugated polymers such as poly(paraphenylene) [102].

The critical role of electron correlation for obtaining accurate band structures and DOS has been illustrated by Sun and Bartlett [103] in the case of polyethylene. On the one hand, their study pointed out the limitations of the HF and DFT (with LDA and BLYP exchange-correlation functionals) approaches whereas, on the other hand, it resolved discrepancies among the experiments. Figure 1.7 shows the

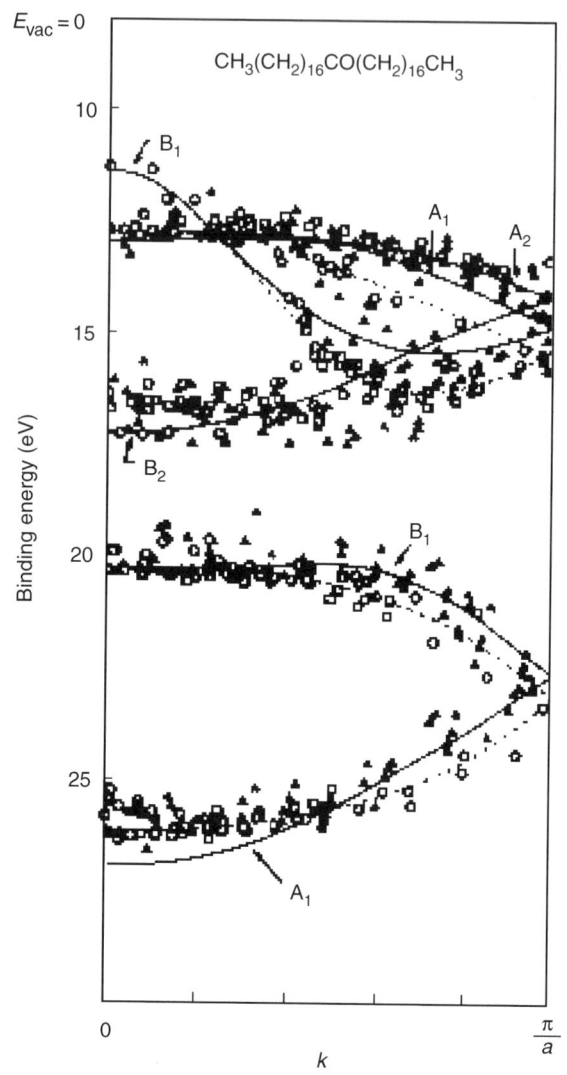

Figure 1.5. Energy band dispersion curves obtained from ARUPS measurements on different models of polyethylene chains in comparison with the theoretical results of Karpfen [46] (solid curve). The theoretical results have been adapted to fit the experimental results by adjusting the correspondence between the experimental and theoretical tops of the lower-lying B_1 band and by contracting the bands 0.8 times. The dotted curves indicate the experimentally deduced dispersion curves for the A and B bands. The symmetry assignments of the energy bands for the ideal chains are due to McCubbin [97]. Reprinted from Seki et al. [94], copyright 1990, by permission of the American Institute of Physics, Woodbury, NY.

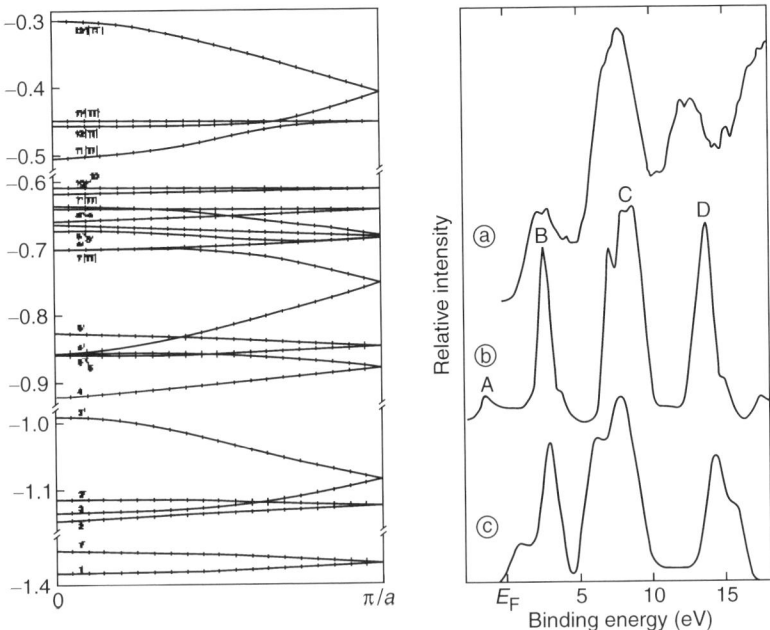

Figure 1.6. Left: *ab initio* band structure of polypyrrole (two rings per unit cell). Right: comparison of the experimental UPS spectrum (a) of polypyrrole with *ab initio* (b) and semiempirical (c) simulated spectra. The spectra have been aligned so that the locations of peak B coincide. Reprinted from André et al. [99], copyright 1984, by permission of the American Institute of Physics, Woodbury, NY.

good agreement between the theoretical and experimental photoelectron spectra as obtained in Sun and Bartlett [103].

1.3.2. Excitation Energies: Band Gaps and Excitons

Excitation energies are key quantities which determine the transport properties (insulator vs semiconductor and conductor behavior) of a polymer as well as its linear and nonlinear optical properties (Section 1.3.3). The spatial localization or delocalization of the excitation defines different physical phenomena [105]. *Photoconduction* is associated with the promotion of an electron from a valence to a conduction band, and its threshold defines the band gap, $E_g = \varepsilon_{LUCO} - \varepsilon_{HOCO}$. The methods for evaluating the corresponding one-particle Hartree–Fock and Kohn–Sham energies are presented in Sections 1.2.2 and 1.2.7, respectively, whereas Section 1.2.6 presents those methods for obtaining the quasi-particle energies within many-body perturbation theory, Green's functions, and coupled cluster schemes. In this description of the excitation processes, the electron–hole interactions are simply assumed to be screened out. Thus, differences between quasi-particle (or one-particle)

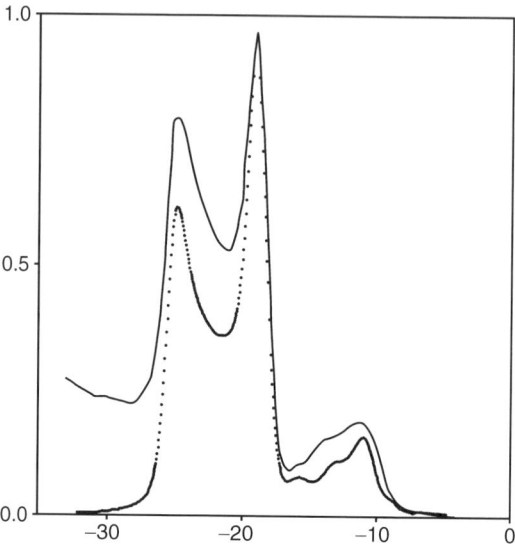

Figure 1.7. Comparison between the experimental (solid line) [104] and the theoretical (dotted lines) MP2/6-31G [103] XPS for polyethylene. Reprinted from Sun and Bartlett [103], copyright 1996, by permission of the American Institute of Physics, Woodbury, NY.

levels cannot reproduce the optical *absorption* and *emission* band edges which are generally associated with excitonic states. *Excitons* are indeed localized excitations. In polymers, they differ from the Frenkel excitons where the electron–hole pair is localized in the same unit cell or from the weakly bound Wannier excitons [106]. Instead, due to the strong covalent interactions between the unit cells, an intermediate (or charge-transfer) exciton theory is needed [107].

Over the last decades several schemes for determining the excitation energies of infinite periodic systems have been presented and applied *ab initio*, in most cases, to simple polymeric structures. Beyond the simple band gap expression, which also constitutes the zero-order approximation to the excitation energies, the first correction consists, for molecular systems, in accounting in an average way for the electron interactions. For polymers, the corresponding expressions for the direct ($\Delta k = 0$) singlet and triplet excitations of closed-shell systems read as:

$$\Delta E^S_{i(k) \to a(k)} = \varepsilon_a(k) - \varepsilon_i(k) - J_{ai}(k) + 2K_{ai}(k) \qquad (1.43)$$

$$\Delta E^T_{i(k) \to a(k)} = \varepsilon_a(k) - \varepsilon_i(k) - J_{ai}(k) \qquad (1.44)$$

where $J_{ai}(k)$ and $K_{ai}(k)$ are Coulomb and exchange integrals between the unoccupied $\phi_a(k)$ and occupied $\phi_i(k)$ COs. These corrections to the SCF orbital energy differences are needed since the electron promoted in a virtual orbital has to feel the influence of an $N-1$ electron potential rather than an N electron potential. However, contrary to molecules, when dealing with periodic infinite systems,

$\Delta E^S_{i(k)\to a(k)} = \Delta E^T_{i(k)\to a(k)} = \varepsilon_a(k) - \varepsilon_i(k)$, because of the zero value of these two-electron integrals defined between COs. One can say that, in this representation, the excitation is so much delocalized that any electron–hole interaction is completely diluted in the immensity of the system or, in other words, that the (first-order) correction brought by this single-configuration approximation evolves as the inverse of the number of interacting unit cells [11, 108, 109]. In the simplest level of approximation to account for the electron–hole interactions, the excited states are described by linear combinations of singly excited configurations obtained by promoting an electron from an occupied to an unoccupied band. An elegant and unified way to present these approaches as well as their higher-order electron correlation versions consists in adopting the theory of propagators [75] and particularly of the excitation propagator or particle–hole propagator. At the random phase approximation (RPA) level, the ground state is chosen to be the Hartree–Fock ground state, obtained by solving equation (1.8), while the operator manifold is restricted to single excitations and de-excitations. Following the usual treatments of propagator theory, the RPA expression for the excitation propagator reads:

$$P(\omega) = 2 \begin{pmatrix} \mathbf{A} + \omega \mathbf{1} & -\mathbf{B} \\ -\mathbf{B}^* & \mathbf{A}^* - \omega \mathbf{1} \end{pmatrix}^{-1} \quad (1.45)$$

The matrix elements of **A** and **B** are defined as:

$$[A(k, k')]_{ai, bj} = (\varepsilon_a(k + \Delta k) - \varepsilon_i(k))\delta_{ab}\delta_{ij}\delta_{kk'}$$
$$+ g(\phi_a(k + \Delta k)\phi_i(k)|\phi_j(k')\phi_b(k' + \Delta k))$$
$$- (\phi_a(k + \Delta k)\phi_b(k' + \Delta k)|\phi_j(k')\phi_i(k)) \quad (1.46)$$

$$[B(k, k')]_{ai, bj} = (\phi_i(k)\phi_b(k' + \Delta k)|\phi_j(k')\phi_a(k + \Delta k))$$
$$- g(\phi_i(k)\phi_a(k + \Delta k)|\phi_j(k')\phi_b(k' + \Delta k)) \quad (1.47)$$

where Δk is the exciton quasi-momentum. To describe singlet (triplet) excitations, $g = 2$ ($g = 0$) while the notation for the two-electron integrals between COs is:

$$(\phi_a(k)\phi_i(k)|\phi_j(k')\phi_b(k')) = \iint d\mathbf{r}\, d\mathbf{r}'\, \phi_a^*(k, \mathbf{r})\phi_i(k, \mathbf{r})\frac{1}{|\mathbf{r} - \mathbf{r}'|}\phi_j^*(k', \mathbf{r}')\phi_b(k', \mathbf{r}') \quad (1.48)$$

In optical excitations, the momentum is small with respect to π/a so that the vertical approximation ($\Delta k = 0$) is justified whereas, for electron energy loss spectra, the change in quasi-momentum has to be considered explicitly. The excitation energies correspond to the poles of $P(\omega)$. Different schemes have been proposed for their search including (i) solving a non-Hermitian eigenvalue problem [110] and (ii) diagonalizing two matrices either symmetric or hermitian [111]. The configuration interaction singles (CIS) scheme or Tamm–Dancoff approximation (TDA) is obtained by setting all the **B** matrix elements to zero. The RPA and TDA procedures have been implemented by Vračko and co-workers [112–114] for studying hydrogen fluoride chains, polyethylene, polyacetylene, and also cytosine and guanine stacks. For example, using the 3–21G basis set, the lowest excitation energy ($\Delta k = 0$) of polyethylene goes from 17.34 eV at the Hartree–Fock level to 13.09 eV in TDA, whereas the difference between TDA and RPA (12.64 eV) is very small [114]. By enlarging the basis set to 6–31G**, Hirata et al. [110] calculated a TDA value of 11.75 eV, whereas the experimental value is 7.6 eV. This demonstrates the importance of including electron–hole interactions but also that additional electron correlation is needed.

In fact, the first *ab initio* evaluations of the excitation energies of stereoregular polymers already incorporated dynamical correlation effects by using quasi-particle energies generated from a MP treatment (Toyozawa's electronic polaron formalism [74]) instead of the HF orbital energies [115, 116]. In addition, as for the evaluation of the correlation energy [60, 61], Wannier functions have been employed. For instance, in the case of polydiacetylenes, such polaron correction to the scheme including only single excitations between the valence and conduction bands leads to a reduction of the excitation energy by about 2 eV [115]. Within this model, Suhai also addressed the limitations of representing the excitations in conjugated organic compounds by Frenkel (on-site) excitons. Further aspects discussed by Suhai encompass the inclusion of the polarization-screening effects during the formation of the electron–hole pair and their description by using phenomenological expressions or perturbation theory. Liegener and Ladik [117] rewrote the Bethe–Salpeter equation (BSE) deriving from the particle–hole propagator under a matrix form and proposed a consistent treatment of all virtual double-excitations, which, in fact, is similar to the second-order excitation propagator approximation. For a model alternating hydrogen chain, they emphasized the role of the double excitations on both the excitation energies and exciton binding energies (defined as the difference between the band gap and the excitation energy). This procedure was later extended by including third-order electron correlation corrections to the quasi-particle energies and applied, for example, to hydrogen fluoride chains and polyacetylene [118].

In addition to developing an efficient direct trial-vector algorithm for searching the poles of equation (1.45), which avoids the explicit evaluation of two-electron integrals over CO basis, Hirata et al. [110] investigated the convergence of the excitation energies with respect to the number of unit cells in the lattice summations, the number of k-points in the first Brillouin zone for integrations, and the size of the basis set. Employing the Namur cutoff criterion [26], the excitation energies converge with the cutoff on the lattice summations slower than the HF total energies and band gap. The same behavior is obtained when considering the evolution of the excitation energies as a function of the oligomer size. Typically, twice as large a value of N as that required to obtain converged total energies is needed to get excitation energies converged up to two decimal places in eV. The convergence with respect to the number of k-points in the integration is very fast. These conclusions were drawn from calculating the lowest singlet and triplet excitation energies of polyethylene. In a subsequent work [119], following the approaches of Suhai [115, 116] and Liegener and Ladik [117, 118], they introduced dynamical electron correlation effects and used even larger basis sets in a way that the calculated excitation energies of polyethylene were in reasonable agreement with experiment. Moreover, they designed an approximate scheme which does not suffer from possible divergences associated with the high- and low-energy bands.

Alternatively, electron correlation effects on the excitation energies can be included using DFT-based approaches [120–122]. It consists first in obtaining the one-particle energies from solving the Kohn–Sham equations and then in adding the self-energy correction to get quasi-particle energies. This can be achieved either by adding a constant shift to the conduction band [122] or from solving the Dyson equation of the one-particle Green's function where the electron self-energy [cf. equation (1.37)] is evaluated in the GW approximation [120, 121]. Then, to determine the excited states, the Bethe–Salpeter equation of the two-particle Green's function describing the coupling between the electron–hole pairs is solved. As shown in Ethridge et al. [123], which employs HF wave-functions, this scheme is not restricted to the DFT framework while it is also well-suited for including screening effects (generally within the random phase approximation) originating from the strongly anisotropic interchain interactions. In the first applications to polymers the wave-functions were described by plane-wave basis sets, the local density approximation was applied, and the Bethe–Salpeter equation was reduced to a few excitations [120–122]. In many cases, the agreement between theory and experiment is rather good. Nevertheless, further developments are needed, for instance, in order to correctly predict the relative position of the $2\,^1A_g$ and $1\,^1B_u$ singlet excitons of PA and PDA [121].

A DFT alternative to GW-BSE-DFT is the time-dependent DFT (TDDFT) scheme. To our knowledge, its first adaptation to infinite one-dimensional periodic systems is due to Hirata and Bartlett [110], who also proposed its Tamm–Dancoff approximation. This scheme is equivalent to the RPA/CIS HF-based approaches described in the same reference [110], where the HF-exchange is appropriately replaced by the exchange-correlation term. From prototypical investigations on polyethylene, it was pointed out that the convergence of the excitation energies with respect to the number of k-points for integration is slower at the TDDFT and TDDFT/TDA levels than with RPA or TDA. This was attributed to an incomplete cancellation of the self-interaction energy (SIE). The TDDFT and TDDFT/TDA excitation energies to the lowest singlet exciton state are smaller than their HF-based analogs and in apparent good agreement with experiment. Indeed, using the 6–31G* basis set, the RPA value amounts to 11.75 eV, whereas the TDDFT/SVWN, TDDFT/BLYP, and TDDFT/B3LYP values are 8.01, 8.04, and 9.14 eV, respectively. The TDA analogs are, in the same order, 11.77, 8.01, 8.04, and 9.14 eV, whereas the experimental value is 7.6 eV. Nevertheless, these TDDFT and TDDFT/TDA excitation energies—using nonhybrid exchange-correlation functionals—collapse to the fundamental Kohn–Sham band gaps and predict zero exciton binding energies. This was also attributed to incomplete cancellation of the SIE and the ultralocality of the conventional exchange-correlation functionals [124]. Since it has been demonstrated to correct the optical linear responses of polymer chains for their substantial overestimation [125] and to describe excitonic effects in crystalline insulators [126], time-dependent current density functional theory (TDCDFT) appears to be an interesting scheme to apply for estimating the excitation energies of stereoregular polymers.

Other applications of *ab initio* band structure theory to excitation processes, of which the essence of the applied methodologies is described above, can be found in Tobita et al. [127] and Liegener [128] and encompass a detailed study of polydiacetylenes as well as the simulation of Auger spectra.

1.3.3. Electronic Polarizabilities and Hyperpolarizabilities

The straightforward generalization to 1-D infinite periodic systems of molecular methods for computing the linear and nonlinear electric field responses employs the scalar potential representation of the interactions between the charges and the field. Adopting the electric dipolar approximation of a clamped nucleus lattice, the perturbation potential reads (in a.u.):

$$V = \mathbf{E}(t) \cdot \sum_{i}^{\text{electrons}} \mathbf{r}_i \qquad (1.49)$$

where $\mathbf{r}_i$ is the position operator associated with electron i and $\mathbf{E}(t)$ is the external or applied time-dependent electric field. In other words, equation (1.49) neglects the quadrupolar and magnetic terms as well as the field-induced nuclear relaxation effects. When the electric field is applied along the periodicity axis of the polymer, the potential which is unbound from below destroys the periodicity. This is not compatible with Bloch's theorem [18] or with Born–von Kàrmàn cyclic boundary conditions [19] and, consequently, no ground state exists for an electron in the presence of a uniform electric field. Several approaches have been proposed to overcome these problems in 1-D but also 3-D periodic systems. Most of these, adopting equation (1.49), have been elaborated with the aim of describing the linear and nonlinear responses of inorganic crystals and have been reviewed in Champagne and Bishop [129].

The first CO-based scheme for determining the longitudinal (hyper)polarizabilities per unit cell is due to Genkin and Mednis [130], who formulated an appropriate uncoupled Hartree–Fock (UCHF) approach for infinite periodic systems. It was applied at the Hückel level of approximation for determining the (linear) polarizability (α) and the second hyperpolarizability (γ) of polyacetylene and polydiacetylenes, as well as their dimers when studying chain-pairing effects [131, 132] as well as for evaluating the first hyperpolarizability (β) of polymers presenting an asymmetric unit cell [133]. *Ab initio* extensions and the corresponding analytical procedures for evaluating the dipole moment integrals have been elaborated and applied by Barbier, Champagne, and co-workers [134, 135], but were restricted to the polarizability. In this scheme, the dynamic longitudinal (denoted L while z is the longitudinal axis) polarizability per unit cell is given under the form of a sum-over-states (SOS) expression:

$$\frac{\alpha_L(\omega)}{2N+1} = \frac{4a}{\pi} \sum_i^{N_d} \sum_a^{N_{du}} \int_0^{\pi/a} \frac{|\Omega_{ia}(k)|^2 [\varepsilon_a(k) - \varepsilon_i(k)]}{\{[\varepsilon_a(k) - \varepsilon_i(k)]^2 - \omega^2\}} dk \quad (1.50)$$

where the $\Omega_{ia}(k)$ are the elements of the antihermitian dipole matrix. Owing to the neglect of field-induced electron reorganization effects on the average electron interactions, the best this approach can achieve is a qualitative prediction of the polymer polarizabilities and their interpretation in terms of the most contributing bands and COs. Indeed, the UCHF method is known to underestimate the polarizability.

A coupled Hartree–Fock scheme has then been elaborated within the RPA level [96, 136, 137]. Contrary to the coupled-perturbed Hartree–Fock (CPHF) or time-dependent Hartree–Fock (TDHF) schemes using the scalar potential representation, the RPA approach, like other polarization propagator schemes, takes advantage of the fact that the field-dependent energies and wave-functions—or their field-derivatives—do not need to be known explicitly to evaluate the polarizability. At this level, the dynamic longitudinal polarizability reads:

$$\alpha_L(\omega) = 2(\Omega^*, \Omega) \begin{pmatrix} \mathbf{A} + \omega \mathbf{1} & \mathbf{B} \\ \mathbf{B}^* & \mathbf{A}^* - \omega \mathbf{1} \end{pmatrix}^{-1} \begin{pmatrix} \Omega \\ \Omega^* \end{pmatrix} \quad (1.51)$$

where the **A** and **B** matrix elements are defined by equations (1.46)–(1.47) using $\Delta k = 0$ and $g = 2$. The corresponding polarizability per unit cell is straightforwardly obtained when rewriting equation (1.51) in an integral form [136]. By setting the **B** matrix elements to zero and keeping only the diagonal part of **A**, the resolvent matrix becomes diagonal, the coupling between the different COs disappears and the polarizability is given by equation (1.50), (i.e. it corresponds to the UCHF expression). This RPA approach has been applied to a set of simple π- and σ-conjugated polymers, including polyacetylene, polyyne, polydiacetylenes, and polysilane, of which the polarizability was decomposed into its band contributions and analyzed with respect to their topology [96, 137]. The convergence of the polarizability value as a function of the number of k-points for integration and the number of unit cells in the lattice summations was investigated in Champagne et al. [136, 138]. In particular, the convergence of the sums generating a logarithmically divergent $1/|m|$ series is ensured either by the orthonormalization condition or by the integration procedure over the reciprocal space. To our knowledge polymeric adaptations of the polarization propagator approaches have not yet been proposed either for the evaluation of nonlinear optical responses or for correlated levels of approximation.

Another method to estimate the asymptotic coupled Hartree–Fock polarizability per unit cell of polymers was proposed by Otto in 1992 [139]. It is based on expressing the scalar potential as a sum of two terms [140]:

$$\mathbf{E}(t) \cdot \mathbf{r} = i\mathbf{E}(t) \cdot e^{i\mathbf{k}\cdot\mathbf{r}} \nabla_{\mathbf{k}} e^{-i\mathbf{k}\cdot\mathbf{r}} - i\mathbf{E}(t) \cdot \nabla_{\mathbf{k}} \qquad (1.52)$$

where the second term of the right-hand side, which describes the polarization current and is nonperiodic, is dropped because for insulating systems with completely filled bands such a current cannot exist. The first term, corresponding to the polarization, is employed to obtain, using Frenkel's variational principle, a set of coupled Hartree–Fock equations that provide an expression for the field-dependent dipole moment. By differentiating the latter with respect to the external electric field amplitude, the static and dynamic polarizability and hyperpolarizabilities are evaluated [139, 141]. The application of this method, however, raised several questions regarding the equivalence between the polymeric and large oligomeric values and the satisfaction of the (hyper)polarizability tensor symmetry [142]. In a more recent investigation [143], the Erlangen group reformulated the TDHF equations by using the intermediate orthonormalization conditions to remove the arbitrary phase factor dependence of the perturbed wave-functions. This enabled to obtain polymeric (hyper)polarizability values commensurate with the large oligomer limit and to generalize the method for computing MP2 longitudinal polarizability per unit cell [144].

In fact, the real breakthrough for computing static and dynamic, linear and nonlinear, responses of stereoregular polymers was due to Kirtman et al. [142], who derived a fully self-consistent procedure by using a noncanonical form of time-dependent perturbation theory due to Karna and Dupuis [145]. They used the vector potential representation of the interaction between the charges and the field:

$$V = \frac{1}{c} \mathbf{A}(t) \cdot \sum_{i}^{\text{electrons}} \mathbf{p}_i \qquad (1.53)$$

where c is the speed of light, $\mathbf{A}(t)$ the vector potential, and $\mathbf{p}_i$ the momentum operator of electron i because it is translationally invariant whereas the commonly used scalar potential is not. In addition, following Genkin and Mednis [130], they replaced the quasi-momentum by:

$$\mathbf{\kappa} = \mathbf{k} + \frac{1}{c}\mathbf{A}(t) \qquad (1.54)$$

When the order-by-order solution of the corresponding TDHF equations are obtained, they are substituted into an appropriate CO formula for the polarization, and fully analytical expressions for the static and dynamic (hyper)polarizabilities are proposed. In these expressions, known as 'iterative' formulas, the first hyperpolarizability depends upon the solution of the first- and second-order perturbation equations [142]. Nevertheless, the terms involving the solution of the second-order equation can be substituted by first-order terms so as to satisfy the $2n+1$

rule and to obtain less iterative formulas [146, 147]. The resulting formulas for the dynamic longitudinal α and β are:

$$\alpha_L(-\omega;\omega) = \sum_{l=-N}^{N} \text{Tr}[\mathbf{P}^{0,l;z}(\omega)\mathbf{M}^{0,l}]$$
$$+ \frac{a}{2\pi}i \int_{-\pi/a}^{\pi/a} dk \, \text{Tr}\{\mathbf{n}[\mathbf{Q}(k)\mathbf{U}^z(k,\omega) - \mathbf{U}^z(k,\omega)\mathbf{Q}(k)]\} \quad (1.55)$$

$$\beta_L(-\omega_\sigma;\omega_1,\omega_2) = \frac{-a}{2\pi} \int_{-\pi/a}^{\pi/a} dk \, \text{Tr}$$
$$\times \left(P_{-\sigma,1,2} \left\{ \mathbf{n} \begin{bmatrix} \mathbf{U}^z(k,-\omega_\sigma)\mathbf{G}^z(k,\omega_1)\mathbf{U}^z(k,\omega_2) \\ -\mathbf{U}^z(k,-\omega_\sigma)\mathbf{U}^z(k,\omega_1)\varepsilon^z(k,\omega_2) \\ +i\mathbf{U}^z(k,-\omega_\sigma)\dfrac{d}{dk}\mathbf{U}^z(k,\omega_2) \end{bmatrix} \right\} \right)$$
(1.56)

where $\mathbf{n}$ is the diagonal occupation number matrix, $\mathbf{M}$ is the cell-dependent dipole moment matrix, and $\mathbf{P}$ is given by equation (1.11). $\mathbf{Q}(k)$ is the transformation matrix to express the k-derivative of the LCAO coefficient matrix $\mathbf{C}(k)$, whereas the $\mathbf{U}^z(k,\omega_1)$ matrix is the usual transformation matrix for obtaining the first-order derivative of the LCAO coefficients with respect to an electric field oscillating at a pulsation ω_1 aligned along the z-axis. The other matrix definitions are standard [145]. This scheme has been applied to determine the polarizability, first and second hyperpolarizabilities of several polymers. Bishop et al. [146] employed a minimal basis set of Gaussian lobe orbitals whereas conventional Gaussian-type atomic orbitals are used in Gu et al. [147] and subsequent studies [148], which compare the NLO properties of different polydiacetylenes and address the crystal packing effects by considering a pair of interacting polyacetylene chains.

The effects of *pseudo*linear-dependencies (cf. Section 1.2.3.3) and long-range interactions in calculating the polarizability and hyperpolarizabilities of stereoregular polymers have been addressed recently [149]. As in field-free band structure calculations, the Coulomb terms in the time-dependent Hartree–Fock (TDHF) equations can be accurately determined using a multipole expansion for the long-range contribution. From studying polyacetylene, polymethineimine, and polysilane, it is found that (i) the exchange term converges rapidly with respect to the size of the short-range region $[-N, N]$ for which it is evaluated, (ii) the inclusion of long-range Coulomb contributions (with N fixed) ensures fast convergence with respect to the size of the medium-range region (Figure 1.2), thus demonstrating the reliability of the multipole expansion approach for the TDHF procedure as well, and (iii) the removal of long range Coulomb contributions leads to small errors (<2 percent), although these long-range terms are often compulsory for convergence in the SCF procedures as well as for matching of the oligomeric and polymeric values.

In addition, it was found in Champagne et al. [149] that (hyper)polarizabilities can be significantly underestimated by eliminating the columns of the transformation matrix associated with small eigenvalues in the Löwdin canonical orthogonalization procedure which has been used to remove *pseudo*linear-dependencies. This is shown to be particularly true for a polysilane where the removed basis set combinations of σ-symmetry are required for a proper description of the (hyper)-polarization. Further developments are now required for generalizing this approach by adding electron correlation effects, either by applying perturbation theory or by adopting DFT schemes.

Other approaches for evaluating linear and nonlinear polymeric responses have also been proposed (i) by Schmidt and Springborg [150] using a dipole moment operator which presents the same periodicity as the Born–von Kàrmàn zone, (ii) by Kudin and Scuseria [151], as well as (iii) by Darrigan et al. [152] using a sawtooth potential in combination with a supercell approach.

1.3.4. Structural (Geometries), Vibrational, and Mechanical Properties

Generally, a geometry optimization proceeds first by the evaluation of the energy gradients and the subsequent atomic displacements on the multidimensional potential energy surface. Two CO strategies have been developed for determining the energy gradients of stereoregular polymers. On the one hand, the numerical strategy consists of using a finite distortion procedure for estimating the forces on the atoms. This simply requires evaluation of the energy for different polymer geometries. This approach is therefore limited by the availability of methods (HF, MPn, CC, DFT) for determining the energy and by the large number of nuclear displacements for each optimization step, which grows rapidly with the size of the unit cell. To our knowledge, the first numerical geometry optimization was carried out at the HF level on linear fluorhydric acid by Karpfen and Schuster [153]. Then, the HF-optimized geometries of more realistic systems including polyacetylene, polymethineimine, polyethylene, and polydiacetylenes [46, 154] were numerically determined. In some cases, to reduce the computational cost, only the bond lengths and valence angles between atoms of the backbone were allowed to be displaced. Numerical evaluations of the forces were also performed at the MP2 level by Suhai [61], Sun and Bartlett [155], and Poulsen et al. [156], or by Knab and co-workers [71] at the CC level. On the other hand, an analytical scheme can be adopted for calculating the energy gradients. Dewar and co-workers presented analytical expressions for the energy first derivatives with respect to atomic Cartesian coordinates in stereoregular periodic systems [14, 157] but their approach was restricted to semiempirical wave-functions and did not consider explicitly the derivative with respect to the unit cell length (referred to as cell pressure or stress tensor). The first *ab initio* formula, including the cell pressure calculation, and implementation, is due to Teramae and co-workers [158, 159]. This was achieved at the Hartree–Fock level and illustrated on a study of the polyacetylene conformers and of all-*trans* polymethineimine. Hirata and Iwata [160] generalized these to the density functional theory level, using both pure and hybrid DFT exchange-correlation functionals. Then, they extended the method to second-order Møller–Plesset perturbation theory [70]. On the other hand, when using plane-wave basis sets, the DFT determination of forces, and subsequently the geometry optimization procedure are simplified because the basis functions are not attached to the nuclei [87]. Adopting the direct space fast multipole method for calculating the long-range Coulomb term, Kudin and Scuseria [93] also implemented energy gradient expressions for pure DFT exchange-correlation functionals using atomic basis sets. The very good accuracy on the Fock matrix elements enabled to avoid the numerical instabilities generally associated with basis sets containing diffuse functions. This was certainly favored by the lack of HF exchange terms, which is known to converge very slowly (see Section 1.2.3.1). Doll and co-workers [161, 162] later presented general analytical gradient expressions for 1-D, 2-D, and 3-D periodic systems at the Hartree–Fock level.

As for the ground state and excitation energies and for the (hyper)polarizabilities, the accuracy on the forces is a function of the number of k-points (N_k)—here for integrating the energy-weighted density matrix (Filon procedure)—of the definition of the short- and medium-range regions (Figure 1.2) as well as the inclusion of long-range contributions. A detailed assessment of the relations between these parameters and the analytical HF energy gradient accuracy was carried out by Jacquemin and co-workers [27–29, 163]. From model calculations on hydrogen fluoride chains [28], it turns out that (i) gradients require smaller N_k values than the energy, (ii) good accuracy is achieved by setting the integral discarding threshold at 10^{-8} a.u. but imposing a tight convergence threshold on the energy (10^{-12} a.u.), and (iii) the inclusion of long-range corrections is essential for performing tight optimizations (small residual forces on the atoms) as well as to speed-up the gradient calculation (smaller medium-range regions). Following the multiple Taylor series expansion scheme presented in Jacquemin et al. [27] for the energy and its gradients, these long-range corrections are added to the Hermite integrals and all subsequent integral building operations take the long-range effects into account automatically. The cell pressure is different from the other nuclear derivatives. Indeed, since it contains many individually diverging terms which cancel each other—the long-range corrections are more important. Moreover, equivalent atoms must be considered in the evaluation of the cell pressure, not for the other gradients. Another important geometrical parameter is the helical angle (see Section 1.2.4) and, for its optimization, its energy gradient. It was shown in Jacquemin and Champagne [163] that the convergence of the helical gradient is very slow and requires the inclusion of long-range corrections.

Geometry optimization procedures also necessitate the selection of parameters to optimize and the choice of a technique to move on the potential energy surface. In what concerns the first issue, internal coordinates consisting of the proper combination of bonds, angles, and torsions allow geometry optimization to be found in significantly fewer steps than a Cartesian coordinate system. Different schemes adopting this idea have been proposed and implemented, either using redundant internal coordinates [164] or delocalized internal coordinates [165]. With respect to the second issue, in addition to the energy gradients, rapidly converging geometry optimization procedures use the Hessian. However, in many cases, the Hessian is not calculated at each optimization step. On the other hand, using gradients at successive steps, quasi-Newton techniques have been implemented: they consist of starting with a crude estimate of the Hessian and improving it during the course of the optimization [164, 166]. The approach of Scuseria and co-workers [93, 164], which presents a close connection with molecular orbital techniques (fast multipole method, redundant internal coordinate), has been applied to optimize, using pure DFT exchange-correlation functionals, the geometry of different polymers ranging from polyacetylene and polyparaphenylenevinylene [93], to carbon nanotubes [93] and polypeptides [167]. This geometry optimization algorithm [164] avoids the explicit evaluation of the cell stress. Indeed, the unit cell length is optimized indirectly through the evaluation of the internal coordinates that extend between adjacent unit cells. Since hybrid exchange-correlation functionals like B3LYP and PBE0 present superior accuracy—in particular when dealing with π-conjugated systems displaying electron conjugation—further developments should encompass the mixing of these pure DFT approaches with the HF-based schemes.

Several applications of these analytical gradient schemes have been reported, including a series of studies on polyacetylene [70, 89, 93, 160]. For all-*trans* polyacetylene, at the HF/6-31G level, the bond length alternation (BLA)—the difference between the lengths of the single and double CC bonds—amounts to 0.112 Å, while it is considerably reduced when including electron correlation within the MP2/6-31G scheme (0.072 Å). This mostly results from an elongation of the C=C bond. Adopting DFT, the BLA is further reduced and attains 0.008, 0.015, 0.013, 0.018, and 0.057 Å with the SVWN, BLYP, PBE, VSXC, and B3LYP exchange-correlation functionals, respectively (6–31G* basis set). Thus, whereas pure DFT exchange-correlation functionals underestimate the BLA, the hybrid B3LYP provides suitable values. To enable the assessment of the different levels of calculation, it should be stated that the experimental values range from 0.08 to 0.10 Å with an uncertainty of about 20 percent.

Simulating vibrational spectra consists of determining the transition energies and corresponding (IR, Raman, etc.) intensities. So far, little has been achieved for the latter, whereas the former have been evaluated within the harmonic approximation from the Hessian or force constant matrix. The harmonic phonon dispersion curves, $\omega(k)$, can be calculated by following Piseri and Zerbi [168] who extended Wilson's GF method, that is, by solving the eigenvalue problem:

$$[\mathbf{M}^{-1/2}\mathbf{F}_X(k)\mathbf{M}^{-1/2} - \omega^2(k)\mathbf{1}]\mathbf{L}(k) = 0 \qquad (1.57)$$

where $\mathbf{M}^{-1/2}\mathbf{F}_X(k)\mathbf{M}^{-1/2}$ is the Hermitian dynamical matrix built from the k-dependent force constant matrix in terms of atomic Cartesian coordinates, $\mathbf{F}_X(k)$. $\mathbf{M}$ is the diagonal matrix of the atomic masses and $\mathbf{L}(k)$ describes the vibrational normal modes. Like the k-dependent Fock matrix elements [equation (1.9)], the $\mathbf{F}_X(k)$ matrix is expressed by a lattice sum:

$$\mathbf{F}_X(k) = \mathbf{B}^\dagger(k)\mathbf{F}_R(k)\mathbf{B}(k)$$
$$= \mathbf{B}^\dagger(k)\left[\sum_{j=-N}^{N} e^{ikja}\mathbf{F}_R^{0,j}\right]\mathbf{B}(k) \qquad (1.58)$$

where the $\mathbf{B}(k)$ matrix accomplishes the transformation from Cartesian to internal coordinates and $\mathbf{F}_R^{0,j}$ is the matrix of the force constants between the atoms of unit cells 0 and j. For $k = 0$, all the equivalent nuclei in the different unit cells move in phase in such a way that the corresponding vibrational normal modes can be active in IR or Raman. The $\omega(k=0)$ values are called the fundamental frequencies. As for polyethylene at $k = \pi/a$, when considering the helical symmetry of polymers, vibrational normal modes with specific $k \neq 0$ values can also be active. In the other cases, the modes are not active because the displacements in the different unit cells are out of phase.

Most of the CO determinations of the vibrational frequencies focus on the fundamental frequencies because the others are inactive in IR and Raman. Simulating the phonon dispersion curves would, on the other hand, require the use of nearest-neighbor approximation [$j = 0$ in equation (1.58)] [168], elaborating a method for displacements which destroy the translational periodicity [169], or using a computationally expensive supercell approach [170].

Again, numerical and analytical procedures have been elaborated to calculate the force constants, that is, the second-order derivatives of the energy with respect to the coordinates. These numerical procedures can, however, be further distinguished by the order of the numerical differentiation because the force constants correspond both to the first-order derivatives of the gradients or to the second-order derivatives of the energy. Karpfen and co-workers [153, 154] used second-order derivatives at the HF level while Sun and Bartlett [155] and Poulsen et al. [156] at the MP2 level. When analytical gradients became available, vibrational frequency calculations proceeded via the evaluation of the first-order derivatives of the gradients either at the HF [14, 70, 157–159], DFT [93, 160, 170], or MP2 [70] levels. In addition, an analytical second-order derivative scheme was also elaborated and implemented by Hirata and Iwata [171] at the HF level of approximation.

At the exception of hybrid oligomer/polymer approaches which have been applied on a large variety of polymers [172], little has been realized in what concerns the IR and Raman intensities. In the double harmonic oscillator approximation, the IR and Raman intensities correspond to (combinations of) first-order derivatives of the dipole moment and dynamic polarizability with respect to vibrational normal coordinates, respectively. Kudin and Scuseria [93] report IR intensities of polyparaphenylene obtained from numerical differentiation of the dipole moment, which is probably evaluated following Kudin and Scuseria [151]. In fact, the evaluation of the longitudinal component of the dipole moment—and therefore of its normal coordinate derivatives—requires a specific treatment. It has been provided by Bishop et al. [146], who highlighted that the dipole moment is only found to a modulus of the unit cell length and who proposed a tractable computational scheme. On these grounds, an analytical procedure for evaluating the dipole moment derivatives has been elaborated and implemented [173]. It opens the way for evaluating the IR intensities as well as static and dynamic vibrational polarizabilities. A finite field approach was also presented by Otto et al. [174] to evaluate the static vibrational polarizabilities. Extended finite field nuclear relaxation schemes and analytical procedures for polarizability and hyperpolarizability derivatives are now needed to envisage the simulation of Raman and hyper-Raman spectra as well as the prediction of vibrational hyperpolarizabilities.

The elastic or Young's modulus for a 1-D periodic system is related to the second-order derivative of the energy with respect to the unit cell length,

$$Y = \frac{a}{S} \times \left(\frac{\partial^2 E_{\mathrm{UC}}}{\partial a^2}\right)_{a_0} \qquad (1.59)$$

where S is the polymer cross section and a_0 is the equilibrium unit cell length. In the numerical procedure, calculating Y consists of optimizing the geometry of the polymer for different (fixed) values of a and differentiating twice the corresponding energy with respect to a. Adopting this method for polyethylene, Suhai determined Y values of 334 and 276 GPa at the HF/6–31G** and MP2/6–31G** levels of approximation, respectively [175]. Alternatively, the Young modulus can be evaluated by following the expressions derived by Hong and Kertész [176],

$$Y = \frac{a}{S} \times (\mathbf{c}^\dagger \mathbf{F}_R^{-1} \mathbf{c})^{-1} \qquad (1.60)$$

where $\mathbf{F}_R$ is the Hessian matrix and $\mathbf{c}$ describes the relation between the internal coordinates and the unit cell length. Thus, the full Hessian is required to compute Y. This scheme enables the geometrical degrees of freedom which contribute to Y [176, 177] to be unraveled. DFT CO calculations of the Young's modulus of polymers, including a comparison of the elongation–relaxation scheme with the Hessian procedure, have been reported by Van Doren and co-workers [177, 178]. For polyethylene, they obtained a Y value of 323 GPa [177]. On the other hand, the Hessian can also be evaluated analytically adopting the approach of Hirata and Iwata [171]. Jacquemin et al. [179] presents such analytical scheme which considers only second-order derivatives with respect to internal coordinates inside the unit cell as well as with respect to the unit cell length. Details on the corresponding coupled-perturbed Hartree–Fock procedure for determining the first-order derivatives of the density matrix and on the computation of the long-range effects are provided in Jacquemin et al. [179].

1.4. FURTHER ASPECTS AND OUTLOOK

The last 40 years have witnessed substantial achievements in the field of *ab initio* polymer quantum theory which enable the simulation of many structural and electronic properties of stereoregular polymers. Nowadays, these *ab initio*

FURTHER ASPECTS AND OUTLOOK

crystalline orbital methods can be used for interpreting experimental data but also for simulating an experiment before it is carried out. Examples of applications have been described in Section 1.3. Nevertheless, by analyzing the breadth of the markets where polymeric materials are used—as well as by considering the considerable achievements that have been made for characterizing theoretically (small) isolated molecules—there is still plenty of room for further developments and their applications in view of creating new polymeric materials with targeted properties.

Among these domains to investigate, one can cite (i) the electronic and vibrational circular dichroism as well as the vibrational Raman optical activity which are specific to chiral, including helical, structures [180], (ii) the effects of the surroundings—which differ from the interactions in 2-D or 3-D periodic structures—in isotropic as well as anisotropic media, for which a first treatment was recently proposed by Cossi within the conductor-like polarizable continuum model [181], (iii) the vibrational (nuclear relaxation as well as curvature) counterpart to the hyperpolarizabilities for which molecular methods were recently presented [182], (iv) the energy and charge transport phenomena which are of paramount importance in electronics [183], (v) the magnetic properties of polymers [54–56], (vi) the simulation of nuclear magnetic resonance spectra in relation to assigning the stereochemistry of polymers [184, 185], and (vii) the relativistic effects.

At several places in this review, it was shown that many approaches initially developed to treat 3-D periodic systems have already been applied to study polymeric materials [80, 86, 87, 120–123]. Further considerations of such methods [186]—both with respect to the tuning of the algorithms and to the applications [187]—for polymer structures would certainly bring *another dimension* to their investigations. Other aspects that have been tackled during these last 40 years and which are not described in details here encompass the study of metallic chains for which the pathological aspects of the restricted HF have been highlighted [188].

Many macromolecular systems are, however, aperiodic with respect to their chemical composition or to their structure and conformation. Among these, DNA and polypeptides, conjugated polymers which have been doped in order to modify their electrical conductivity, and random co-polymers require a specific treatment of aperiodicity or randomness. Different strategies have been proposed including the negative factor counting method [189] and Green's matrix formalism for treating local impurities and interface states [190]. Other approaches to studying periodic and aperiodic structures present some similarities with molecular methods although they differ from the oligomer approach [16]. The local space approximation (LSA) developed by Kirtman and co-workers [191] may be viewed either as an embedding technique or as a procedure for combining different fragments of a system to generate the whole. In particular, the LSA can be applied to infinite systems, it allows electronic charge to be transferred between the local region and its surroundings, and different levels of theory can be adopted for different regions. LSA treatments have been worked out and applied at the HF as well as correlated levels, for the density matrix as well as for its derivatives. Another approach, the elongation method, has been proposed by Imamura and co-workers [192] and has been extensively applied to periodic and aperiodic polymers. The successive steps of the elongation method are (i) solving the SCF equation for the so-called starting cluster, (ii) localizing the corresponding molecular orbitals and classifying these as active or frozen, (iii) adding a monomer unit to the cluster and solving the SCF equation of the whole extended system by considering only the monomer and the active MOs, (iv) localizing the MOs, and so on. Thus, by adding stepwise monomer units and localizing the MOs, the size of the SCF problem is kept constant. The method has been applied with semiempirical Hamiltonians [192] as well as *ab initio* at the HF and DFT levels [193]. In addition to total energies, the elongation method has been developed to perform geometry optimizations, to calculate excitation energies, and, more recently, to evaluate the static electric polarizability and hyperpolarizabilities [194]. The richness of polymers being in their structure diversity, methods capable of treating aperiodicity and randomness are expected to focus an increased interest in the near future.

ACKNOWLEDGMENTS

B.C. thanks the Belgian National Fund for Scientific Research for his Senior Research Associate position. This work is the result of numerous and fruitful interactions on many aspects of the *Quantum Theory of Polymers* the author has had during the last 15 years with J.M. André, D.M. Bishop, F. Castet, J. Delhalle, E.B. Deumens, J.G. Fripiat, F.L. Gu, D. Jacquemin, B. Kirtman, D.H. Mosley, Y. Öhrn, T.D. Poulsen, J.Q. Sun, and M.G. Vračko. The author also thanks J.M. André, R.J. Bartlett, and N. Ueno for allowing him to reproduce figures from References [94, 99, 103].

REFERENCES

[1] J. Ladik, *Acta Phys. Hung.* **18**, 173 (1965); *Acta Phys. Hung.* **18**, 185 (1965).
[2] J.M. André, L. Gouverneur, and G. Leroy, *Int. J. Quantum Chem.* **1**, 427 (1967); *Int. J. Quantum Chem.* **1**, 451 (1967).
[3] J.M. André, *Adv. Quantum Chem.* **12**, 65 (1980).
[4] M. Kertész, *Adv. Quantum Chem.* **15**, 161 (1982).
[5] M. Kertész, *Int. Rev. Phys. Chem.* **4**, 125 (1985).
[6] M. Springborg, *Int. Rev. Phys. Chem.* **12**, 241 (1993).
[7] J. Ladik, *Phys. Rep.* **313**, 171 (1999).
[8] J. Ladik, *Quantum Theory of Polymers as Solids* (Plenum Press, New York, 1988).
[9] J.M. André, J. Delhalle, and J.L. Brédas, *Quantum Chemistry Aided Design of Organic Polymers for Molecular Electronics* (World Scientific Publishing Company, London, 1991).

[10] T.K. Rebane, in *Methods of Quantum Chemistry*, M.G. Veselov (ed.) (Academic Press, New York, 1965), p. 147.
[11] M. Kertész, *Chem. Phys.* **44**, 349 (1979).
[12] W.L. McCubbin and R. Manne, *Chem. Phys. Lett.* **2**, 230 (1968); M.H. Wangbo, R. Hoffmann, and R.B. Woodward, *Proc. R. Soc. Ser. A* **366**, 23 (1979).
[13] K. Morokuma, *Chem. Phys. Lett.* **6**, 186 (1970); D.L. Beveridge, I. Jano, and J. Ladik, *J. Chem. Phys.* **56**, 4744 (1972).
[14] M.J.S. Dewar, Y. Yamaguchi, and S.H. Suck, *Chem. Phys. Lett.* **51**, 175 (1977).
[15] J.M. André, L.A. Burke, J. Delhalle, G. Nicolas, and Ph. Durand, *Int. J. Quantum Chem.* **S13**, 283 (1979).
[16] K. Müllen and G. Wegner (eds), *Electronic Materials: the Oligomer Approach* (Wiley-VCH, Weinheim, 1998).
[17] B. Champagne, D. Jacquemin, J.M. André, and B. Kirtman, *J. Phys. Chem. A* **101**, 3158 (1997); E.J. Weniger and B. Kirtman, in T.E. Simas, G. Avdelas, and J. Vigo-Aguiar special issue "Numerical Methods in Physics, Chemistry and Engineering" Computers and Mathematics with Applications **45**, 189 (2003).
[18] F. Bloch, *Z. Phys.* **52**, 555 (1928).
[19] M. Born and T. von Kàrmàn, *Z. Phys.* **13**, 297 (1912); *Z. Phys.* **14**, 15 (1913).
[20] D.H. Mosley, J.G. Fripiat, B. Champagne, and J.M. André, *Int. J. Quantum Chem.* **S27**, 793 (1993).
[21] L.E. McMurchie and E.R. Davidson, *J. Comput. Chem.* **26**, 218 (1978).
[22] D. Jacquemin, B. Champagne, J.M. André, E. Deumens, and Y. Öhrn, *J. Comput. Chem.* **23**, 1430 (2002).
[23] P.M.W. Gill, *Adv. Chem. Phys.* **25**, 141 (1994).
[24] C.M. Wilson-Zicovitch and R. Dovesi, *Int. J. Quantum Chem.* **67**, 299 (1998); *Int. J. Quantum Chem.* **67**, 311 (1998).
[25] L. Piela and J. Delhalle, *Int. J. Quantum Chem.* **13**, 605 (1978).
[26] J. Delhalle, L. Piela, J.L. Brédas, and J.M. André, *Phys. Rev. B* **22**, 6254 (1980); L. Piela, J.M. André, J.L. Brédas, and J. Delhalle, *Int. J. Quantum Chem.* **S14**, 405 (1980).
[27] D. Jacquemin, B. Champagne, and J.M. André, *J. Chem. Phys.* **111**, 5306 (1999).
[28] D. Jacquemin, B. Champagne, and J.M. André, *J. Chem. Phys.* **111**, 5324 (1999).
[29] D. Jacquemin and B. Champagne, *Int. J. Quantum Chem.* **80**, 863 (2000).
[30] D. Jacquemin, J.M. André, and B. Champagne, *J. Chem. Phys.* **118**, 373 (2003).
[31] J. Delhalle and J.L. Calais, *J. Chem. Phys.* **85**, 5286 (1986); L.Z. Stolarczyk, M. Jeziorska, and H.J. Monkhorst, *Phys. Rev. B* **37**, 10646 (1988); M. Jeziorska, L.Z. Stolarczyk, J. Paldus, and H.J. Monkhorst, *Phys. Rev. B* **41**, 12473 (1990); C. Pisani, M. Causa, and R. Orlando, *Int. J. Quantum Chem.* **38**, 419 (1990).
[32] D. Jacquemin, J.G. Fripiat, and B. Champagne, *Int. J. Quantum Chem.* **89**, 452 (2002).
[33] J. Des Cloizeaux, *Phys. Rev. A* **135**, 685 (1964); L. Piela, J.M. André, J.G. Fripiat, and J. Delhalle, *Chem. Phys. Lett.* **77**, 143 (1981).
[34] L. He and D. Vanderbilt, *Phys. Rev. Lett.* **86**, 5341 (2001); S.N. Taraskin, D.A. Drabold, and S.R. Elliott, *Phys. Rev. Lett.* **88**, 196405 (2002).

[35] J. Delhalle, J. Cizek, I. Flamant, J.L. Calais, and J.G. Fripiat, *J. Chem. Phys.* **101**, 10717 (1994); I. Flamant, J.G. Fripiat, J. Delhalle, and F.E. Harris, *Theor. Chem. Acc.* **104**, 350 (2000); J. Delhalle, J.G. Fripiat, and F.E. Harris, *Int. J. Quantum Chem.* **90**, 587 (2002).
[36] J.Q. Sun and R.J. Bartlett, *J. Chem. Phys.* **106**, 5554 (1997).
[37] W.J. Schneider and J. Ladik, *J. Comput. Chem.* **2**, 376 (1981).
[38] M. Abramowitz and I. Stegun, *Handbook of Mathematical Functions* (Dover, New York, 1968).
[39] G. Evans, *Practical Numerical Integration* (Wiley, New York, 1993).
[40] L.N.G. Filon, *Proc. R. Soc. Edinburgh* **49**, 38 (1928–1929).
[41] S.M. Chase and L.D. Fosdick, *Commun. ACM* **12**, 453 (1969).
[42] J.G. Fripiat, J.M. André, J. Delhalle, and J.L. Calais, *Int. J. Quantum Chem.* **S24**, 593 (1990).
[43] J.G. Fripiat, J.M. André, J. Delhalle, and J.L. Calais, *Int. J. Quantum Chem.* **S25**, 603 (1991).
[44] P.O. Löwdin, *J. Chem. Phys.* **18**, 365 (1950); P.O. Löwdin, *Adv. Phys.* **5**, 1 (1956); P.O. Löwdin, *J. Appl. Phys.* **33** (suppl.), 251 (1962).
[45] S. Suhai, P.S. Bagus, and J. Ladik, *Chem. Phys.* **68**, 467 (1982).
[46] A. Karpfen, *J. Chem. Phys.* **75**, 238 (1981).
[47] A. Imamura and H. Fujita, *J. Chem. Phys.* **61**, 115 (1974).
[48] I.I. Ukrainskii, *Theor. Chim. Acta* **38**, 139 (1975).
[49] A. Blumen and C. Merkel, *Phys. Status Solidi* **B83**, 425 (1977).
[50] T. Shimanouchi and S.I. Mizushima, *J. Chem. Phys.* **23**, 707 (1955); R.E. Hugues and J.L. Lauer, *J. Chem. Phys.* **30**, 1165 (1959); C.X. Cui and M. Kertész, *J. Am. Chem. Soc.* **111**, 4216 (1989).
[51] M. Springborg and R.O. Jones, *J. Chem. Phys.* **88**, 2652 (1988).
[52] J.M. André, D.P. Vercauteren, V.P. Bodart, and J.G. Fripiat, *J. Comput. Chem.* **5**, 535 (1984).
[53] T.D. Poulsen, K.V. Mikkelsen, J.G. Fripiat, and B. Champagne, *J. Chem. Phys.* **113**, 5958 (2000).
[54] M. Mitani, Y. Takano, Y. Yoshioka, and K. Yamaguchi, *J. Chem. Phys.* **111**, 1309 (1999);
[55] M. Mitani, H. Mori, Y. Takano, D. Yamaki, Y. Yoshioka, and K. Yamaguchi, *J. Chem. Phys.* **113**, 4035 (2000).
[56] N. Tyutyulkov, M. Baumgarten, and F. Dietz, *Chem. Phys. Lett.* **353**, 231 (2002).
[57] A.B. Kunz, *Phys. Rev. B* **6**, 606 (1972); S.T. Pantelides, A.J. Mickish, and A.B. Kunz, *Phys. Rev. B* **10**, 2602 (1974).
[58] J.Q. Sun and R.J. Bartlett, *J. Chem. Phys.* **104**, 8553 (1996).
[59] J.Q. Sun and R.J. Bartlett, *J. Chem. Phys.* **107**, 5058 (1997); J.Q. Sun and R.J. Bartlett, *Phys. Rev. Lett.* **80**, 319 (1998).
[60] S. Suhai, *Int. J. Quantum Chem.* **23**, 1239 (1983).
[61] S. Suhai, *Chem. Phys. Lett.* **96**, 619 (1983); S. Suhai, *Phys. Rev. B* **27**, 3506 (1983).
[62] G.H. Wannier, *Phys. Rev.* **52**, 191 (1937).

REFERENCES

[63] E.I. Blount, *Solid State Phys.* **13**, 305 (1963); see also R. Knab, W. Förner, J. Cizek, and J. Ladik, *J. Mol. Struct. (THEOCHEM)* **366**, 11 (1996) for optimal localization of the Wannier functions.

[64] W. Förner, *Int. J. Quantum Chem.* **43**, 221 (1992).

[65] C.M. Liegener, *J. Phys. C: Solid State Phys.* **18**, 6011 (1985); S. Suhai, *Int. J. Quantum Chem.* **S27**, 131 (1993).

[66] S. Suhai, *J. Chem. Phys.* **101**, 9766 (1994); S. Suhai, *Phys. Rev. B* **51**, 16553 (1995).

[67] P.Y. Ayala, K.N. Kudin, and G.E. Scuseria, *J. Chem. Phys.* **115**, 9698 (2001).

[68] J. Delhalle, J.G. Fripiat, and F.E. Harris, *Int. J. Quantum Chem.* **90**, 1326 (2002).

[69] M. Yu, S. Kalvoda, and M. Dolg, *Chem. Phys.* **224**, 121 (1997).

[70] S. Hirata and S. Iwata, *J. Chem. Phys.* **109**, 4147 (1998).

[71] W. Förner, R. Knab, J. Cizek, and J. Ladik, *J. Chem. Phys.* **106**, 10248 (1997); R. Knab, W. Förner, and J. Ladik, *J. Phys.: Condens. Matter* **9**, 2043 (1997).

[72] S. Hirata, I. Grabowski, M. Tobita, and R.J. Bartlett, *Chem. Phys. Lett.* **345**, 475 (2001).

[73] S. Hirata, R. Podeszwa, M. Tobita, and R.J. Bartlett, *J. Chem. Phys.* **120**, 2581 (2004).

[74] Y. Toyozawa, *Prog. Theor. Phys.* **12**, 422 (1954); J.T. Devreese, A.B. Kunz, J. Mintmire, and T.C. Collins, *Solid State Commun.* **11**, 679 (1993).

[75] J. Linderberg and Y. Öhrn, *Propagators in Quantum Chemistry*, Second Edition (Wiley-Interscience, Hoboken, 2004).

[76] B.T. Pickup and O. Goscinski, *Mol. Phys.* **26**, 1013 (1973); Y. Öhrn and G. Born, *Adv. Quantum Chem.* **13**, 1 (1981); L.S. Cederbaum, W. Domcke, J. Schirmer, and W. von Niessen, *Adv. Chem. Phys.* **65**, 115 (1986).

[77] C.M. Liegener, *J. Chem. Phys.* **88**, 6999 (1988); I.J. Palmer and J. Ladik, *J. Comput. Chem.* **15**, 814 (1994).

[78] J.Q. Sun and R.J. Bartlett, *Topics in Current Chemistry* (Springer, Berlin, 1999), Vol. 203, p. 121.

[79] M. Deleuze, J. Delhalle, B.T. Pickup, and J.L. Calais, *Phys. Rev. B* **46**, 15668 (1992); M. Deleuze and B.T. Pickup, *J. Chem. Phys.* **102**, 8967 (1995); M. Deleuze, M.K. Scheller, and L.S. Cederbaum, *J. Chem. Phys.* **103**, 3578 (1995); M. Deleuze, J. Delhalle, B.T. Pickup, and J.L. Calais, *Adv. Quantum Chem.* **26**, 35 (1995); M. Nooijen and R.J. Bartlett, *Int. J. Quantum Chem.* **63**, 601 (1996).

[80] G. Stollhoff and P. Fulde, *J. Chem. Phys.* **73**, 4548 (1980); K. Becker and P. Fulde, *J. Chem. Phys.* **91**, 4223 (1989); G. König and G. Stollhoff, *J. Chem. Phys.* **91**, 2993 (1989); P. Fulde and G. Stollhoff, *Int. J. Quantum Chem.* **42**, 103 (1992).

[81] J.W. Mintmire and C.T. White, *Phys. Rev. Lett.* **50**, 101 (1983); J.W. Mintmire and C.T. White, *Phys. Rev. B* **28**, 3283 (1983); J.W. Mintmire, *Phys. Rev. B* **39**, 13350 (1989).

[82] J.W. Mintmire, J.R. Sabin, and S.B. Trickey, *Phys. Rev. B* **26**, 1743 (1982).

[83] J.W. Mintmire, *Density Functional Methods in Chemistry*, J. Labanowski and J. Andzelm (eds) (Springer, New York, 1991), p. 125; M.S. Miao, P.E. Van Camp, V.E. Van Doren, J.J. Ladik, and J.W. Mintmire, *Phys. Rev. B* **54**, 10430 (1996); M.S. Miao, P.E. Van Camp, V.E. Van Doren, J.J. Ladik, and J.W. Mintmire, *J. Chem. Phys.* **109**, 9623 (1998).

[84] M. Springborg and O.K. Andersen, *J. Chem. Phys.* **87**, 7125 (1987); M. Springborg, J.L. Calais, O. Goscinski, and L.A. Eriksson, *Phys. Rev. B* **44**, 12713 (1991).

[85] H. Meider and M. Springborg, *J. Phys. Chem. B* **101**, 6949 (1997); M. Springborg, *J. Am. Chem. Soc.* **121**, 11211 (1999); M. Springborg, *Int. J. Quantum Chem.* **77**, 843 (2000).

[86] A. Borrmann, B. Montanari, and R.O. Jones, *J. Chem. Phys.* **106**, 8545 (1997); B. Montanari and R.O. Jones, *Chem. Phys. Lett.* **272**, 347 (1997). See also R.O. Jones and O. Gunnarsson, *Rev. Mod. Phys.* **61**, 689 (1989) for general aspects of DFT.

[87] P. Blaha, K. Schwarz, P. Sorantin, and S.B. Trickey, *Comput. Phys. Commun.* **59**, 399 (1990); P. Vogl and D.K. Campbell, *Phys. Rev. B* **41**, 12797 (1990); C. Ambrosch-Draxl, J.A. Majewski, P. Vogl, and G. Leising, *Phys. Rev. B* **51**, 9668 (1995).

[88] G. Te Velde and E.J. Baerends, *Phys. Rev. B* **44**, 7888 (1991).

[89] S. Hirata, H. Torii, and M. Tasumi, *Phys. Rev. B* **57**, 11994 (1998).

[90] M. Causa, R. Dovesi, C. Pisani, R. Colle, and A. Fortunelli, *Phys. Rev. B* **36**, 891 (1987); M. Causa and J. Zupan, *Chem. Phys. Lett.* **220**, 145 (1994); A. Zupan and M. Causa, *Int. J. Quantum Chem.* **56**, 337 (1995).

[91] C. Pisani, R. Dovesi, C. Roetti, M. Causa, R. Orlando, S. Casassa, and V.R. Saunders, *Int. J. Quantum Chem.* **77**, 1032 (2000).

[92] K.N. Kudin and G.E. Scuseria, *Chem. Phys. Lett.* **289**, 611 (1998).

[93] K.N. Kudin and G.E. Scuseria, *Phys. Rev. B* **61**, 16440 (2000).

[94] K. Seki, U.O. Karlsson, R. Engelhardt, and E.E. Koch, *Chem. Phys. Lett.* **103**, 343 (1984); K. Seki, N. Ueno, U.O. Karlsson, R. Engelhardt, and E.E. Koch, *Chem. Phys.* **105**, 247 (1986); N. Ueno, K. Seki, H. Fujimoto, T. Kuramochi, K. Sigita, and H. Inokuchi, *Phys. Rev. B* **41**, 1176 (1990).

[95] K. Seki, U.O. Karlsson, R. Engelhardt, E.E. Koch, and W. Schmidt, *Chem. Phys.* **91**, 459 (1984); S. Narioka, H. Ishii, K. Edamatsu, K. Kamiya, S. Hasegawa, T. Ohta, N. Ueno, and K. Seki, *Phys. Rev. B* **52**, 2362 (1995).

[96] B. Champagne and J.M. André, *Nonlin. Optics.* **9**, 25 (1995).

[97] W.L. McCubbin, *Electronic Structure of Polymers as Molecular Crystals*, J.M. André and J. Ladik (eds) (Plenum Press, London, 1974), p. 171.

[98] J. Delhalle and S. Delhalle, *Int. J. Quantum Chem.* **11**, 349 (1977); J. Delhalle, D. Thelen, and J.M. André, *Comput. Chem.* **3**, 1 (1979).

[99] J.M. André, D.P. Vercauteren, G.B. Street, and J.L. Brédas, *J. Chem. Phys.* **80**, 5643 (1984).

[100] J. Delhalle, J.M. André, S. Delhalle, J.J. Pireaux, R. Caudano, J.J. Verbist, *J. Chem. Phys.* **60**, 595 (1974).

[101] J. Delhalle, *Chem. Phys.* **5**, 306 (1974).

[102] J.L. Brédas, R.R. Chance, R. Silbey, G. Nicolas, and Ph. Durand, *J. Chem. Phys.* **77**, 371 (1982); B. Champagne, D.H. Mosley, J.G. Fripiat, and J.M. André, *Phys. Rev. B* **54**, 2381 (1996).

[103] J.Q. Sun and R.J. Bartlett, *Phys. Rev. Lett.* **77**, 3669 (1996).

[104] J.J. Pireaux, S. Svensson, E. Basilier, P.A. Malmqvist, U. Gelius, R. Caudano, and K. Siegbahn, *Phys. Rev. A* **14**, 2133 (1976).

[105] J.L. Brédas, J. Cornil, D. Beljonne, D.A. dos Santos, and Z. Shuai, *Acc. Chem. Res.* **32**, 267 (1999).

REFERENCES

[106] J. Callaway, *Quantum Theory of the Solid State* (Academic, San Diego, CA, 1991), Section 6.6.

[107] Y. Takeuti, *Prog. Theor. Phys.* **18**, 421 (1957); Y. Takeuti, Suppl. *Prog. Theor. Phys.* **12**, 75 (1959).

[108] M. Kertész, J. Koller, and A. Ažman, *Croat. Chem. Acta* **55**, 85 (1982).

[109] J.M. André, J.L. Brédas, B. Thémans, and L. Piela, *Int. J. Quantum Chem.* **23**, 1065 (1983).

[110] S. Hirata, M. Head-Gordon, and R.J. Bartlett, *J. Chem. Phys.* **111**, 10774 (1999).

[111] J. Linderberg and Y. Öhrn, *Int. J. Quantum Chem.* **12**, 161 (1977).

[112] M.G. Vračko and M. Zaider, *Int. J. Quantum Chem.* **42**, 321 (1992); M.G. Vračko and M. Zaider, *Int. J. Quantum Chem.* **47**, 119 (1993).

[113] M.G. Vračko and M. Zaider, *Radiat. Res.* **138**, 18 (1994).

[114] M.G. Vračko, B. Champagne, D.H. Mosley, and J.M. André, *J. Chem. Phys.* **102**, 6831 (1995).

[115] S. Suhai, *Phys. Rev. B* **29**, 4570 (1984).

[116] S. Suhai, *J. Chem. Phys.* **85**, 611 (1986); S. Suhai, *Int. J. Quantum Chem.* **29**, 469 (1986).

[117] C.M. Liegener and J. Ladik, *Chem. Phys.* **106**, 339 (1986).

[118] C.M. Liegener and J. Ladik, *Phys. Rev. B* **35**, 6403 (1987); C.M. Liegener, *J. Chem. Phys.* **88**, 6999 (1988).

[119] S. Hirata and R.J. Bartlett, *J. Chem. Phys.* **112**, 7339 (2000).

[120] M. Rohlfing and S.G. Louie, *Phys. Rev. Lett.* **82**, 1959 (1999); M. Rohlfing and S.G. Louie, *Phys. Rev. B* **62**, 4927 (2000).

[121] J.W. van der Horst, P.A. Bobbert, M.A.J. Michels, and H. Bässler, *J. Chem. Phys.* **114**, 6950 (2001); J.W. van der Horst, P.A. Bobbert, and M.A.J. Michels, *Phys. Rev. B* **66**, 35206 (2002).

[122] A. Ruini, M.J. Caldas, G. Bussi, and E. Molinari, *Phys. Rev. Lett.* **88**, 206403 (2002).

[123] E.C. Ethridge, J.L. Fry, and M. Zaider, *Phys. Rev. B* **53**, 3662 (1996).

[124] S.J.A. van Gisbergen, P.R.T. Schipper, O.V. Gritsenko, E.J. Baerends, J.G. Snijders, B. Champagne, and B. Kirtman, *Phys. Rev. Lett.* **83**, 694 (1999); S. Grimme and M. Parac, *Chem. Phys. Chem.* **3**, 292 (2003).

[125] M. van Faassen, P.L. de Boeij, R. van Leeuwen, J.A. Berger, and J.G. Snijders, *J. Chem. Phys.* **118**, 1044 (2003).

[126] P.L. de Boeij, F. Kootstra, J.A. Berger, R. van Leeuwen, and J.G. Snijders, *J. Chem. Phys.* **115**, 1995 (2001).

[127] M. Tobita, S. Hirata, and R.J. Bartlett, *J. Chem. Phys.* **114**, 9130 (2001).

[128] C.M. Liegener, *Phys. Rev. B* **43**, 7561 (1993); H. Ågren, F. Gel'mukhanov, and C.M. Liegener, *Int. J. Quantum Chem.* **63**, 313 (1997).

[129] B. Champagne and D.M. Bishop, *Adv. Chem. Phys.* **126**, 41 (2003).

[130] V.N. Genkin and P.M. Mednis, *Sov. Phys. JETP* **27**, 609 (1968); [*Zh. Eksp. Teor. Fiz.* **54**, 1137 (1968)].

[131] C. Cojan, G.P. Agrawal, and C. Flytzanis, *Phys. Rev. B* **15**, 909 (1977).

[132] G.P. Agrawal, C. Cojan, and C. Flytzanis, *Phys. Rev. B* **17**, 776 (1978).

[133] B. Champagne, D. Jacquemin, and J.M. André, *Nonlinear Optical Properties of Organic Materials VIII*, G.R. Möhlmann (ed.), SPIE Proceedings Vol. 2527, 1995, p. 71.

[134] C. Barbier, *Chem. Phys. Lett.* **142**, 53 (1987); C. Barbier, J. Delhalle, and J.M. André, *J. Mol. Struct. (THEOCHEM)* **188**, 299 (1989).

[135] B. Champagne and J.M. André, *Int. J. Quantum Chem.* **42**, 1009 (1992); B. Champagne, D.H. Mosley, and J.M. André, *J. Chem. Phys.* **100**, 2034 (1994).

[136] B. Champagne, D.H. Mosley, J.G. Fripiat, and J.M. André, *Int. J. Quantum Chem.* **46**, 1 (1993); B. Champagne, D.H. Mosley, J.G. Fripiat, and J.M. André, *Adv. Quantum Chem.* **35**, 95 (1999).

[137] B. Champagne, D.H. Mosley, and J.M. André, *Int. J. Quantum Chem.* **S27**, 667 (1993); B. Champagne and Y. Öhrn, *Chem. Phys. Lett.* **217**, 551 (1994); B. Champagne, J.M. André, and Y. Öhrn, *Int. J. Quantum Chem.* **57**, 811 (1996).

[138] B. Champagne, D.H. Mosley, and J.M. André, *Chem. Phys. Lett.* **210**, 232 (1993); B. Champagne, J.G. Fripiat, D.H. Mosley, and J.M. André, *Int. J. Quantum Chem.* **S29**, 429 (1995).

[139] P. Otto, *Phys. Rev. B* **46**, 10876 (1992).

[140] J. Ladik and P. Otto, *Int. J. Quantum Chem.* **S27**, 111 (1993); J. Ladik, *Nonlinear Optical Materials: Theory and Modeling*, S.P. Karna and A.T. Yeates (eds) (American Chemical Society, Washington, DC, 1996), Chap. 10, p. 174.

[141] P. Otto, F.L. Gu, and J. Ladik, *J. Chem. Phys.* **110**, 2717 (1999).

[142] B. Kirtman, F.L. Gu, and D.M. Bishop, *J. Chem. Phys.* **113**, 1294 (2000).

[143] A. Martinez, P. Otto, and J. Ladik, *Int. J. Quantum Chem.* **94**, 251 (2003).

[144] P. Otto, A. Martinez, A. Czaja, and J. Ladik, *J. Chem. Phys.* **117**, 1908 (2002).

[145] S.P. Karna and M. Dupuis, *J. Comput. Chem.* **12**, 487 (1991).

[146] D.M. Bishop, F.L. Gu, and B. Kirtman, *J. Chem. Phys.* **114**, 7633 (2001).

[147] F.L. Gu, D.M. Bishop, and B. Kirtman, *J. Chem. Phys.* **115**, 10548 (2001).

[148] F.L. Gu, Y. Aoki, and D.M. Bishop, *J. Chem. Phys.* **117**, 385 (2002); B. Kirtman, B. Champagne, F.L. Gu, and D.M. Bishop, *Int. J. Quantum Chem.* **90**, 709 (2002).

[149] B. Champagne, D. Jacquemin, F.L. Gu, Y. Aoki, B. Kirtman, and D.M. Bishop, *Chem. Phys. Lett.* **373**, 539 (2003).

[150] K. Schmidt and M. Springborg, *Phys. Chem. Chem. Phys.* **1**, 1743 (1999).

[151] K.N. Kudin and G.E. Scuseria, *J. Chem. Phys.* **113**, 7779 (2000).

[152] C. Darrigan, M. Rérat, G. Mallia, and R. Dovesi, *J. Comput. Chem.* **24**, 1305 (2003).

[153] A. Karpfen and P. Schuster, *Chem. Phys. Lett.* **44**, 459 (1976).

[154] A. Karpfen, *Chem. Phys. Lett.* **64**, 299 (1979); A. Karpfen and J. Petkov, *Theor. Chim. Acta* **53**, 65 (1979); A. Karpfen, *J. Phys. C* **13**, 5673 (1980); A. Karpfen and A. Beyer, *J. Comput. Chem.* **5**, 11 (1984).

[155] J.Q. Sun and R.J. Bartlett, *J. Chem. Phys.* **108**, 301 (1998).

[156] T.D. Poulsen, K.V. Mikkelsen, J.G. Fripiat, D. Jacquemin, and B. Champagne, *J. Chem. Phys.* **114**, 5917 (2001).

[157] M.J.S. Dewar, Y. Yamaguchi, and S.H. Suck, *Chem. Phys.* **43**, 145 (1979).

[158] H. Teramae, T. Yamabe, C. Satoko, and A. Imamura, *Chem. Phys. Lett.* **101**, 49 (1983).

[159] H. Teramae, T. Yamabe, and A. Imamura, *J. Chem. Phys.* **81**, 3564 (1984).

REFERENCES

[160] S. Hirata and S. Iwata, *J. Chem. Phys.* **107**, 10075 (1997).

[161] K. Doll, V.R. Saunders, and N.M. Harrison, *Int. J. Quantum Chem.* **82**, 1 (2001).

[162] K. Doll, *Comp. Phys. Commun.* **137**, 74 (2001).

[163] D. Jacquemin and B. Champagne, *Int. J. Quantum Chem.* **85**, 539 (2001).

[164] K.N. Kudin, G.E. Scuseria, and H.B. Schlegel, *J. Chem. Phys.* **114**, 2919 (2001).

[165] J. Andzelm, R.D. King-Smith, and G. Fitzgerald, *Chem. Phys. Lett.* **335**, 321 (2001).

[166] B. Civalleri, Ph. D'Arco, R. Orlando, V.R. Saunders, and R. Dovesi, *Chem. Phys. Lett.* **348**, 131 (2001).

[167] R. Improta, V. Barone, K.N. Kudin, and G.E. Scuseria, *J. Am. Chem. Soc.* **123**, 3311 (2001).

[168] L. Piseri and G. Zerbi, *J. Mol. Spectrosc.* **26**, 254 (1961).

[169] J.Q. Sun and R.J. Bartlett, *J. Chem. Phys.* **109**, 4209 (1998).

[170] S. Hirata and S. Iwata, *J. Chem. Phys.* **108**, 7901 (1998).

[171] S. Hirata and S. Iwata, *J. Mol. Struct. (THEOCHEM)* **451**, 121 (1998).

[172] see for instance C.X. Cui and M. Kertész, *J. Chem. Phys.* **93**, 5257 (1990); C.X. Cui and M. Kertész, *Macromolecules* **25**, (1992); M. Kofranek, H. Lischka, and A. Karpfen, *J. Chem. Phys.* **96**, 982 (1992).

[173] D. Jacquemin, J.M. André, and B. Champagne, *J. Chem. Phys.* **118**, 3956 (2003).

[174] P. Otto, A. Martinez, and J. Ladik, *J. Chem. Phys.* **111**, 6100 (1999).

[175] S. Suhai, *Int. J. Quantum Chem.* **S18**, 161 (1984).

[176] S.Y. Hong and M. Kertész, *Phys. Rev. B* **41**, 11368 (1990).

[177] F. Bartha, F. Bogar, A. Peeters, C. Van Alsenoy, and V. Van Doren, *Phys. Rev. B* **62**, 10142 (2000).

[178] M.L. Zhang, M.S. Miao, A. Peeters, C. Van Alsenoy, J.J. Ladik, and V.E. Van Doren, *Solid State Commun.* **116**, 339 (2000).

[179] D. Jacquemin, J.M. André, and B. Champagne, *J. Chem. Phys.* **118**, 373 (2003).

[180] A.A. Hansen, *Mol. Phys.* **34**, 1473 (1977); W. Hug, *Handbook of Vibrational Spectroscopy*, J.M. Chalmers and P.R. Griffiths (eds) (Wiley, Chichester, 2002) **1**, 745.

[181] M. Cossi, *Chem. Phys. Lett.* **384**, 179 (2004).

[182] D.M. Bishop, M. Hasan, and B. Kirtman, *J. Chem. Phys.* **103**, 4157 (1995); J.M. Luis, M. Duran, B. Champagne, and B. Kirtman, *J. Chem. Phys.* **113**, 5203 (2000); O. Quinet, B. Kirtman, and B. Champagne, *J. Chem. Phys.* **118**, 505 (2003).

[183] J. Hofkens, M. Cotlet, T. Vosch, P. Tinnefeld, K.D. Weston, G. Ego, A. Grimsdale, K. Müllen, D. Beljonne, J.L. Brédas, S. Jordens, G. Schweitzer, M. Sauer, and F. De Schryver, *Proc. Natl Acad. Sci. USA* **100**, 13146 (2003); R. Gronheid, A. Stefan, M. Cotlet, J. Hofkens, J. Qu, K. Müllen, M. Van der Auweraer, J.W. Verhoeven, and F.C. De Schryver, *Angew. Chem. Int. Edn* **42**, 4209 (2003).

[184] P.S. Asirvatham, V. Subramanian, R. Balakrishnan, and T. Ramasami, *Macromolecules* **36**, 921 (2003).

[185] F. Mauri, B.G. Pfrommer, and S.G. Louie, *Phys. Rev. Lett.* **77**, 5300 (1996).

[186] S. Hammes-Schiffer, *J. Chem. Phys.* **101**, 375 (1994); B. Delley, *J. Chem. Phys.* **113**, 7756 (2000); P. Fulde, *Adv. Phys.* **51**, 909 (2002).

[187] E.B. Starikov, *Int. J. Quantum Chem., Quantum Biol. Symp.* **22**, 145 (1995); E.B. Starikov, *Int. J. Quantum Chem.* **68**, 421 (1998).

[188] J. Delhalle and J.L. Calais, *Int. J. Quantum Chem.* **S21**, 115 (1987); J.L. Calais and J. Delhalle, *Int. J. Quantum Chem.* **42**, 35 (1992); I. Flamant, J. Delhalle, and J.G. Fripiat, *Int. J. Quantum Chem.* **63**, 709 (1997).

[189] R.S. Day and F. Martino, *Chem. Phys. Lett.* **84**, 86 (1981).

[190] M. Seel, G. Del Re, and J. Ladik, *J. Comput. Chem.* **3**, 451 (1982); P. Otto, A.K. Bakhshi, and J. Ladik, *J. Mol. Struct. (THEOCHEM)* **135**, 209 (1986); M.A. Abdel-Raouf, P. Otto, J. Ladik, and M. Seel, *Phys. Rev. B* **40**, 1450 (1989); M. Seel, *J. Mol. Struct. (THEOCHEM)* **188**, 381 (1989).

[191] B. Kirtman and C. de Melo, *J. Chem. Phys.* **75**, 4592 (1981); B. Kirtman and C.E. Dykstra, *J. Chem. Phys.* **85**, 2791 (1986); B. Kirtman and C. de Melo, *J. Chem. Phys.* **86**, 1624 (1987); B. Kirtman and C.E. Dykstra, *J. Chem. Phys.* **90**, 7251 (1989); C.E. Dykstra and B. Kirtman, *A. Rev. Phys. Chem.* **41**, 155 (1990); K.A. Robins and B. Kirtman, *J. Chem. Phys.* **99**, 6777 (1993); B. Kirtman, *Topics in Current Chemistry, Localization and Correlation—a Tribute to Ede Kapuy*, P. Surjan (ed.) (Springer, Berlin, 1999), Vol. 203, p. 147.

[192] A. Imamura, Y. Aoki, K. Maekawa, *J. Chem. Phys.* **95**, 5419 (1991).

[193] Y. Aoki and A. Imamura, *J. Chem. Phys.* **97**, 8432 (1992); Y. Aoki, S. Suhai, and A. Imamura, *J. Chem. Phys.* **101**, 10808 (1994).

[194] M. Mitani, Y. Aoki, and A. Imamura, *Int. J. Quantum Chem.* **64**, 301 (1997); Y. Kurihara, Y. Aoki, and A. Imamura, *J. Chem. Phys.* **108**, 10303 (1998); F.L. Gu, Y. Aoki, A. Imamura, D.M. Bishop, and B. Kirtman, *Mol. Phys.* **101**, 1487 (2003).

2

QUANTUM-CHEMISTRY-BASED FORCE FIELDS FOR POLYMERS

GRANT D. SMITH AND OLEG BORODIN
University of Utah, Salt Lake City, Utah, USA

CONTENTS

2.1.	The Classical Potential Function for Polymers	48
	2.1.1. The Form of the Potential	48
	2.1.2. Existing Potentials	51
	2.1.3. Sources of Data for Force Field Parameterization	51
	2.1.4. Determination of Partial Atomic Charges and Dipole Polarizability	54
	2.1.5. Determination of Dispersion and Repulsion Parameters	59
	2.1.6. Parameters for Covalent Bonds	63
	2.1.7. Valence Bends	64
	2.1.8. Dihedral Potential	66
	2.1.9. Improper Torsions	68
	2.1.10. Parameterization Algorithm	68
2.2.	Quantum-Chemistry-Based Force Field and Simulations of 1,4-Polybutadiene	70
	2.2.1. Force Field Needs for PBD	70
	2.2.2. The Quantum Chemistry Data Set for Force Field Parameterization	71
	2.2.3. Force Field Parameterization	73
	2.2.4. A Comment on the Transferability of the Dihedral Potential	76
	2.2.5. Potential Function Validation	77
	2.2.6. The Need for Accurate Potentials	82

Molecular Simulation Methods for Predicting Polymer Properties, edited by Vassilios Galiatsatos.
ISBN 0-471-46481-3   Copyright © 2005 John Wiley & Sons, Inc.

2.3. Quantum-Chemistry-Based Polarizable Force Fields for PEO and
    PEO–LiBF$_4$ 83
    2.3.1. Development of the Many-Body Polarizable PEO
           Force Field   83
    2.3.2. Effect of Polarizability on PEO Structural and Dynamic Properties   84
    2.3.3. Force Field Development Field and MD Simulations
           of PEO–LiBF$_4$ 87
    2.3.4. MD Simulations of PEO–LiBF$_4$ 88

2.1. THE CLASSICAL POTENTIAL FUNCTION FOR POLYMERS

The ability to accurately represent the potential energy of an ensemble of atoms is central to simulations of real materials. Various forms of classical potentials (force fields) for polymers can be found in the literature [1–3]. The form of potential most appropriate for a given polymer depends largely upon the properties of interest to the simulator. Our interest lies in reproducing the static, thermodynamic and dynamic (transport and relaxational) properties of nonreactive polymer systems and, as a consequence, we have extensive experience in parameterization of potentials that accurately reproduce the molecular geometry, nonbonded interactions, and conformational energetics of the polymers we study. As demonstrated in two detailed examples provided in Sections 2.2 and 2.3, we have found that a relatively simple representation of the classical potential energy works remarkably well for these polymer properties. However, parameterization of even simple potentials is a challenging task.

2.1.1. The Form of the Potential

Potentials that can handle chemical reactions [4] are beyond the scope of our discussion. For a nonreactive polymer the classical force field represents the total potential energy of an ensemble of atoms $V(\mathbf{r})$ with positions given by the vector $\mathbf{r}$ as a sum of nonbonded interactions $V^{NB}(\mathbf{r})$ and contributions due to all bond, valence bend, and dihedral interactions:

$$V(\mathbf{r}) = V^{NB}(\mathbf{r}) + \sum_{\text{bonds}} V^{\text{bond}}(r_{ij}) + \sum_{\text{bends}} V^{\text{bend}}(\theta_{ijk}) + \sum_{\text{dihedrals}} V^{\text{tors}}(\phi_{ijkl}) \quad (2.1)$$

The various interactions are illustrated in Figure 2.1. The dihedral (torsional) term also includes improper torsion or out-of-plane bending interactions. This simple force field is generally applicable to organic polymers and is capable of accurately reproducing conformations, structure, thermodynamic properties, and dynamics of polymers as illustrated in Sections 2.2 and 2.3. If very accurate representation of vibrational spectra of the polymer is of primary interest, it will likely be necessary to include cross-terms between bonded interactions [5].

THE CLASSICAL POTENTIAL FUNCTION FOR POLYMERS   49

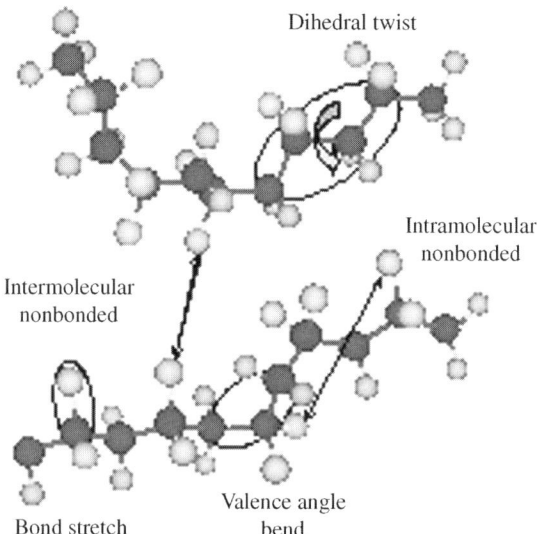

Figure 2.1. Bonded and nonbonded (two-body) interactions arising in a polymer system as represented by a simple classical potential.

Commonly, the nonbonded energy $V^{NB}(\mathbf{r})$ consists of the sum of the two-body repulsion and dispersion energy terms between atoms i and j represented by the Buckingham (exponential-6) potential, the energy due to the interactions between fixed partial atomic or ionic charges (Coulomb interaction), and the energy due to many-body polarization effects:

$$V^{NB}(\mathbf{r}) = V^{pol}(\mathbf{r}) + \frac{1}{2}\sum_{i,j=1}^{N} A_{ij}\exp(-B_{ij}r_{ij}) - \frac{C_{ij}}{r_{ij}^6} + \frac{q_i q_j}{4\pi\varepsilon_0 r_{ij}} \qquad (2.2)$$

Here, q_i is the partial charge on atom i and r_{ij} is the separation between atoms i and j. Nonbonded interactions are typically included between all atoms of different molecules and between atoms of the same molecule separated by more than two bonds. It is not uncommon, however, to scale (reduce in strength) intramolecular nonbonded interactions between atoms separated by three bonds or to excluded them entirely. Therefore, care must be taken in implementing potential functions to ensure that these interactions are treated as intended. Repulsion interactions have the shortest range and typically become negligible at 1.5 σ (atomic radius). We have found that the exponential form more accurately reproduces repulsive interactions than the inverse 12 form of the standard Lennard–Jones potential. Dispersion interactions (forces) are longer range, being significant in magnitude out to distances of around 2.5 σ. The Coulomb interaction is long-range, necessitating use of special summing methods for polar polymers [6, 7].

A further complication arises in polar materials when many-body dipole polarization is taken into account explicitly. In contrast to dispersion–repulsion and Coulomb interactions, the potential energy due to dipole polarization is not pairwise additive and is given by a sum of the interaction energy between the induced dipoles μ_i and the electric field $\mathbf{E}_i^0$ at atom i generated by the permanent charges in the system (q_i), the interaction energy between the induced dipoles, and the energy required to induce the dipole moments [8]:

$$V^{\text{pol}}(\mathbf{r}) = -\sum_{i=1}^{N} \mu_i \cdot \mathbf{E}_i^0 - \frac{1}{2}\sum_{i,j}^{N} \mu_i \cdot \bar{T}_{ij} \cdot \mu_j + \sum_{i=1}^{N} \frac{\mu_i \cdot \mu_i}{2\alpha_i} \quad (2.3)$$

where $\mu_i = \alpha_i \mathbf{E}_i^{\text{tot}}$, α_i is the isotropic atomic polarizability, $\mathbf{E}_i^{\text{tot}}$ is the total electrostatic field at the atomic site i due to permanent charges and induced dipoles, and the second-order dipole tensor is given by

$$\bar{T}_{ij} = \nabla_i \nabla_j \frac{1}{4\pi\varepsilon_0 r_{ij}} = \frac{1}{4\pi\varepsilon_0 r_{ij}^3}\left[\frac{3\mathbf{r}_{ij}\mathbf{r}_{ij}}{r_{ij}^2} - 1\right] \quad (2.4)$$

where $\mathbf{r}_{ij}$ is the vector from atom i to atom j. Because of the expense involved in simulations with explicit inclusion of many-body dipole polarization, we have utilized where possible a two-body approximation for these interactions, as described in Section 2.3.

The contributions due to bonded interactions are represented as

$$V^{\text{bond}}(r_{ij}) = \frac{1}{2}k_{ij}^{\text{bond}}\left(r_{ij} - r_{ij}^0\right)^2 \quad (2.5)$$

$$V^{\text{bend}}(\theta_{ijk}) = \frac{1}{2}k_{ijk}^{\text{bend}}\left(\theta_{ijk} - \theta_{ijk}^0\right)^2 \quad (2.6)$$

$$V^{\text{tors}}(\phi_{ijkl}) = \frac{1}{2}\sum_{n=1,2,\ldots} k_{ijkl}^{\text{tors}}(n)\left[1 - \cos(n\phi_{ijkl})\right]$$

or

$$V^{\text{tors}}(\phi_{ijkl}) = \frac{1}{2}k_{ijkl}^{\text{oop}}(\phi_{ijk})^2 \quad (2.7)$$

Here, r_{ij}^0 is an equilibrium bond length,* θ_{ijl}^0 is an equilibrium valence bend angle and ϕ_{ijkl} is a torsional angle while k_{ij}^{bond}, k_{ijk}^{bend}, $k_{ijkl}^{\text{tors}}(n)$ and k_{ijkl}^{oop} are the bond, bend, torsion, and improper torsion (out-of-plane bending) force constants, respectively. The indices indicate which (bonded) atoms are involved in the interaction.

---

*The equilibrium bond length or bond angle does not necessarily correspond to the bond length or angle of the molecular mechanics optimized geometry. Nonbonded interactions can result in an optimized bond length or angle that differs from the equilibrium value.

THE CLASSICAL POTENTIAL FUNCTION FOR POLYMERS

These geometric parameters and force constants, combined with the nonbonded parameters $q_i, \alpha_i, A_{ij}, B_{ij}$ and C_{ij}, constitute the classical force field for a particular polymer.

2.1.2. Existing Potentials

By far the most convenient way to obtain a force field for any polymer is to utilize an extant one. In general, force fields can be divided into three categories: (a) force fields parametrized based upon a broad training set of molecules such as small organic molecules, peptides, or amino acids, including AMBER [1, 9, 10], COMPASS [11], OPLS-AA [3] and CHARMM [12]; (b) generic potentials such as DREIDING [13] and UNIVERSAL [14] that are not parameterized to reproduce properties of any particular set of molecules; and (c) specialized force fields carefully parametrized to reproduce properties of a specific compound such as water [15–17], polymers [18, 19], polymer aqueous solutions [20] and polymer electrolytes [21]. A procedure for parameterizing the latter class of potential is described below. A summary of the data used in the parameterization of some of the most common force fields is presented in Table 2.1.

In choosing a potential, two major issues must be faced—the quality of the potential and the transferability of the potential. The quality of a potential can be estimated by examining the quality and quantity of data used in parameterizing the potential. For example, AMBER ff99 uses a much higher level of quantum chemistry calculation for determination of dihedral parameters than AMBER ff94 (see Table 2.1). The ability of the force fields to describe the molecular and condensed-phase properties of the training set is another indicator of the force field quality. For example, the OPLS-AA force field for perfluoroalkanes does not correctly predict the splitting of *gauche* and *trans* conformational states obtained from high-level quantum chemistry, despite the fact that the force field has been specifically designed for perfluoroalkanes [22]. The issue of transferability of a potential is faced when a high-quality force field, adequately validated for compounds similar to the one of interest, is used in modeling related compounds not in the training set, or in modeling entirely new classes of materials. Transferability depends tremendously upon the particular potential function parameter, with some parameters being in general quite transferable between similar compounds and others being much less so. In practice, parameterized force fields (AMBER, OPLS, and CHARMM) can work well within the class of molecules they have been parameterized upon. However, when the force field parameters are utilized for compounds similar to those in the original training set but not contained in the training set, significant errors can appear and the quality of force field predictions is often comparable to those found using generic force fields [23]. Much of this difficulty arises from the nontransferability of partial charges and torsional parameters as described below.

2.1.3. Sources of Data for Force Field Parameterization

When the conclusion is reached that existing force fields are inadequate for a given polymer, either because a complete force field for the polymer of interest does not exist, or that existing potentials have not been adequately validated for the polymer

TABLE 2.1. Summary of the Primary Data Used in Force Field Parameterization

Interactions	AMBER [ff94, ff99, ff02]	OPLS-AA	CHARMM	DREIDING
Repulsion–dispersion	PVT, ΔH^{vap}	PVT, ΔH^{vap}	PVT, ΔH^{vap}, crystal structure, QC [scaled HF/6–31G(d)]	Crystal structures and sublimation energies
Electrostatic	QC [HF/6–31G*, HF/6–31G*, B3LYP/cc-pVTZ]	PVT, ΔH^{vap}	QC [HF/6–31G(d)], exp. dipoles	Predictive method [Gasteiger et al.]
Polarization	[N/A, N/A, exp.]	N/A	N/A	N/A
Bond/bend	X-ray structure, IR, Raman	Most values were taken from AMBER[ff94] with some values from CHARMM	IR, Raman, microwave and electron diffraction, X-ray crystal data, QC [HF/6–31G(d) scaled by 0.9] to supplement experimental data on vibrational frequency	All bond and bend force constants were assumed the same for single bonds (700 kcal/mol/Å^2) and 100 (kcal/mol/rad^2, equilibrium bond length and bending angles were obtained from the structures of hydrides Based on hybridization and independent of the particular types involved
Torsion	exp., QC [MP2/6–31G*, MP4/TZP,GVB-LMP2, MP4/TZP,GVB-LMP2]	QC [HF/6–31G*]	Microwave and electron diffraction to determine dihedral term phase and multiplicity, experimental and QC (various levels) gas-phase data on relative conformational energies to determine torsional parameters	
Training set	Peptides, nucleic acids, organics	Organic liquids	Peptides. Special attention was paid to peptide interaction with TIP3P water	Generic

J. Gastriger, M. Marsili, *Tetrahedron*, **36**, 3219 (1980).

of interest, one must undertake the challenging task of parameterizing, or at least partially parameterizing, a new potential function. In order to carry out this task, one requires data against which the force field parameters (or subset thereof) given in Section 2.1.1 can be fit. As can be seen in Table 2.1, there are two primary sources for such data: experiment and *ab initio* quantum chemistry calculations. Whereas *molecular* properties useful for force field parameterization, for example conformational populations, equilibrium molecular geometries, rotational energy barriers, dipole moments, intermolecular binding energies, and molecular polarizabilities, are available from experiment for some polymers and model molecules, experimentally measured structural, thermodynamic, and dynamic data for *condensed phases* (melt and/or crystal) of the polymer of interest or closely related polymers are particularly useful in force field parameterization and validation. In particular, experimentally measured thermodynamic properties of polymer melts and crystals are crucial in parameterizations of dispersion interactions, which, as described below, are difficult to obtain from quantum chemistry.

The second source of information for force field parameterization is quantum chemistry studies of single molecules or molecular clusters. While high-level quantum chemistry calculations are not yet possible on high polymers and condensed phases (particularly amorphous phases), such calculations are feasible on small molecules representative of polymer repeat units and oligomers as well as molecular clusters that reproduce interactions between polymer segments or the interaction of a polymer segment with surfaces, solvents, ions, and so on. These calculations can provide the molecular level information, particularly regarding molecular geometry, electrostatic potential and polarizability, the conformational energy surface, and nonbonded interactions that must be reproduced by a force field that will accurately predict structural, thermodynamic, and dynamic properties of a polymer.

Of key importance in utilizing quantum chemistry calculations on oligomers and molecular clusters for polymer force field parameterization is use of an adequate level of theory. The level of theory must allow for accurate determination of geometries, energetics, electrostatic potential and polarizabilities while remaining computationally viable. As a rule of thumb for compounds consisting of first and second row atoms, we have found that augmented correlation-consistent polarizable basis sets (e.g. aug-cc-pvDz) [24, 25] with geometry optimization at the DFT level (e.g. B3LYP) [26, 27] and determination of energies, dipole moments, polarizabilities and electrostatic potentials using Møller–Plesset second-order perturbation theory (MP2) [28, 29], work quite well. This level of theory often provides molecular dipole moments within a few percent,* molecular polarizabilities within 10 percent, energies of important conformers within ± 0.3 kcal/mol, rotational energy barriers between important conformations within ± 0.5 kcal/mol, and intermolecular binding energies after basis set superposition error (BSSE) correction within 1 kcal/mol. However, whenever force field parameterization for any new class of polymer for which extensive quantum chemistry studies do not exist is undertaken, a comprehensive study of the influence of basis set and

---

*Determined relative to experiment or higher levels of theory.

electron correlation on molecular geometries, conformational energies, cluster energies, dipole moments, molecular polarizabilities, and electrostatic potential should be systematically carried out.

Our philosophy has been to rely heavily on high-level quantum chemistry studies of model compounds and molecular clusters for force field parameterization, and to utilize available condensed-phase experimental data for force field validation and empirical adjustment of parameters, when necessary. In Section 2.1.10 we provide a generic algorithm for force field parameterization for a new polymer. In Sections 2.2 and 2.3 we give examples of parameterization of quantum-chemistry-based potentials as well as pertinent results from simulations utilizing those potentials that serve to illustrate the utility of the quantum-chemistry-based potentials and the need for accurate potentials in simulations of polymers.

2.1.4. Determination of Partial Atomic Charges and Dipole Polarizability

Most polymers are sufficiently polar that Coulomb interactions must be accurately represented in order to adequately reproduce intra- and intermolecular interactions. Important exceptions are polyolefins and perfluoroalkanes. For such largely nonpolar polymers it is often possible to 'subsume' atoms (i.e. hydrogen or fluorine atoms) into their attached carbons, resulting in a very computationally efficient 'united atom' potential. Parameterization of a united atom potential for a nonpolar polymer is considered in Section 2.2. Unfortunately, there have been several attempts to utilize united atom potentials in simulations of polymers in which Coulomb interactions involving hydrogen atoms strongly influence intramolecular interactions (e.g. local conformations) and intermolecular interactions. We are strongly of the opinion that this is dangerous without first carrying out simulations with an accurate all-atom, polar potential and extensively comparing polymer properties obtained with the united atom and all-atom potentials.

For polar polymers it is often sufficient to treat Coulomb interactions with fixed partial atomic charges and neglect explicit inclusion of many-body polarizability. However, for highly polar polymers, or particularly in the presence of charged species (counter ions or salts), it is critical that dipole polarizability be considered either explicitly or through a mean-field two-body approximation. Derivation of potentials for a polar polymer with explicit consideration of dipole polarizability, and the role of polarizability in determining structural, thermodynamic, and dynamic properties of polar polymers, is considered in Section 2.3.

2.1.4.1. Determination of Partial Charges. For nonpolarizable potentials, Coulomb interactions are represented through fixed partial atomic charges, as given by the last term of equation (2.2). Partial charges are single molecule property and therefore can be determined utilizing quantum chemistry and experimental data on single molecules (as opposed to molecular complexes or condensed phases). In parameterization of partial atomic charges, we attempt to represent as accurately as possible the electrostatic potential in the vicinity of model molecules as determined from high-level quantum chemistry calculations. Typically we utilize electrostatic potentials for the most important (highest population) conformations of oligomers that represent one to several repeat units of the polymer. A typical example is shown in Figure 2.2 for 1,2-dimethoxyethane, a model molecule (dimer) for poly(ethylene oxide) (PEO).

The electrostatic potential at a grid of points, typically consisting of 5000–150,000 evenly spaced points spanning distances of 2.5–4.0 Å for C atoms, 1.9–4.0 Å for O, and 1.9–4.0 Å for H, is determined typically at the MP2/aug-cc-pvDz or B3LYP/aug-cc-pvDz level. These distances correspond to the first coordination (or solvation) shell of the corresponding atoms in a typical polymer melt. The electrostatic potential on this grid is then reproduced as accurately as possible by adjusting the partial charges of the various atoms in order to minimize $\chi^2(\mathbf{q})$

$$\chi^2(\mathbf{q}) = \sum_{i=1}^{N_{CONF}} \sum_{j=1}^{N_{GRID}} \omega_{ij} \left[\phi_{ij}^{QC} - \phi_{ij}^{MM}(\mathbf{q}) \right]^2 \qquad (2.8)$$

where $\mathbf{q}$ represents the partial atomic charges, ϕ_{ij}^{QC} and ϕ_{ij}^{MM} are the electrostatic potential for the ith conformer of the model molecule at the jth grid point from quantum chemistry and molecular mechanics, respectively, and ω_{ij} is the weight given this point in the optimization. The weighting of the electrostatic grid depends upon whether it is more desirable to have a better representation of

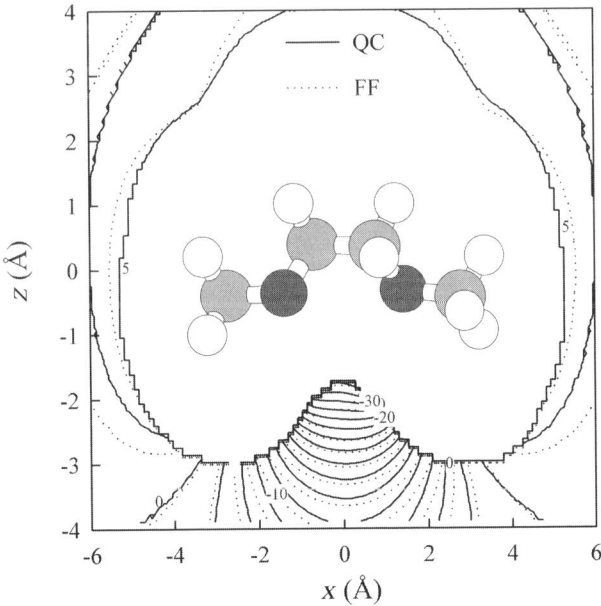

Figure 2.2. Electrostatic potential in the plane of a 1,2-dimethoxyethane molecule as obtained from *ab initio* wave functions at the B3LYP/aug-cc-pvDz level (QC) and partial atomic charges fit to reproduce the quantum chemistry potential (FF). Energies are in kcal/mol.

the electrostatic potential very near the molecule (important when ions are present) or a better representation over the entire nearest-neighbor packing distance. For example, large weight given to points in the proximity of ether oxygen atom improves the description of the electrostatic potential very near the oxygen atoms, which is required for accurate description of PEO–Li^+ interactions (Section 2.3). The optimization is carried out with constraints of fixed total charge for the molecule (usually charge neutral), accurate representation of the molecular dipole moment, that like atoms (e.g. methyl hydrogen atoms of a given methyl group) have the same charge, and that the polymer repeat unit be charge-neutral. Figure 2.2 illustrates the quality of agreement that can be achieved in representing the electrostatic potential with partial atomic charges. Such fits usually yield a root-mean-square deviation of the electrostatic potential between the force field and quantum chemistry of less than 1.0 kcal/mol. Note that in application of the resulting charges in polymer simulations the charge–charge interactions between neighbor and next-neighbor atoms are excluded and that third-neighbor interactions are frequently scaled.

2.1.4.2. Parameterization of the Many-Body Polarizable Model. In some cases, for example in the presence of small charged species (Li^+, Na^+, etc.), a polymer can be highly polarized by its environment. The electrostatic interactions can then no longer be adequately described only by fixed partial charges, and the force field needs to be augmented with additional terms describing polarization of a molecule [equations (2.3) and (2.4)]. The most commonly used polarizable models are: (a) the isotropic atomic polarizability model, in which electrons are assumed to be localized on atoms which are polarized by the external electric field [21, 30]; (b) the bond polarizability model, in which electrons are assumed to be localized between the atoms, comprising a bond which is polarized by the external electric field; and (c) the charge fluctuating model [31], in which partial charges are redistributed among the atoms in response to the electric field.

The bond polarizability model is more computationally expensive than the atomic polarizability model. The computationally efficient charge fluctuating model suffers from artifacts such as the inability to induce out-of-plane dipole moments in planar molecules. The atomic polarizability model does not suffer from these drawbacks and is our model of choice. In implementation of the atomic polarizability model, the issue of intramolecular polarization must be faced. The model has been implemented allowing full intramolecular polarization [32] where polarization interactions between all atoms are considered. Other implementations completely neglect intramolecular polarization interactions [33, 34], that is, the electric field produced by other atoms in the same molecule is not allowed to polarize an atom in that molecule. For polymers the latter approach is not appropriate due to the nonlocal nature of most intramolecular interactions, while the former approach can lead to instabilities resulting from interactions between bonded atoms. In our implementation of the atomic polarization model an atom in a molecule is not allowed to be polarized by the other atoms of the same molecule if it is part of the same bond and bend, while all remaining intramolecular polarization interactions are considered explicitly.

A general problem with parameterization of the atomic polarizability model is the partitioning of molecular polarizability into atomic contributions. Experiments and quantum chemistry calculations yield molecular polarizabilities, whereas atomic polarizabilities are required for the polarizable model. Two approaches have been suggested to solve this problem. In the first approach, quantum chemistry calculations are performed on a training set of molecules containing the atom types of interest and the molecular polarizabilities of these molecules are fitted in order to find a set of atomic polarizabilities that yields the best fit for *all* of these molecular polarizabilities. Transferability of the atomic polarizabilities between chemically similar molecules is assumed. In the second approach, the polarization potential energy around a model molecule is calculated from quantum chemistry at a number of points in the vicinity of the molecule as follows. First, a unit test charge is placed at a point of interest and quantum chemical calculations are performed in order to determine the electrostatic potential energy of the complex, given as the difference between the total energy of the complex (molecule + charge) and the total energy of the molecule alone. This procedure is repeated, this time without allowing polarization of the molecule by the test charge. This consists simply of calculating the electrostatic potential at the point of interest due to the molecule (without the test charge), precisely as is done in determination of partial charges described in Section 2.1.4.2. The difference between these two electrostatic potential energies (with and without test-charged induced polarization) is due to polarization of the molecule by the test charge, $V_{\text{test}}^{\text{QC}}(\mathbf{r})$, which can be straightforwardly represented in the classical potential through application of equation (2.3). For the particularly simple case of negligible intramolecular polarization,

$$V_{\text{test}}^{\text{MM}}(\mathbf{r}) = -\sum_{i=1}^{N} \boldsymbol{\mu}_i \cdot \mathbf{E}_i - \frac{1}{2} \sum_{i,j}^{N} \boldsymbol{\mu}_i \cdot \bar{T}_{ij} \cdot \boldsymbol{\mu}_j + \sum_{i=1}^{N} \frac{\boldsymbol{\mu}_i \cdot \boldsymbol{\mu}_i}{2\alpha_i} \qquad (2.9)$$

where $\mathbf{E}_i$ is the electric field at atom i due to the test charge and $\boldsymbol{\mu}_i = \alpha_i \mathbf{E}_i$. Analogous to the procedure for determining partial charges described above, the atomic polarizabilities are then adjusted to reproduce $V_{\text{test}}^{\text{QC}}(\mathbf{r})$ as accurately as possible for all points of interest by minimizing $\chi^2(\boldsymbol{\alpha})$

$$\chi^2(\boldsymbol{\alpha}) = \sum_{i=1}^{N_{\text{CONF}}} \sum_{j=1}^{N_{\text{GRID}}} \omega_{ij} \left(V_{\text{test}ij}^{\text{QC}} - V_{\text{test}ij}^{\text{MM}}(\boldsymbol{\alpha}) \right)^2 \qquad (2.10)$$

where $\boldsymbol{\alpha}$ represents the set of atomic polarizabilities.

The first approach described above assumes atomic polarizabilities are transferable between similar molecules and hence requires no effort for parameterization for a molecule similar to those in the original training set. However, this approach can yield errors of 10–20 percent in description of the polarization energy, as seen for

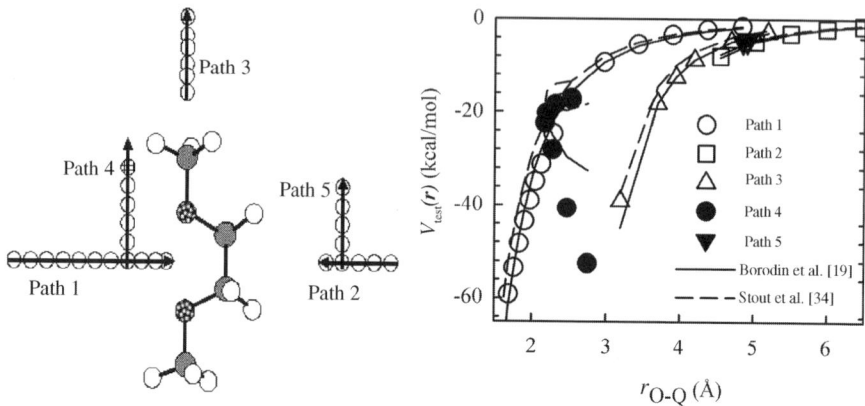

Figure 2.3. Polarization energy along important paths near a 1,2-dimethoxyethane molecule from quantum chemistry calculations at B3LYP/aug-cc-pvDz level (symbols) and from atomic polarizabilities (lines) as a function of distance between the test charge and the nearest oxygen atom.

1,2-dimethoxyethane in Figure 2.3. Note in particular that the description along path 1 at $r_{O-Q} \approx 2$ Å is especially important for PEO–Li-salt systems discussed in Section 2.3, and that using atomic polarizabilities parameterized for molecules similar to 1,2-dimethoxyethane [34] results in an underestimation of the polarization energy along this path. In contrast, the second approach [21] yields a much more accurate description of the polarization energy.

One should also keep in mind that in the polarizable model the electrostatic potential around a molecule is no longer only a function of partial charges. The partial charges on the atoms separated by more than a certain number of bonds (typically two in our model) from a given polarizable atom create an electric field at that atom, resulting in an induction of an atomic dipole moment (intramolecular polarization) that contributes to the electrostatic field around the molecule that should be considered during partial charge fitting. Thus, atomic polarizabilities should be determined before fitting partial charges and the contribution of self-polarization to the electrostatic potential should be included in the fit of partial charges. If significant intramolecular polarization is present, this procedure will need to be performed iteratively as the partial charges will in turn influence the polarization potential energy.

2.1.4.3. Transferability of Partial Atomic Charges and Dipole Polarizabilities.
We have found that atomic dipole polarizabilities are much less sensitive to the identity of neighboring atoms and the particular structure of the chemical repeat unit than partial atomic charges. Hence, atomic dipole polarizabilities parameterized for a given polymer are often transferable to related polymers (however, see discussion in Section 2.1.4.2). Unfortunately, the same does not hold for partial atomic charges. Partial atomic charges are highly dependent upon the bonding environment.

Accurate representation of Coulomb interactions in polar polymers requires parameterization of partial atomic charges for every polymer of interest. This can be accomplished to some degree by use of 'charge recipes' based upon iterative partial equalization of orbital electronegativity [35], but a much more accurate set of partial charges can be obtained from fitting the electrostatic potential obtained from quantum chemistry studies of appropriate model compounds as described above and in Section 2.3. Since the same compounds will most likely be utilized for investigation of conformational energetics (see Section 2.4.8), and parameterization of partial atomic charges is relatively straightforward, there is usually no reason not to obtain and utilize the more accurate quantum-chemistry-based partial atomic charges.

2.1.5. Determination of Dispersion and Repulsion Parameters

Dispersion and repulsion parameters for most modern force fields (AMBER, OPLS-AA, etc.) have been obtained by fitting to experimental PVT data and heat of vaporization of oligomer liquids or structure and sublimation energies of molecular crystals. This approach has been successful in describing the thermodynamic and structural properties of many organic liquids and peptides. If no experimental data are available or are of questionable accuracy, it may be impossible to determine repulsion and dispersion parameters completely based upon empirical data alone. In these cases approximate relations such as the London formula (see below) or quantum chemistry calculations may be the best source of data for completing parameterization of dispersion and repulsion parameters.

Determining accurate dispersion and repulsion parameters from quantum chemistry is challenging. This is particularly true for dispersion parameters, whose parameterization requires very accurate treatment of electron correlation in determining the binding energy of molecular clusters. Carrying out quantum chemistry studies of molecular clusters of sufficient accuracy to allow for final determination of dispersion parameters for any substance more complex than a noble gas or molecules consisting of more than about three heavy atoms is very computationally expensive. In addition, dispersion parameters utilized in condensed phase simulations do not represent true two-body interactions but rather a mean-field (or potential of mean force) representation of two body and higher-body interactions that can alter the value of C_{ij} by 10–20 percent from the true two-body value. Our approach has been to utilize empirical values for repulsion and dispersion parameters where we believe high-quality, validated values exist. Otherwise we have utilized quantum chemistry to establish repulsion parameters and initial values for dispersion parameters that have been subsequently adjusted to reproduce experimental thermodynamic data for condensed phases of oligomers and polymers. Where such data are not available and quantum chemistry calculations are prohibitively expensive, we have utilized the London formula with the assumption that the average excitation energies can be approximated by the ionization potentials in order to obtain a rough estimate (within 30–40 percent) of the dispersion parameters.

Under these assumptions the dispersion parameters are given by:

$$C_{ij} = -\frac{3}{2(4\pi\varepsilon_0)} \frac{IP_i IP_j}{IP_i + IP_j} \alpha_i \alpha_j \quad (2.11)$$

where IP_i and α_i are the ionization potential and polarizability of atom i.

2.1.5.1. Repulsion Parameters from Quantum Chemistry. The binding energy of molecular clusters as a function of configuration (e.g. intermolecular separation), particularly for clusters comprised of two molecules, is very useful for determining repulsion parameters. Using the exponential representation of the interatomic repulsion [equation (2.2)], we can establish repulsion parameters (A_{ij} and B_{ij}) for various atomic pairs based upon the BSSE-corrected Hartree–Fock (HF) energy for such clusters as a function of separation provided that accurate charges and atomic polarizabilities have been previously determined, as illustrated in Figure 2.4. For a cluster

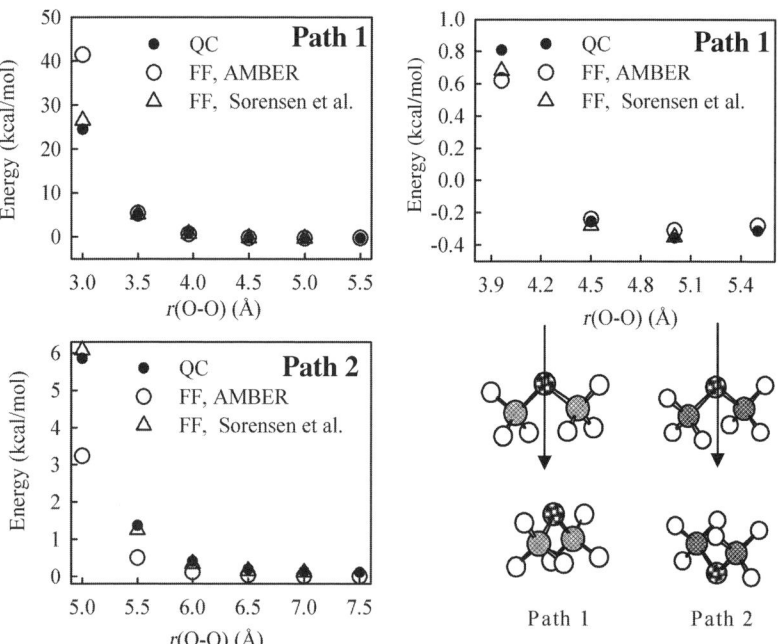

Figure 2.4. The total energy minus dispersion energy of a dimethyl ether cluster as a function of distance between ether oxygen atoms from HF/aug-cc-pvDz quantum chemistry calculations (QC) and two classical potentials. The force field energy is due to electrostatic contributions described by charges and polarizabilities obtained as described in the text and in Borodin and Smith [19] and empirical repulsion parameters from the AMBER [1] and Sorensen et al. [37] force fields. Two scales are shown for path 1.

of two molecules the repulsion energy contribution to the binding energy is given as

$$V^{REP}(\mathbf{r}) = V^{HF}(\mathbf{r}) - V^{COUL}(\mathbf{r}) \tag{2.12}$$

Here $\mathbf{r}$ represents a (fixed) configuration of the cluster and $V^{HF}(\mathbf{r})$ is the HF contribution to the binding energy. $V^{COUL}(\mathbf{r})$ is the molecular mechanics contribution of partial atomic charges and dipole polarizability (if treated explicitly) to the binding energy. Both $V^{COUL}(\mathbf{r})$ and $V^{HF}(\mathbf{r})$ are given as the difference in the respective quantity between the cluster and the sum of the individual molecules with the same molecular geometry as in the cluster. $V^{HF}(\mathbf{r})$ should be corrected for BSSE using the counterpoise method [36]. The corresponding objective function for determination of the repulsion parameters is

$$\chi^2(\mathbf{A}_{ij}, \mathbf{B}_{ij}) = \sum_{k=1}^{N_{CONF}} \omega_k \left[V^{REP}(\mathbf{r}_{CONF}) - \sum_{i=1}^{N_1} \sum_{j=1}^{N_2} A_{ij} \exp(-B_{ij} r_{ij}) \right]^2 \tag{2.13}$$

The double sum includes interactions between intermolecular atomic pairs only and is taken over the N_1 and N_2 atoms of the two molecules. All N_{CONF} configurations of the cluster are considered with corresponding weight ω_k. Typically, the lowest energy configurations are given greater weight in the fitting procedure. We have found that the HF energy for the molecular clusters we have utilized to establish repulsion parameters typically saturates quickly with basis set size, with the aug-cc-pvDz basis set often providing HF binding energies within 0.1 kcal/mol or better of the estimated complete basis limit [19].

We have found that quantum-chemistry-based repulsion parameters for carbon, oxygen and hydrogen atoms obtained as described above agree well with the empirical exp-6 repulsion parameters obtained by fitting polymer crystal structure and energies [37] and from fits to thermodynamic properties of molecular liquids [1], as suggested by Figure 2.4. Some deviation of the AMBER force field from the quantum chemistry data and the empirical exp-6 force field indicates a well-known problem of the inverse-12 potential predicting too steep a repulsion. We conclude that repulsion parameters can be determined with reasonable effort from quantum chemistry calculations, at least as accurately as from fitting to experimental data.

2.1.5.2. Dispersion Parameters from Quantum Chemistry. In principle, the contribution of dispersion (London forces) to the binding energy of a molecular cluster is given as the difference between the BSSE-corrected binding energy determined from quantum chemistry with electron correction, $V^{CORR}(\mathbf{r})$, and that determined at the HF level, that is:

$$V^{DIS}(\mathbf{r}) = V^{CORR}(\mathbf{r}) - V^{HF}(\mathbf{r}) \tag{2.14}$$

with the corresponding objective function

$$\chi^2(\mathbf{C}_{ij}) = \sum_{k=1}^{N_{CONF}} \omega_k \left(V^{DIS}(\mathbf{r}) - \sum_{i=1}^{N_1} \sum_{j=1}^{N_2} C_{ij} r_{ij}^{-6} \right)^2 \quad (2.15)$$

We have found that good estimates (within 10–20 percent) of the dispersion energy contribution to the binding of many molecular clusters can be obtained by making a basis set extrapolation of the dispersion contribution as illustrated in Figure 2.5a and scaling the complete basis set extrapolation energies by the ratio of MP4 energies to MP2 energies using the aug-cc-pvDz basis set. Good agreement of the dispersion energy obtained from the scaled complete basis extrapolated energies with that pre-

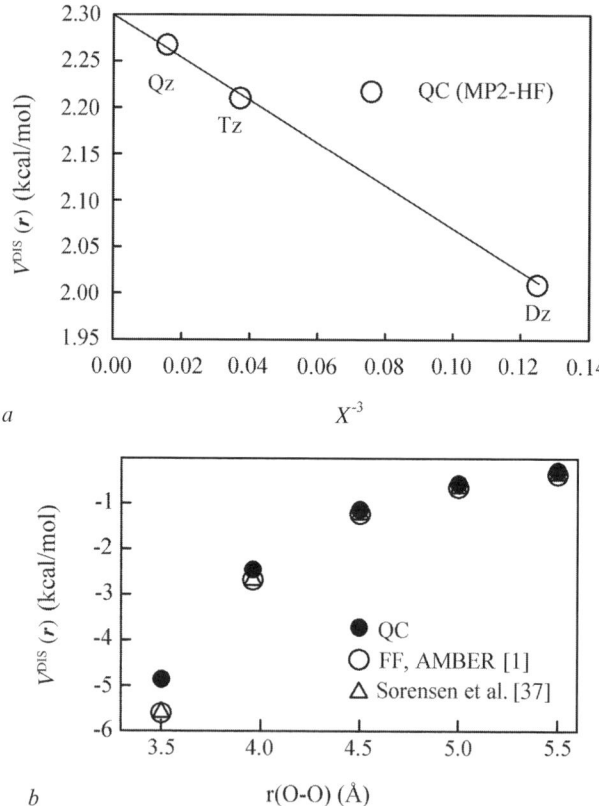

Figure 2.5. (a) X^{-3} extrapolation of the correlation energy $V^{DIS}(\mathbf{r})$ [equation (2.14)] for a dimethyl ether cluster at the minimum energy along path 1 from Figure 2.4. Aug-cc-pvXz basis sets are denoted as Xz, for X = D, T, Q. (b) The dispersion energy for the dimethyl ether cluster along path 1 from quantum chemistry calculations using partial charges and polarizabilities given in Borodin and Smith [19] to determine the electrostatic energy.

dicted by empirical dispersion parameters is shown in Figure 2.5b. The quantum chemistry predictions of the dispersion energy are around 10 percent lower than the dispersion contribution from the empirical force fields. These results agree well with the findings of benchmark calculations on He_2, Ne_2, and Ar_2 that indicated that the BSSE-corrected complete basis set extrapolation of the MP4 complex energies was approximately 10 percent below empirical values [38].

2.1.5.3. Empirical Adjustment of Dispersion Parameters. Regardless of the source of data utilized to parameterize dispersion interactions (experimental thermodynamic and structural data on molecular crystals or liquids, quantum chemistry data on molecular clusters, or polarizabilities, ionization energies and electron affinities for use in the London Formula), it will likely be necessary to make (hopefully) minor empirical adjustments to the dispersion parameters so as to yield highly accurate thermodynamic properties for the condensed-phase polymer. Adjustments as large as ± 10 percent can be anticipated. Density for polymer melts or oligomer liquids and heats of vaporization are particularly useful for this purpose. The cohesive energy density, specific gravity, thermal expansion coefficient, compressibility, heat of vaporization, and surface/interfacial tension are also sensitive to dispersion interactions. In contrast, liquid-phase conformations, structure and dynamics are relatively insensitive to dispersion interactions.

2.1.5.4. Transferability of Dispersion and Repulsion Parameters. As with atomic polarizabilities, we have found that dispersion and repulsion parameters are in general not highly sensitive to the particular bonding environment and molecular structure of the repeat unit, and hence are highly transferable between similar atom types in different polymers. Note that this transferability assumes that the hybridization of the atom has not changed. For example, the dispersion/repulsion parameters (or atomic polarizability) used for an sp^2 carbon atom will not be applicable to an sp^3 carbon atom. Dramatically different bonding environments can also result in the need to reparameterize dispersion and repulsion parameters.

2.1.6. Parameters for Covalent Bonds

For reproduction of most polymer properties of interest, it is important to represent the atomic bond geometry (bond lengths) accurately. However, as long as reasonably large bond force constants are employed (i.e. the bonds are reasonable stiff), the precise value of the bond force constant influences only the frequencies (and amplitudes) of the highest frequency polymer normal modes. Note that because these modes are treated classically each makes a contribution of kT to the heat capacity irrespective of the value of the force constant that would not be realized in a real polymer at reasonable temperature. Additionally, the lack of zero-point vibration in the classical simulation influences the polymer structure factor, but only on length scales comparable to the bond length. Beyond these minor effects, bond stretching does not make an important contribution to the thermodynamics, structure, or dynamics of the polymer. It is quite common to constrain bonds to their equilibrium length during simulations in order to allow for use of a larger integration time step. Care must be taken, however, when bond force constants are artificially lowered in order to allow larger time steps. Here, allowing too much flexibility in the bond lengths can influence conformational energetics and hence chain configurations and conformational dynamics. Similarly, the presence of a (naturally) very flexible bond would necessitate that additional attention be paid to the value of the stretching force constant.

2.1.6.1. Parameterization of Covalent Bonds. In practice, we normally constrain bonds, eliminating the need for parameterization of bond force constants. Otherwise, we take reasonable values for the stretching force constants from the literature. We parameterize equilibrium bond lengths for each type of bond (e.g. $C^{sp^2}-O^{sp^2}$ or $C^{sp^3}-H$), represented by $\mathbf{r}^0$, so as to accurately reproduce the bond lengths of model compounds determined from high-level quantum chemistry. The corresponding objective function is

$$\chi^2(\mathbf{r}^0) = \sum_{i=1}^{N_{\text{CONF}}} \sum_{j=1}^{N_{\text{BONDS}}} \omega_{ij} \left[r_{ij}^{\text{QC}} - r_{ij}^{\text{MM}}(\mathbf{r}^0) \right]^2 \quad (2.16)$$

where the sum is over all bonds of each conformation of the model molecule(s). Here r_{ij}^{QC} and r_{ij}^{MM} are the bond lengths of the jth bond of the ith conformation from quantum chemistry and molecular mechanics, respectively. Note that the latter is obtained from the molecular mechanics-optimized (lowest energy) geometry of the given conformation. There is some flexibility in choosing in the number of bond types in a given molecule. As a rule of thumb, it is simpler to utilize as few bond types as is reasonable (based upon the atoms participating in the bond and their hybridization) and introduce additional refinement of bond types (e.g. based upon next-neighbors) as needed to provide sufficient flexibility to accurately represent the molecular geometry. It should be possible to reproduce all bond lengths within ± 0.01 Å.

2.1.6.2. Transferability of Covalent Bond Parameters. Equilibrium bond lengths are transferable within a few percent [22] between polymers of similar structure and bonding, indicating that, if accuracy of better than a few percent is required, the equilibrium bond length should be reparameterized based upon high-level quantum chemistry calculations. One should also keep in mind that an error of a few percent in equilibrium bond lengths might result in errors of a few percent in the density of a liquid.

2.1.7. Valence Bends

As with covalent bonds, the precise value of the valence bend force constant will not significantly influence properties beyond those associated with the high-frequency bend vibrations. However, the same bending force constants used in parameterization of the dihedral potential (Section 2.1.8 below) must be utilized in subsequent simulations. The bending of backbone valence angles in particular can strongly influence conformational energetics, and hence stiffening or softening valence bends after final parameterization of the dihedral potential can result in conformational energies and particularly rotational energy barriers that are too high or too low. To reiterate, it is critical that values of valence bend force constants and equilibrium valence bend angles be firmly established before final parameterization of the dihedral potential is undertaken, and that these values be used in conjunction with the dihedral parameters in any simulation.

2.1.7.1. Parameterization of Valence Angles. As with covalent bonds, we have taken bending force constants from the literature where available. Where these are not available, we have taken the bending force constants directly from quantum chemistry normal mode frequencies determined for representative model molecules with appropriate scaling of the force constants, ranging from 80 to 98 percent, depending upon the level of theory. In other cases we calculated the energy associated with distortion of various valence bends of representative model compounds from quantum chemistry and have adjusted the force constant so as to reproduce this distortion energy with the classical potential. We parameterize equilibrium bend angles for each type of valence bend (e.g. $C^{sp^3}-C^{sp^2}-O^{sp^2}$ or $H-C^{sp^3}-H$), represented by θ^0, so as to accurately reproduce the bend angles in model compounds as determined from high-level quantum chemistry. The corresponding objective function is:

$$\chi^2(\theta^0) = \sum_{i=1}^{N_{CONF}} \sum_{j=1}^{N_{BENDS}} \omega_{ij} \left[\theta_{ij}^{QC} - \theta_{ij}^{MM}(\theta^0) \right]^2 \quad (2.17)$$

where the sum is over all bends of each conformation of the model molecule(s). Here θ_{ij}^{QC} and θ_{ij}^{MM} are the bend angles of the jth valence bend of the ith conformation of the molecule from quantum chemistry and molecular mechanics, respectively. As with bond lengths, the latter are obtained from the molecular mechanics-optimized geometry of the given conformation. Typically we have found it sufficient to base bond types solely on the identity and hybridization of the atoms participating in the bend. It should be possible to reproduce all important valence angles within $\pm 2°$.

2.1.7.2. Transferability of Valence Angle Parameters. Valence bend force constants and equilibrium angles are generally transferable to like interactions in related polymers, assuming that *the dihedral potential is subsequently parameterized* to accurately reproduce conformational energetics for the polymer of interest. However, relatively small changes in equilibrium angles can influence the dipole moment and electrostatic potential around the molecule. Since the quantum chemistry geometry of the model molecule(s) is likely to have been determined for utilization in parameterization of the partial charges (above) and the dihedral potential (below), it is internally consistent and will provide a better representation of the dipole moment and electrostatic potential if equilibrium bend angles are obtained by fitting to the quantum chemistry geometry of the model molecule(s).

2.1.8. Dihedral Potential

It is crucial that the conformational energies, specifically the relative energies of important conformations and the rotational energy barriers between them, be accurately represented in any simulation of a real polymer. Conformational energetics, in addition to segmental geometry and nonbonded interactions, determine the properties of a specific polymer and differentiate one polymer from another. Hence, without accurate representation of the conformational energetics, one cannot honestly claim to be simulating any particular polymer, nor can one expect to accurately predict the properties of the polymer.

We have found that, as a minimum, a force field must be able to reproduce the relative energies of the important conformations of single dihedrals and dihedral pairs (dyad) in model oligomers, and that where possible the ability of a potential so parameterized to reproduce the conformational energies of larger oligomers should be investigated. When specific intramolecular interactions (e.g. hydrogen bonding) occur, it is necessary to investigate model compounds sufficiently long to include such interactions. As with dispersion energies, the conformational energies and rotational energy barriers obtained from quantum chemistry for model molecules are quite sensitive to the level of theory utilized, both basis set size and treatment of electron correlation. Fortunately, we have found that it is typically not necessary to conduct geometry optimizations with electron correlation (e.g. MP2 level)—for many compounds SCF or DFT geometries are sufficient. Unfortunately, relative conformational energies and rotational energy barriers obtained at the SCF and DFT level are usually not sufficiently accurate, necessitating the calculation of MP2 energies at SCF or DFT geometries.

2.1.8.1. Parameterization of Dihedral Potential. In fitting the dihedral potential, it is sometimes possible to utilize only 1-, 2- and 3-fold dihedral terms [$n = 1, 2$ and 3 in equation (2.7)]. However, we have often found it necessary to utilize up to 6-fold dihedral terms. One must be cognizant of possible artifacts (e.g. spurious minima and conformational energy barriers) that can be introduced into the conformational energy surface when higher-fold terms ($n > 3$) with large amplitudes are utilized. It is best to generate and examine the entire conformational energy surface for single dihedrals and dihedral pairs for model molecules from molecular mechanics after (and perhaps during) parameterization in order to ascertain that such artifacts have not been introduced.

We parameterize the dihedral force constants for each term (typically $n = 1-3$ or $n = 1-6$) of each dihedral type (e.g. $C^{sp^3}-C^{sp^3}-O^{sp^3}-C^{sp^3}$), represented by $\mathbf{k}^{tors}(n)$, so as to accurately reproduce the geometries and energies of the important conformers and rotational energy barriers of the model molecules (oligomers). In addition to these stationary points, we often map out additional points on the conformational energy surface by constraining selected dihedrals to certain angles and optimizing the remaining degrees of freedom. The corresponding objective function is:

$$\chi^2\big[\mathbf{k}^{\text{tors}}(n)\big] = \sum_{i=1}^{N_{\text{CONF}}} w_i \big\{V_i^{\text{QC}} - V_i^{\text{MM}}\big[\mathbf{k}^{\text{tors}}(n)\big]\big\}^2 \\ + \sum_{i=1}^{N_{\text{CONF}}} \sum_{j=1}^{N_{\text{TORS}}} \omega_{ij} \big\{\phi_{ij}^{\text{QC}} - \phi_{ij}^{\text{MM}}\big[\mathbf{k}^{\text{tors}}(n)\big]\big\}^2 \qquad (2.18)$$

Here N_{CONF} represents all low-energy conformers, rotational energy barriers and constrained conformations of the model molecule(s) investigated, V_i^{QC} and V_i^{MM} are the quantum chemistry energy and molecular mechanics energy (from the molecular mechanics optimized geometry) of the ith conformer, respectively, relative to the lowest energy conformer of the molecule, and ϕ_{ij}^{QC} and ϕ_{ij}^{MM} are the quantum chemistry and molecular mechanics (optimized) torsional angle of the jth dihedral of the ith conformer. It is not always necessary (or desirable) to include all possible torsions in the potential function representation of a molecule. Often, omitting 'duplicate' terms that involve the same central bond but difference pendent atoms, for example, representing rotations about a C-CH$_2$-CH$_2$-C bond with a single C-C-C-C dihedral type, reduces the number of unknown parameters, thereby simplifying the fitting procedure without sacrificing the ability of the potential to fit the quantum chemistry potential surface. Reproducing the conformational energy surface (geometry and energies of important conformers and rotational energy barriers) of the model compounds with a classical potential is typically much more difficult than reproducing bond and valence bend geometries. Depending upon the polymer and conformation, dihedral angles for some important conformers can be expected to vary by as much as $\pm 10°$ from quantum chemistry, although for most conformers the fit should be much better. Similarly, the energies for some conformers may vary by as much as ± 0.5 kcal/mol from quantum chemistry, with most conformers and barriers being represented within ± 0.2 kcal/mol.

2.1.8.2. Transferability of Dihedral Potentials.
Dihedral potentials are in our experience *not* transferable between polymers, even closely related polymers. When even minor changes in the chemical structure are made, a new set of partial charges is required, and can result in incorrect conformational energies if dihedral parameters from the original polymer are utilized. Furthermore, differences in charges, bend force constants, geometries and dispersion/repulsion parameters make it impossible to import dihedral parameters from another potential even for an identical polymer. In all cases, the importance of conformational energies and rotational energy barriers and their sensitivity to many of the force field parameters necessitates comparison of the conformational energy surface from quantum chemistry and molecular mechanics for appropriate model compounds. Since such an investigation requires the same quantum chemistry data needed for parameterization of the dihedral potentials, it is recommended to parameterize (or reparameterize) the dihedral potential based upon these data. In summary, it is our philosophy that, once

all other force field parameters that influence conformational energies are established (see the fitting algorithm in Section 2.1.9), the torsional potential should be adjusted to fit the conformational energy surface, and the torsional parameters have little or no intrinsic physical significance (or transferability) in and of themselves.

2.1.9. Improper Torsions

Improper torsions, or out-of-plane bending interactions, occur with planar sp^2 hybridization. For a given sp^2 center there are four (equivalent) out-of-plane interactions. We typically set the force constants for three of these to zero and represent the out-of-plane bending interaction with a single term. Where high-quality values are available we utilize out-of-plane bending force constants from the literature. When necessary, we parameterize force constants for out-of-plane bending interactions based upon quantum chemistry energies for distortion of the equilibrium planar geometry of the molecule at the sp^2 center of interest. As with bends, the precise value of the out-of-plane bending constant is not of critical importance, but needs to be established before final parameterization of the dihedral potential. Our experience indicates that out-of-plane bending force constants are highly transferable.

2.1.10. Parameterization Algorithm

We recommend the following algorithm for systematic parameterization and validation of potential functions for polymers. As can be seen in Sections 2.2 and 2.3, the idiosyncrasies of real materials make it expedient to vary the procedure depending upon needs, computational resources, and available experimental and quantum chemistry data.

2.1.10.1. Determine Force Field Parameterization Needs. First, one should establish the quality of existing potentials. Have they been parameterized for the polymer of interest? Have they been parametrized for closely related polymers? What data (quantum chemistry and experiment) were used in the parameterization? How well are these data reproduced by the potential? What validation steps have been carried by the originators of the potential or by others who have utilized the potential?

The second step involves determining what force field parameters are missing or need reparameterization. Parameters that are not highly transferable and are likely to need parameterization are partial charges and dihedral parameters. Equilibrium bond lengths and angles are more transferable but should be parameterized based upon quantum chemistry studies of model compounds if these are needed for other uses (validation or parameterization). While atomic polarizabilities (for many-body polarizable potentials) and dispersion–repulsion parameters are generally transferable, they may not exist (or be validated) for the system of interest. For bond, bend and improper torsion force constants, reasonable existing values are generally applicable, but it is possible that new interaction types not previously parameterized will be encountered.

2.1.10.2. Establish the Quantum Chemistry Data Set for Force Field Parameterization. The next step is to determine the set of model molecules(s) to be utilized in the force field parameterization. If dispersion–repulsion parameters are needed, this will include molecular complexes containing the intermolecular interactions of interest. Oligomers containing all single conformations and conformational pairs extant in the polymer should be included. Larger oligomers should be investigated if computationally feasible. A search for existing quantum chemistry studies of these and related molecules should be conducted before undertaking quantum chemistry calculations. When a new class of polymer/model molecule (one for which extensive quantum chemistry studies have not yet been conducted) is being investigated, the influence of basis set and level of theory on conformational energies, geometries, dipole moment and electrostatic potential for single molecules and binding energies for clusters should be systematically investigated. Comparison with experiment (binding energies, molecular geometry, conformational energies, etc.) can also help establish what level of theory is adequate.

Initial geometry optimization should be performed at a modest level of theory to determine important conformations/configurations of single molecules/clusters as needed. These initial geometries provide insight into the conformational energy surface of the polymer as well as starting geometries for higher-level calculations. Once the level of theory is established, all important conformers and rotational energy barriers for the model molecule(s), as well as dipole moments and the electrostatic potential around the lowest energy conformers should be determined. BSSE-corrected binding energies for important configurations of the molecular clusters should also be determined.

2.1.10.3. Parameterize and Validate Force Field. Parameterization should be carried out in the following order (refer to Sections 2.1.4 to 2.1.9 for detailed procedures). First, partial atomic charges and atomic polarizabilities should be established, followed by parameterization of dispersion–repulsion parameters as needed. Equilibrium bond lengths and angles, and corresponding force constants (bend, bond, and improper torsion, as needed) should be determined using the best current estimate of the nonbonded and torsional parameters. Next, the torsional parameters should be determined with the best current estimate of bend, bond, improper torsion, and nonbonded parameters. The bonded parameters (bends, bonds, torsions, improper torsions) should be determined iteratively using this procedure until no further improvement (agreement with quantum chemistry) is possible.

Simulations utilizing the current version of the potential (resulting from the above iterative procedure) should be conducted in order to investigate the quality of the dispersion parameters by comparing thermodynamic properties (e.g. liquid density as a function of temperature, heats of vaporization) of the model molecules or longer oligomers with available experimental data. Empirical adjustments (hopefully 10 percent or smaller) to the dispersion parameters should be made as needed.

The bonded potential should then be reparameterized using the new values of the dispersion parameters if these have been empirically adjusted. Finally, the potential should be validated through extensive comparison of structural, thermodynamic, and dynamic properties obtained from simulations of oligomers and polymers with available experimental data.

2.2. QUANTUM-CHEMISTRY-BASED FORCE FIELD AND SIMULATIONS OF 1,4-POLYBUTADIENE

The dynamics and relaxational behavior of 1,4-polybutadiene (PBD) have been the subject of extensive experimental study in recent years. PBD is a good glass former, and its simple chemical structure, narrow molecular weight distribution, and wide variety of available microstructures make it ideal for investigations of the glass transition as well as subglass and high-temperature dynamics. These properties, and an extensive experimental data base, make PBD attractive for simulation studies. We have performed extensive simulation and experimental studies of the relaxation behavior of PBD above the glass transition temperature (PBD melt) [39–45] using a quantum-chemistry-based potential, emphasizing comparison with experiment for local, segmental and chain dynamics, and the elucidation of relaxational mechanisms. Here we describe the parameterization of a potential for PBD as an example of a quantum-chemistry-based potential for a nonpolar polymer, and present salient results that demonstrate the quality of the quantum chemistry based potential and emphasize the need for high-quality potentials.

2.2.1. Force Field Needs for PBD

Following the procedure outlined in Section 2.1.10, we begin by establishing the force field needs for PBD. Because PBD is essentially nonpolar, Coulomb interactions contribute insignificantly to intermolecular and intramolecular nonbonded interactions. Hence, it is not necessary to utilize partial atomic charges in the classical force field representation of PBD and it should be possible to utilize a united atom potential for PBD where the hydrogen atoms are subsumed into their attached carbons. Our goal was to parameterize a united atom potential for PBD that accurately reproduces the conformational energetics of the polymer. While several united atom potentials exist for PBD [46, 47], none of them have been extensively validated against high-level quantum chemistry data. As shown below, the extant potentials do not provide a good representation of the conformational energy surface for PBD.

Where possible, we utilized existing bonded and nonbonded parameters in our PBD potential. Bond lengths were constrained in our simulations, eliminating the need for covalent bond force constants. Valence angle bending force constants, specifically force constants for the CH_2-CH_2-CH and $CH_2-CH-CH$ bends, were taken from an extant united atom potential for PBD [46]. The CH_2-CH_2 dispersion–repulsion parameters were taken from our united atom potential for polyethylene [48]. Note that a Lennard–Jones form of the dispersion–repulsion interactions was utilized in these simulations. These and all quantum-chemistry-based potential parameters, parameterized as described below, can be found in Smith and co-workers [39, 40]. Remaining to be parameterized for the PBD united atom potential were CH_2-CH and $CH-CH$ dispersion–repulsion parameters, equilibrium bond lengths and angles, and dihedral parameters for the α (*cis*), α (*trans*) and β dihedrals, as illustrated in Figure 2.6. Because of the subsuming of hydrogen atoms into their attached carbons, improper torsion (out-of-plane bending) interactions centered at C^{sp^2} atoms are eliminated.

2.2.2. The Quantum Chemistry Data Set for Force Field Parameterization

2.2.2.1. Model Molecules and Level of Theory.
In PBD, conformational dynamics involve rotations about allyl CH=CH–CH_2–CH_2 (α) bonds and CH-CH_2–CH_2-CH (β) bonds, illustrated in Figure 2.6. A major component of our force field parameterization effort involves obtaining accurate energies for conformers and rotational energy barriers for these dihedrals, and then representing these energies with the classical potential function. As model compounds for investigating rotations about the α bond we have chosen *cis*-2-pentene and *trans*-2-pentene. For rotations about the β bonds, and for correlations between the α and β bonds, we have investigated 1,5-hexadiene. For possible coupling effects across a *cis* double bond we have investigated *cis*-3-hexene. Note that the allyl (α) dihedrals have low energy

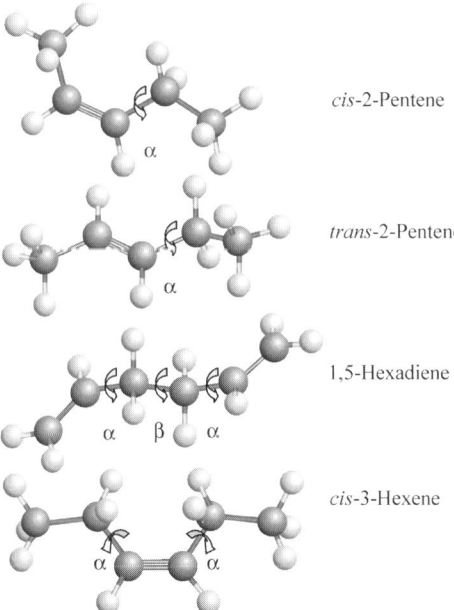

Figure 2.6. Model compounds for 1,4-polybutadiene. The backbone torsions are labeled.

conformations in the *cis* (*cis* = 0°) and *skew* ($s^\pm = \pm 120°$) states as opposed to the *trans* ($t = 180°$) and *gauche* ($g^\pm = \pm 60°$) states for alkyl dihedrals (e.g. butane).

Previous quantum chemistry studies of the energetics of allyl bond rotations have concentrated on 1-butene [49–51] and substituted 1-butene [50–52], including *trans*-2-pentene [50]. For the latter, only the *skew–cis* energy difference was determined. In these studies, geometry optimizations were performed at modest levels of theory. We could find no studies of the conformational energies in *cis*-2-pentene, 1,5-hexadiene or *cis*-3-hexene.

In a quantum chemistry study of *n*-butane and *n*-hexane [53] we demonstrated that inclusion of electron correlation effects in geometry optimizations is important in obtaining accurate geometries and energies for conformers and rotational energy barriers in simple hydrocarbons. While we found that reasonably accurate values of conformer energies could be obtained with modest basis sets, such as 6–31G*, accurate values for the rotational energy barriers required additional diffuse and polarization functions in the atomic basis sets. We showed that quite accurate geometries and energies for conformers and rotational energy barriers in *n*-alkanes (conservatively, ± 0.2 and ± 0.5 kcal/mol for conformers and barriers, respectively) are obtained at the MP2/6–311G** level. Consequently, we have performed our quantum chemistry energy and geometry calculations on the PBD model compounds at the MP2/6–311G** level.

2.2.2.2. Quantum Chemistry Calculations on Model Molecules. A complete set of geometries, conformational energies and rotational energy barriers for the PBD model compounds (Figure 2.6) can be found in Smith and co-workers [39, 40]. The conformational energy as a function of the α dihedral for the 2-pentene compounds and for the β dihedral in $s^+\beta s^+$ conformations of 1,5-hexadiene is shown in Figure 2.7. In addition to the conformational minima and rotational energy barriers, these figures also show results of quantum chemistry optimizations obtained with constrained dihedral angles for the α (2-pentene) and β (1,5-hexadiene) bonds.

Comparison of calculated relative conformer energies and rotational energy barriers with experiment is useful for establishing the accuracy of the quantum chemistry calculations. In contrast to *n*-alkanes, the conformational energetics of simple alkenes have been the subject of only limited experimental investigation. From the temperature dependence of the IR spectrum of 1,5-hexadiene [54], the *gauche* state of the β bond has been estimated to be about 0.2 kcal/mol higher in energy than the *trans* state. This is in quite good agreement with our quantum chemistry energies for the $s^+g^\pm s^+$ conformers of 1,5-hexadiene. We could find no experimental data for 2-pentene. For 1-butene, values of 1.60 kcal/mol [55] and 2.12 kcal/mol [56] have been reported for the *t* (or $s^+\_s^-$) barrier relative to the skew conformer. Experimental estimates of 0.15 kcal/mol [55] and −0.43 kcal/mol [56] have been reported for the energy of the *cis* conformer in 1-butene. For the purpose of comparison with these data, we performed MP2/6–311G**-level calculations on 1-butene. We obtain energies of 0.41 and 2.36 kcal/mol for the *cis* conformer and *t* (or $s^+\_s^-$) barrier, respectively. Given the paucity of experimental data, and the large

2.2.3. Force Field Parameterization

At this point we are prepared to parameterize a quantum-chemistry-based united atom potential for PBD. The question remains as to whether a united atom model can accurately represent conformational energetics and intermolecular interactions that fundamentally determine the static and dynamic properties of a polymer melt. In previous work, we have clearly demonstrated the ability of a united atom

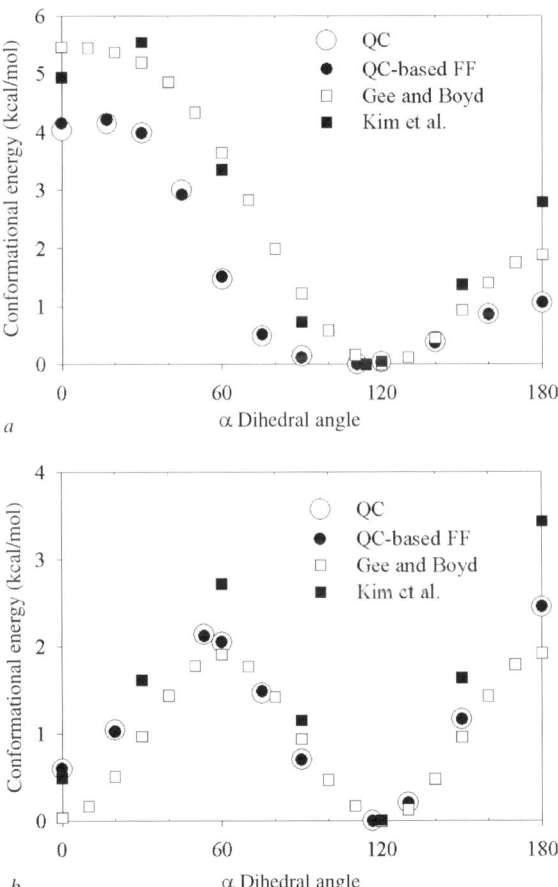

Figure 2.7. Conformational energies of (*a*) *cis*-2-pentene as a function of the α dihedral angle, (*b*) *trans*-2-pentene as a function of the α dihedral angle and (*c*) $s^+\beta s^+$ 1,5-hexadiene as a function of the β dihedral angle as obtained from quantum chemistry and various force fields. Gee & Boyd ref. 46, Kim et al. ref. 47.

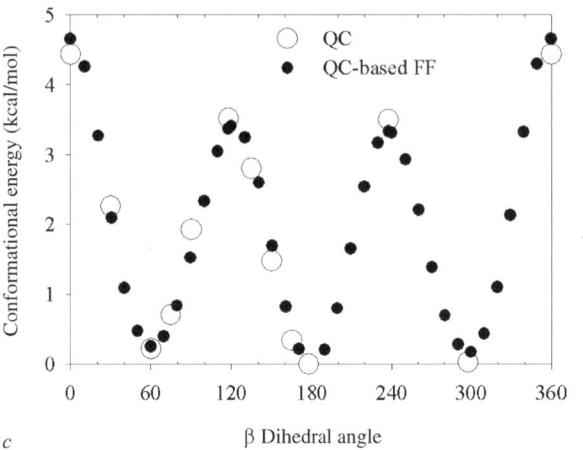

Figure 2.7. *Continued.*

force field for polyethylene to accurately reproduce a wide variety of static and dynamic properties of this polymer [48]. On the other hand, we have shown that for poly(ethylene oxide) an all-atom model is required in order to reproduce the conformations of the polymer and important short-range intermolecular interactions [57]. The primary difference in these systems lies in the fact that important specific nonbonded, primarily electrostatic, interactions in PEO between hydrogen atoms and oxygen atoms strongly influence the intermolecular structure and the conformational energies. However, hydrocarbons such PBD are essentially nonpolar and therefore we do not expect important, specific intermolecular interactions to be manifest in either polymer which would be subsequently lost in a united atom representation, nor have we seen evidence of this in the case of polyethylene.

2.2.3.1. Parameterization of Dispersion–Repulsion Parameters. In our united atom representation of PBD we exclude all 1,4 intramolecular nonbonded interactions. As a consequence, CH_2-CH nonbonded interactions do not arise in the model compounds and hence can be parameterized independent of the bonded interactions (and vice versa). Determination of the CH_2-CH dispersion–repulsion parameters is considered in Section 2.2.3.4. As an initial guess we utilized the CH_2-CH_2 nonbonded potential to describe $CH-CH$ dispersion–repulsion interactions. The $CH-CH$ potential was subsequently adjusted to optimize representation of conformational energies as discussed in Section 2.2.3.4.

2.2.3.2. Equilibrium Bond Lengths and Valence Bond Angles. The geometry parameters for bonds and angles were determined as described in Section 2.1.6 and 2.1.7, utilizing the model compounds shown in Figure 2.6. The classical potential [39, 40] represented well all angles and bond lengths.

2.2.3.3. Dihedral Parameters. The dihedral potential for the α and β bonds were represented by a six-term cosine series [equation (2.7)] as described in Section 2.1.8. For the double bond, we initially employed parameters taken from the united atom force field [46]. However, the barrier for rotation of 14.0 kcal/mol yielded by this model is insufficient to preclude *cis/trans* isomerization at higher melt temperatures. Therefore, we performed quantum chemistry calculations on 2-butene to obtain a reasonable estimate of the energy of rotation about the double bond, which was subsequently fit by a 2-fold potential sufficient to preclude *cis/trans* isomerization.

The torsional potentials for the α and β dihedral were parametrized so as to best reproduce the conformational energies of *cis*-2-pentene, *trans*-2-pentene and 1,5-hexadiene as obtained from quantum chemistry, as illustrated in Figure 2.7. As we consider intramolecular nonbonded interactions only between centers separated by four or more bonds, the only nonbonded interaction involved in the 2-pentene compounds is that between the end methyl groups. In a polymer, these will be methylene groups, and are represented as such in our united atom model for 2-pentene. A comparison between the classical potential and quantum chemistry for the energies and geometries of all conformers and barriers for the PBD model compounds can be found in Smith and co-workers [39, 40]. Agreement between the force field and quantum chemistry is excellent. Also shown in Figure 2.7 are the conformational energies of 2-pentene yielded by the force fields used in previous simulations of PBD. While qualitatively in agreement with the quantum chemistry results, these force fields show important quantitative variations between themselves and are in quite poor agreement quantitatively with quantum chemistry. These discrepancies clearly illustrate the need for a new potential for use in simulations of PBD if we desire to make quantitative comparisons with experiment.

The energies and geometries of the $s^+\beta s^+$ conformations of 1,5-hexadiene allow us to parameterize the torsional potential for the β bond. While these conformational energies, plotted in Figure 2.7c, do depend upon the $CH-CH$ nonbonded potential, the geometry of these conformations is such that the dependence is fairly weak. As an initial guess we used the CH_2-CH_2 nonbonded potential for $CH-CH$ nonbonded interactions, then parameterized the β torsional potential to best match quantum chemistry energies and geometries of 1,5-hexadiene.

2.2.3.4. CH–CH and CH_2–CH Nonbonded Parameters. The $(\beta\alpha)\ g^{\pm} cis$ conformations involve important second-order effects. The ability of our force field to reproduce these effects is reflected in its ability to reproduce the energies of the s^+g^+cis and s^+g^-cis conformers of 1,5-hexadiene. These conformers involve steric interference between hydrogen atoms, and the conformer geometries reflect distortions of the dihedral so as to minimize these interactions. However, by adjusting the $CH-CH$ nonbonded potential, we can do a credible job in reproducing these effects in a united atom representation. The torsional parameters for the β bond were subsequently reparameterized to reflect the updated $CH-CH$ nonbonded potential. The agreement between quantum chemistry and the final force field energies for all calculated stationary points of the model compounds is good for all important conformers [39, 40], and differences are within the uncertainties in the quantum chemistry energies.

Finally, we parametrized the CH_2–CH nonbonded potential so as to reproduce the density of a melt of PBD as a function of temperature. For this purpose, molecular dynamics simulations were performed on a melt of poly($cis_{0.5}$-r-$trans_{0.5}$-butadiene) at 298, 373 and 413 K. Experimental [58] and simulation densities are shown in Figure 2.8 as a function of temperature. Good agreement between experiment and simulation can be seen.

2.2.4. A Comment on the Transferability of the Dihedral Potential

Based upon our experiences in parameterizing the dihedral potential for the allyl bond in *cis*-2-pentene and *trans*-2-pentene, a comment on the transferability of the force field is in order. Previous efforts to parameterize force fields for the allyl bond based on quantum chemistry calculations have employed relatively low level quantum chemistry calculations on 1-butene [51, 52]. For these force fields, and for those employed previously in simulations of PBD, the same torsional potential is employed for allyl bonds adjacent to *cis* double bonds as is used to describe allyl bonds adjacent to *trans* double bonds. Our quantum chemistry calculations show that the energies of the $s^{\pm}tcis$ conformer of 1,5-hexadiene and the *cis* conformer of 1-butene are nearly the same as that for the *cis* conformer of *trans*-2-pentene, indicating that the former two molecules are reasonable model compounds for the allyl bond adjacent to a *trans* double bond. However, the parameterized dihedral potential for the allyl bond in *cis*-2-pentene is significantly different from that for *trans*-2-pentene. Hence, for the united atom model, a dihedral potential parametrized for an allyl bond based upon *trans*-2-pentene (or 1-butene) cannot be used to describe an allyl bond adjacent to a *cis* double bond. Hence, our conclusion is

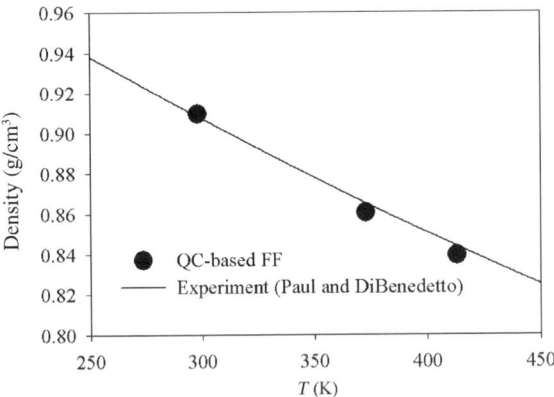

Figure 2.8. Density of a 1,4-polybutadiene melt as a function of temperature from experiment and simulation. Simulation values include a 1 percent increase in density over the actual simulation densities in order to account for molecular weight effects.

that the conformational characteristics of the allyl bond are fundamentally different, depending upon whether the adjacent double bond is *trans* or *cis*. This difference cannot be represented simply through differences in the intramolecular nonbonded interactions, as is assumed implicitly in the other force fields discussed here.

2.2.5. Potential Function Validation

2.2.5.1. Details of the Molecular Dynamics Simulations. We have utilized the quantum chemistry based united atom potential described above in simulations of a PBD melt over wide range of temperatures from those well above the glass transition temperature to those approaching the glass transition temperature. We have investigated thermodynamic properties, structural properties, transport properties and dynamics (relaxations) on length scales ranging from those of the individual dihedral to overall chain dynamics. Relaxation phenomena investigated include dihedral dynamics, dielectric relaxation, normal mode relaxation, viscoelastic response, NMR spin–lattice relaxation, and the single-chain dynamic structure factor. These properties have been extensively compared with experiments, many of which were performed for the purpose of head-to-head comparisons with our simulations. These comparisons have served to validate the quantum chemistry based potential and have provided important insights into relaxation mechanisms in PBD that would have been possible from experiment alone. Furthermore, confidence in the accuracy of our potential provided by this extensive validation allows us to extract mechanistic information from simulations that cannot be verified directly through experiment and utilize this information in developing a picture of relaxation phenomena and the interrelationship of polymer dynamics, structure and heterogeneity on various length and time scales. These studies have also led to important insights into why popular theories of polymer dynamics fail to accurately describe dynamics and relaxation in PBD.

Detailed descriptions and results of our extensive simulations of the PBD melt can be found in Smith and co-workers [39–45]. Here we present a few salient results that serve to illustrate the accuracy of the quantum-chemistry-based potential and the need for accurate potentials in simulations. All simulations were conducted on an ensemble of 40 PBD chains. The chains are statistical copolymers of 1,4-*cis*, 1,4-*trans* and 1,2-vinyl units.* The copolymer chains were generated to reproduce the copolymer statistics and molecular weight of a sample of anionically polymerized PBD that was utilized for most of the experiments considered below. This sample, as well as the simulation chains generated to represent the material, will be referred to as PBD-1600, reflecting the polymer molecular weight. Hence, the simulation chains represent as accurately as possible the molecular structure, molecular weight, geometry, conformational energetics, and nonbonded interactions manifested in the real polymer melt. Simulations were performed using the molecular simulation package Lucretius (www.che.utah.edu/~gdsmith/mdcode/main.html). The simulations were performed at 1 atm pressure over temperatures

---

*The quantum chemistry based potential for the 1,2-vinyl unit is presented in Smith et al. [40].

ranging from 253 to 500 K, with simulation times of 30 up to 200 ns, depending upon temperature.

2.2.5.2. Static Structure Factor and Radius of Gyration.

The static structure factor for a deuterated PBD melt obtained from neutron scatting [59] and simulation using our quantum-chemistry-based potential are compared in Figure 2.9. From simulation the static structure factor is determined using the relationship:

$$S(q) = \frac{\left\langle \sum_{i,j=1}^{N} b_i b_j \frac{\sin q |\mathbf{r}_j - \mathbf{r}_i|}{q |\mathbf{r}_j - \mathbf{r}_i|} \right\rangle}{\sum_{i,j=1}^{N} b_i b_j} \qquad (2.19)$$

where $|\mathbf{r}_j - \mathbf{r}_i|$ is the distance between atoms i and j, b_i and b_j are the scattering lengths of atoms i and j, respectively, and the double sum is over all atoms in the system. For the purpose of calculating the static structure factor from simulations, the hydrogen (deuterium) atoms were geometrically placed on the saved trajectories of the PBD chains as yielded by the positions of the united atoms. Good agreement can be seen between experiment and simulation. The first peak (amorphous halo) is particularly dependent on intermolecular interactions, indicating that nonbonded dispersion and repulsion interactions well represented in the force field. We can therefore have confidence that the force field reproduces accurately intermolecular packing in the PBD melt.

Neutron scattering studies have also yielded the radius of gyration of the PBD-1600 chains. At 353 K values of 15.4 and 15.1 Å were obtained from experiment and simulation, respectively [40]. This excellent agreement is a strong indication that the conformational populations of the PBD chains are accurately reproduced by the quantum-chemistry-based potential.

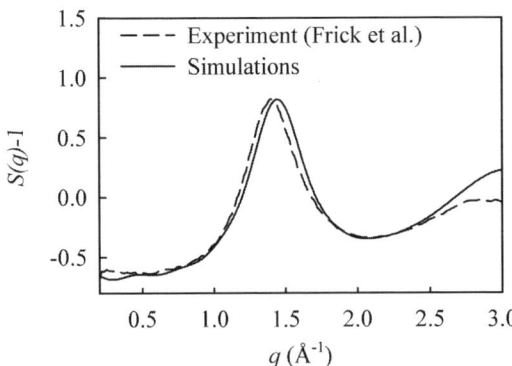

Figure 2.9. The static structure factor of a 1,4-polybutadiene melt from experiment (270 K) and simulation (273 K) using a quantum-chemistry based force field.

2.2.5.3. NMR Spin–Lattice Relaxation Times.

NMR ^{13}C spin–lattice relaxation times are sensitive to the reorientational dynamics of ^{13}C–^{1}H vectors. The motion of the attached proton(s) causes fluctuations in the magnetic field at the ^{13}C nuclei, resulting in decay of their magnetization in NMR experiments. While the time scale for the experimentally measured decay of the magnetization of a ^{13}C nucleus in a polymer melt is typically on the order of seconds, the corresponding decay of the ^{13}C–^{1}H vector autocorrelation function is on the order of nanoseconds, and hence is directly accessible to simulations. The spin–lattice relaxation time can be determined from simulation utilizing the relationships [60]:

$$\frac{1}{nT_1} = K[J(\omega_H - \omega_C) + 3J(\omega_C) + 6J(\omega_H + \omega_C)] \tag{2.20}$$

where $J(\omega)$ is the spectra density function as angular frequency ω given by:

$$J(\omega) = \frac{1}{2}\int_{-\infty}^{\infty} P_2(t)e^{i\omega t}\,dt \tag{2.21}$$

Here n is the number of attached protons, ω_H and ω_C are the proton and ^{13}C resonance frequencies. The constant K assumes values of 2.29×10^9 and 2.42×10^9 s^{-2} for sp^3 and sp^2 nuclei, respectively.

T_1 values can be determined for ^{13}C nuclei in various chemical (bonding) environments due to the different chemical shifts of these nuclei. The various resolvable resonances (neglecting those associated with the vinyl groups) are shown in Figure 2.10. The orientational autocorrelation function $P_2(t)$ is obtained from the simulation trajectory using the relationship

$$P_2(t) = \frac{1}{2}\left\{3\langle[\mathbf{e}_{CH}(t) \cdot \mathbf{e}_{CH}(0)]^2\rangle - 1\right\} \tag{2.22}$$

where $\mathbf{e}_{CH}(t)$ is the unit vector along a C–H bond at time t. Equation (2.22) is an ensemble average over all carbon atoms of each type as denoted in Figure 2.10. For the purpose of calculating the C–H ACF from simulations, the hydrogen atoms were geometrically placed on the saved trajectories of the PBD chains as yielded by the positions of the united atoms.

Experiments and simulations were performed on the PBD-1600 at a temperature of 353 K. Reasonable agreement between experiment and simulation was found, with the simulation T_1 values being approximately a factor of 2 longer than experiment for most resonances, indicating that the C–H vector was relaxing too fast in the simulations. The C–H vector relaxation is strongly correlated with the rate of conformational transitions. Based on our experience with polyethylene, where a comparison of united and explicit atom (with hydrogen atoms) simulations using quantum-chemistry-based potential functions that yielded very similar conformational energetics was made, it appears that use of a united atom potential results in conformational dynamics that are somewhat faster than those yielded by the

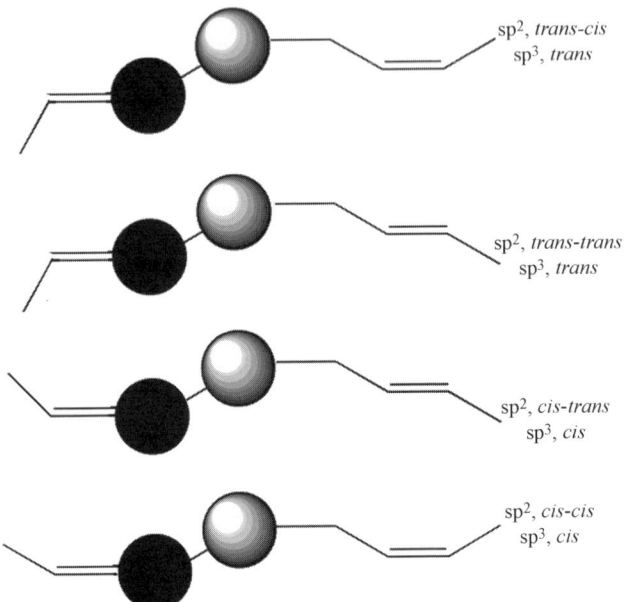

Figure 2.10. Resolvable ^{13}C NMR resonances in 1,4-polybutadiene.

equivalent all-atom potential. Consistent with this experience, we modified the united atom PBD potential in such a way as to slightly raise the rotational energy barriers between important conformers, while at the same time leaving the relative conformer energies (and hence the conformer populations) essentially unchanged. This 'modified' potential [40] has been used in essentially all of our PBD simulations, including those described below.

A comparison of T_1 values for PBD-1600 from experiment and simulation (using the modified potential) was made over the temperature range 273–353 K. Results for the highest and lowest temperatures are shown in Figure 2.11. Excellent agreement is seen for all resolvable resonances over the entire temperature range. The agreement between experiment and simulation for T_1 indicates that the potential is accurately reproducing conformational dynamics in PBD over the entire temperature range investigated. In addition, the C–H vectors have a nonnegligible component parallel to the local chain backbone whose reorientation dynamics also influences T_1. Since relaxation of this component is strongly tied to segmental relaxation, it appears that the simulations are also capturing accurately dynamics on the segmental length scale.

2.2.5.4. Single Chain Dynamic Structure Factor. Neutron spin echo (NSE) measurements of the single chain dynamic structure factor allow one to probe the dynamics of a polymer chain over a wide range of length and time scales. The single chain dynamic structure factor is obtained from simulations of the isotropic

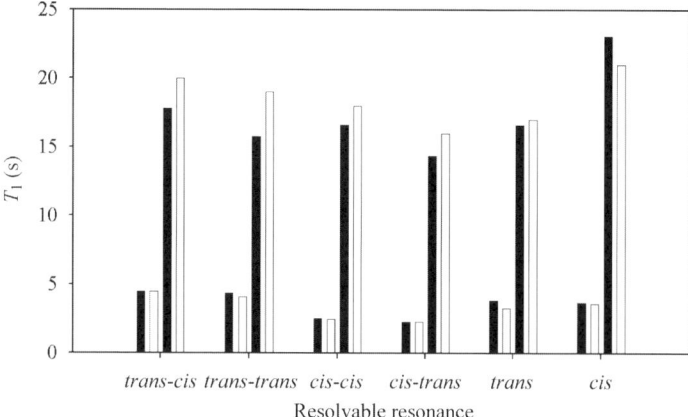

Figure 2.11. A comparison of ^{13}C NMR spin–lattice relaxation times for PBD-1600 from experiment (black) and simulation (white) for the resolvable resonances. Short relaxation times correspond to 273 K, long relaxation times to 353 K.

melt utilizing the simple relationship:

$$S(q,t) = \frac{\left\langle \sum_{i,j=1}^{N} \frac{\sin q|\mathbf{r}_j(t) - \mathbf{r}_i(0)|}{q|\mathbf{r}_j(t) - \mathbf{r}_i(0)|} \right\rangle}{\left\langle \sum_{i,j=1}^{N} \frac{\sin q|\mathbf{r}_j(0) - \mathbf{r}_i(0)|}{q|\mathbf{r}_j(0) - \mathbf{r}_i(0)|} \right\rangle} \qquad (2.23)$$

Here, $|\mathbf{r}_j(t) - \mathbf{r}_i(0)|$ is the distance between (united) atom j at time t and atom i at time 0. The double sum is take over the N atoms *of the same chain*. The ensemble average is taken over all chains and many time origins.

The single chain dynamic structure factor for PBD-1600 at 353 K obtained from NSE measurements is shown in Figure 2.12. Roughly, the length scale associated with a given momentum transfer is $r \approx 2\pi/q$. Hence, the smallest q value measured corresponds to $r \approx 125$ Å, a length scale larger than the mean-square end-to-end distance of the chain, and the largest q corresponds to $r \approx 20$ Å, a length scale a few times larger than the statistical segment length of PBD. Experimentally accessible time scales range from 50 ps to approximately 20 ns, covering the range of dynamics from subsegmental to overall chain relaxation (e.g. the Rouse time). A comparison between experiment and simulation for the single-chain dynamic structure reveals excellent agreement for all times and momentum transfers. The agreement for the smaller q values indicates that the center-of-mass diffusion of the PBD chain is reproduced well in the simulations by the quantum-chemistry-based potential. The agreement for larger q values indicates that the simulations are accurately capturing internal modes of motion of the polymer chain.

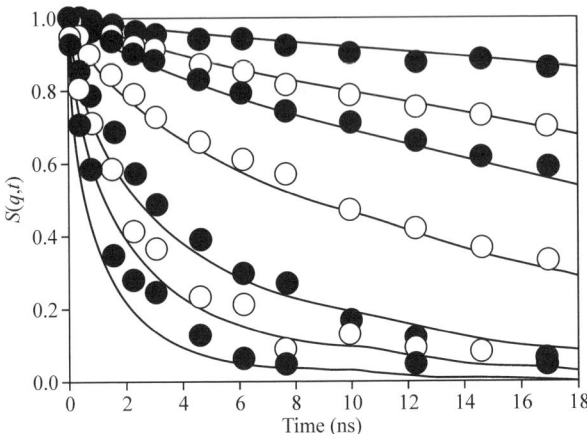

Figure 2.12. A comparison of the single-chain dynamic structure factor for PBD-1600 at 353 K from NSE experiments (symbols) and simulations (lines). q-values (in Å^{-1}, from the slowest relaxation to the fastest) are 0.05, 0.08, 0.10, 0.14, 0.20, 0.24 and 0.30.

2.2.6. The Need for Accurate Potentials

Our experience in parameterizing a quantum-chemistry-based united atom potential for PBD and subsequent experience in performing simulations using this potential for purposes of validation of the potential and for investigation of the dynamics and relaxation behavior of PBD has taught us several important lessons regarding the need for accurate potentials. It is the responsibility of the simulator to clearly demonstrate that his/her simulations, and the conclusions draw from them for a particular polymer, have any relationship to the real behavior of that polymer. The quality of agreement between experiment and simulations for dynamics and relaxation from the length scale of the individual dihedral to the length scale the polymer chain, illustrated in Figures 2.11 and 2.12, provides a difficult-to-refute foundation for the accuracy of our PBD simulations using a quantum-chemistry-based potential, allowing us to predict properties and extract mechanistic interpretations of relaxation behavior of PBD from our simulations with confidence.

We have found that dynamics on all length scales are very sensitive to the details of the conformational energy surface. The differences in conformational energies and rotational energy barriers predicted by several force fields for PBD model compounds (Figure 2.7) can be expected to yield chains of with different conformational statistics and very different conformational dynamics. When one wishes to utilize simulations to predict the behavior of a particular polymer or establish mechanisms for dynamics and relaxations, accurate representation of the conformational surface is paramount. Without high-level quantum chemistry data on conformational energetics of model molecules, it is impossible to evaluate the accuracy of extant potentials, to decide between them, or to parameterize a new potential.

2.3. QUANTUM-CHEMISTRY-BASED POLARIZABLE FORCE FIELDS FOR PEO AND PEO–LiBF$_4$

The need to include many-body polarization interactions in classical force fields in order to accurately predict the structural and dynamic properties of water and aqueous salt solutions has been actively investigated for the last two decades. The inclusion of polarization is considered to be important if both gas-phase and condensed-phase properties of water need to be accurately described [15], while effective two-body potentials have been rather successful in description of condensed-phase densities, heats of vaporization, structure factor, rotational, and translations dynamics [16, 17]. Investigations of aqueous salt solutions have revealed that polarization is important for solutions containing the smallest singly charged ions such as Li$^+$, and for doubly and triply charged cations [61, 62]. In this section, we discuss the importance of including polarization terms into force fields for oligomers of PEO and PEO-based polymer electrolytes.

PEO is a polar polymer with a wide range of technologically important applications that include solid polymer electrolytes [63], novel surfactants [64], modification of natural and artificial membranes [65, 66], aqueous biphasic separations [67], and prevention of protein denaturation [68]. In these applications PEO interacts with salts and/or with water, indicating a need for accurate PEO–water and PEO–salt potentials. Development of an accurate PEO–water potential is described elsewhere [69] together with the results of molecular dynamics (MD) simulations of PEO in aqueous solutions [69–76], whereas development of the PEO–Li$^+$ salt force field is described in detail in Borodin and co-workers [19, 21].

2.3.1. Development of the Many-Body Polarizable PEO Force Field

A quantum-chemistry based many-body polarizable force field for PEO has been developed [19] according to the methodology described in Section 2.1. First, we determined atomic polarizabilities by fitting the polarization potential around the *tgt* conformer of 1,2-dimethoxyethane along the paths shown in Figure 2.3. Then we fitted partial charges to reproduce the electrostatic potential on a grid of points around a dimethyl ether molecule, the important conformers of 1,2-dimethoxyethane and the all-*trans* conformer of diglyme (the PEO trimer). Self-polarization of the molecules was taken into account during charge fitting (see Sections 2.1.4.1 and 2.1.4.2). The root-mean-square deviation of the electrostatic potential around the PEO oligomers between that provided by the force field partial charges and atomic polarizabilities and quantum chemistry calculated at the B3LYP/aug-cc-pvDz level was around 1 kcal/mol. The electrostatic potential around the most important condensed-phase 1,2-dimethoxyethane conformer from the force field and quantum chemistry calculations is shown in Figure 2.2. The force field slightly underestimates the electrostatic potential in the vicinity of ether oxygen atoms compared with the quantum chemistry values, but in general the agreement between the force field and quantum chemistry calculations is quite good.

In Section 2.1, we demonstrated that the repulsion parameters of Sorensen et al. [37] for interactions between carbon, oxygen and hydrogen atoms, obtained by fitting polymer crystal structures and thermodynamic properties, are in excellent agreement with quantum chemistry predictions, whereas the AMBER parameters showed somewhat larger deviations from the quantum chemistry data, as shown in Figure 2.4. We adopted the empirical repulsion parameters of Sorensen et al. [37] without modification for our PEO simulations. The two-body dispersion parameters of Sorensen et al. [37] were reduced by 7 percent to yield an accurate description of the density of diglyme at 293 K in MD simulations using the many-body polarizable PEO potential. A slight reduction in the strength of the (attractive) dispersion interactions is reasonable considering that our potential includes net attractive polarization effects that were not considered in the parameterization of the two-body potential of Sorensen et al. [37]. The dispersion energy from quantum chemistry calculations, shown in Figure 2.5, was around 10 percent lower than the dispersion energy from the empirical force field (see Section 2.1.5).

Equilibrium bond length and bending angles were fit to obtain the best description of the quantum chemistry geometries of the most important 1,2-dimethoxyethane conformers. Bonds were constrained, whereas bending force constants were taken from previous work [77]. Torsional parameters were fit to reproduce relative conformational energies of 1,2-dimethoxyethane and the barriers between then as obtained from quantum chemistry. Torsional parameters determined based on 1,2-dimethoxyethane (the PEO dimer) were found to be transferable to diglyme (the PEO timer), and provided a good description of conformational energies and rotational energy barriers when compared with quantum chemistry values for this larger compound.

2.3.2. Effect of Polarizability on PEO Structural and Dynamic Properties

We performed MD simulations of PEO oligomers such as dimethyl ether, 1,2-dimethoxyethane, diglyme and short polydisperse PEO with molecular number $M_n = 398$ and molecular weight $M_w = 475$ Da, and longer monodisperse PEO with $M_w = 2300$ Da [78]. Both the quantum-chemistry-based many-body polarizable force field described above and a nonpolarizable two-body force field were used. The only difference between the two-body nonpolarizable force field and the many-body polarizable force field was an increase in the dispersion parameters back to the original values of Sorensen et al. [37] and the removal of the polarization in the former. Results of the MD simulations are reported in Borodin et al. [78]. Good agreement between MD simulation predictions and experiment for the enthalpy of vaporization, the static structure factor, the incoherent dynamics structure factor, the diffusion coefficient, the frequency dependent dielectric constant and ^{13}C T_1-spin–lattice relaxation times was found for both force fields. However, some differences in properties of PEO and its oligomers were found between simulations using the many-body polarizable and two-body nonpolarizable force fields, as discussed below.

2.3.2.1. Thermodynamic and Structural Properties. We found that turning off polarizability and compensating for the loss of attractive interactions by an increase in the strength of the dispersion interactions (see above) results in essentially identical thermodynamic properties (heats of vaporization and densities) for PEO and its oligomers, as obtained using the many-body polarizable potential. Furthermore, neglecting the many-body polarizable interactions had no effect on the PEO melt structure factor. This clearly suggests that, for moderately polar polymers, the contribution of many-body polarization to intermolecular interactions can be compensated for by an increase in the strength of the dispersion interactions.

2.3.2.2. Condensed-Phase Conformations. The polarizable and nonpolarizable force fields yielded significantly different liquid phase conformational populations for 1,2-dimethoxyethane as shown in Table 2.2, whereas gas-phase conformational populations were the same within error bars for the two force fields. Conformational populations from MD simulations with the many-body polarizable force field are in better agreement with IR experiments [79] than those from the two-body nonpolarizable force field. This difference can be associated with polarization of the 1,2-dimethoxyethane molecule in the liquid phase, leading to both an increase in dipole moment and changes in conformational populations due to intermolecular and intramolecular interactions between induced dipoles and partial charges. If the fixed partial charges are increased in the nonpolarizable potential in order to effectively account for the missing induced dipoles, the conformer populations are in much better agreement with those yielded by the polarizable potential, as shown in Table 2.2.

2.3.2.3. Dynamics. We also investigated the influence of many-body polarizable interactions on PEO dynamics measured through the decay of the incoherent intermediate scattering function (ISF). For isotropic systems such as liquids, the neutron

TABLE 2.2. Conformer Populations of 1,2-Dimethoxyethane from Simulation and Experiment

Conformer	Nonpolarizable Original Charges	Polarizable Original Charges	Nonpolarizable Increased Charges	Experiment
ttt	0.16 (0.24)[a]	0.10 (0.24)	0.15	0.12
tgt	0.37 (0.24)	0.46 (0.26)	0.49	0.49
tg^+g^-	0.18 (0.27)	0.17 (0.27)	0.15	0.33
tgg	0.12 (0.06)	0.15 (0.06)	0.11	0.11
ttg	0.12 (0.14)	0.07 (0.12)	0.07	0.07

[a]Values in parentheses are for the gas phase. Estimated uncertainties in simulation values are ± 0.02 or less. All values are at 273 K.

ISF, $I(q, t)$ is given by [80]:

$$I(q, t) = \frac{\left\langle \sum_{i=1}^{N} \frac{\sin q|\mathbf{r}_i(t) - \mathbf{r}_i(0)|}{q|\mathbf{r}_i(t) - \mathbf{r}_i(0)|} \right\rangle}{\left\langle \sum_{i=1}^{N} \frac{\sin q|\mathbf{r}_i(0) - \mathbf{r}_i(0)|}{q|\mathbf{r}_i(0) - \mathbf{r}_i(0)|} \right\rangle} \tag{2.24}$$

where $|\mathbf{r}_i(t) - \mathbf{r}_i(0)|$ is the distance between atom i at time t and at time 0, q is the magnitude of the momentum transfer vector, and $< >$ denotes an average over all time origins for atoms (N) with a significant incoherent cross-section (i.e. hydrogen atoms). The decay of the ISF from simulations using the nonpolarizable potential is somewhat faster than that for the polarizable potential as shown in Figure 2.13. Comparison of relaxation times from the time integrals of the ISF revealed that the difference in relaxation time is on the order of 10 percent, indicating that many-body polarization effects have only a small influence on the dynamics of PEO. An increase in partial charges to effectively account for the missing polarization of PEO in simulations with the nonpolarizable potential led to a slowing down of PEO dynamics of the order of 10 percent.

2.3.2.4. Conclusions About the Role of Polarization in Simulations of PEO.
Inclusion of the many-body polarizable interactions into the PEO force field led to additional intermolecular and intermolecular interactions between induced dipole moments and charges, resulting in a significant stabilization of the *tgt* and *tgg* conformers of PEO in the melt and a slowing down of PEO dynamics by approximately 10 percent. Both of these effects can be reproduced by the nonpolarizable potential if the partial charges are increased to effectively account for the missing induced dipole moments in PEO melts.

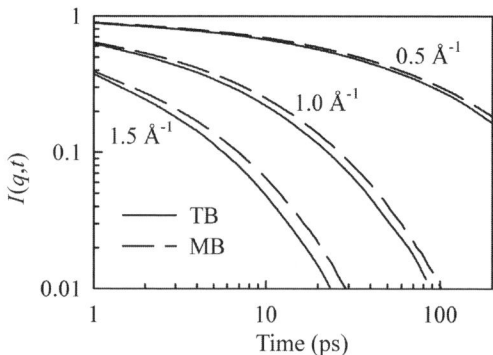

Figure 2.13. Incoherent intermediate dynamic structure factor for PEO ($M_W = 2380$) at 393 K from simulations.

2.3.3. Force Field Development Field and MD Simulations of PEO–LiBF$_4$

2.3.3.1. Development of the Many-Body Polarizable Potential. Our previous quantum chemistry study of Li$^+$–PF$_6^-$ interactions showed that polarization accounts for $\sim$40–50 kcal/mol of the Li$^+$–PF$_6^-$ total complexation energy of about 130 kcal/mol. Investigation of a unit charge interaction with 1,2-dimethoxyethane (Figures 2.2 and 2.3) indicated that the many-body polarization energy is comparable to the electrostatic potential at Li–O separations of $\sim$2 Å, which is approximately the position of the first peak of the Li–O radial distribution function in PEO–lithium salt solutions. Therefore, it is important to include polarization effects in parameterization of PEO–Li$^+$ interactions. The development of our PEO–LiBF$_4$ force fields is discussed in detail in Borodin et al. [81]. Here, we concentrate on development of a two-body potential for PEO–LiBF$_4$ that accounts for polarization and comparison of the properties of PEO–LiBF$_4$ polymer electrolytes obtained using the many-body and two-body polarizable potentials.

2.3.3.2. Development of the Two-Body Polarizable Potential. Because MD simulations with explicit inclusion of many-body polarizable interactions are approximately two to four times more expensive than MD simulations of identical systems with a force field containing only two-body interactions, it is desirable to develop, if possible, a two-body force field for PEO–LiBF$_4$ interactions that takes into account the effects of many-body polarization. We tried several two-body approximations to the many-body force field. We found that increasing the strength of the dispersion interaction to account for the many-body polarization contribution to the 1,2-dimethoxyethane–Li$^+$ complexation energy, analogous to our treatment of PEO discussed in Section 2.3.2, dramatically overestimated the binding energy of PEO–Li$^+$ in the condensed-phase yielding a Li$^+$ self-diffusion coefficient from MD simulations that is orders of magnitude smaller than experimental estimates. Complete omission of the polarization energy from the force field was also unacceptable, resulting in phase-separation of the PEO–LiBF$_4$ system in MD simulations.

Based on the above studies, we conclude that many-body interactions are very important for accurate prediction of properties of polymer electrolytes such as PEO–LiBF$_4$, and that developing a reasonable two-body approximation for the many-body force field is not trivial. We settled on an approach related to that employed in our previous studies of PEO-based polymer electrolytes [82, 83]. Considering only the lowest-order term in $\mathbf{r}_{ij}$ in equations (2.3) and (2.4), the polarization energy between two charged atomic centers with dipolar polarizability is given by:

$$V^{\text{pol}}(r_{ij}) = \frac{1}{4\pi\varepsilon_0}\left(q_i^2 \alpha_j + q_j^2 \alpha_i\right) = -\frac{D_{ij}}{r^4} \tag{2.25}$$

where α_i is the atomic dipole polarizability of atom i. This expression is exact only for two isolated ions. For a system with more than two charged polarizable atoms equation (2.25) is not exact as it neglects many-body induction. In application of equation (2.25), we assume that atoms of PEO and the BF_4^- anions are polarized only by Li^+ in their first coordination shell (3.5–4 Å). At further distance we assume that the system is sufficiently isotropic that net polarization effects are negligible, and hence the D_{ij} parameter is scaled to zero for distances beyond the first coordination shell using a distance dependent dielectric constant.

In previous work [82] we used D_{ij} parameters obtained from fits to the *ab initio* quantum chemistry calculations of Li^+–dimethyl ether binding energies that were uniformly scaled in order to account for condensed phase effects. For the PEO–$LiBF_4$ force field we took a more rigorous approach for obtaining D_{ij} parameters. Specifically, we performed MD simulations of PEO–$LiBF_4$ solutions at EO:Li = 15:1 concentration at 393 K using the many-body polarizable force field. The induced dipole moments from these simulations were used to estimate the net polarization interactions between PEO, $LiBF_4^-$ and the Li^+ cations as a function of separation, allowing us to estimate D_{ij} parameters from equation (2.25).

MD simulations of the PEO–$LiBF_4$ solutions using the two-body polarizable potential parameterized as described above indicated that these simulations yield slower ion dynamics and a lower fraction of cations not complexed by the BF_4^- anions than those from the MD simulations with the many-body polarizable force field. We subsequently made empirical adjustments to the D_{ij} parameters in order to decrease the strength of interaction between Li^+, PEO and BF_4^-, thus increasing the ion dynamics and improving the agreement for the fractions of 'free' Li^+ from MD simulations with the many-body and two-body polarizable potentials within simulation error bars [81].

2.3.4. MD Simulations of PEO–LiBF$_4$

MD simulations of PEO–$LiBF_4$ comparing the two-body polarizable potential and many-body polarizable potential parameterized as described above have been reported elsewhere [81]. The simulations were performed for 20 ns using the many-body polarizable force field and more than 100 ns using the two-body force field. Structural and dynamic properties were calculated for the last 5 and 40 ns of the simulations with the many-body and two-body polarizable force fields, respectively.

2.3.4.1. Structural Properties of PEO–LiBF$_4$. The complexation environment of Li^+ is of great interest in polymer electrolytes as it is expected to be intimately related to the Li^+ transport mechanism. We calculated the radial distribution functions [RDF, $g^{Li-X}(r)$] and distance dependent coordination numbers [$n^{Li-X}(r)$], shown in Figure 2.14, from MD simulations with the many-body and two-body polarizable force fields. The position of the first peak of the Li–O RDF is at slightly larger Li–O separation for the two-body polarizable force field. These most probable distances from MD simulations are similar to those obtained from analysis of the neutron diffraction with isotopic substitution experiments on PEO–LiI [84] and PEO–LiTFSI [85]. Simulations with the two-body polarizable force field predict a markedly larger number of ether oxygen atoms in the first coordination

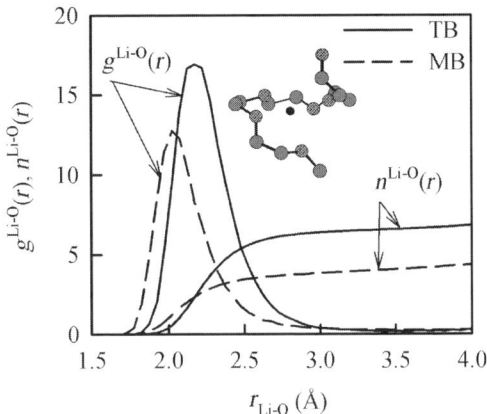

Figure 2.14. Li–O RDF and coordination numbers $n^{\text{Li-O}}(r)$ from simulations. Typical Li$^+$ coordination by a PEO chain is also shown.

shell of a Li$^+$ cation. We tried to match both the Li–O RDF and Li–O coordination numbers obtained from simulations with the many-body and two-body polarizable force fields by changing $D_{\text{Li-X}}$ parameters in the two-body force field, where X = O, C, H. No $D_{\text{Li-X}}$ parameters would allow matching of the position and magnitude of the first maximum of the Li–O RDF and coordination numbers simultaneously, indicating a deficiency of the two-body approximation.

2.3.4.2. Transport Properties of PEO–LiBF$_4$. Ion transport properties such as cation and anion self-diffusion coefficients and ionic conductivity are of great technological importance in polymer electrolytes. Ion mean-square displacements (MSD), self-diffusion coefficients (D) and ionic conductivity (λ) were calculated from MD simulations using the relationship [81]:

$$\text{MSD}(t) = \frac{1}{N}\sum_i^N [\mathbf{r}_i(t) - \mathbf{r}_i(0)]^2 \tag{2.26}$$

$$D = \lim_{t \to \infty} \frac{\text{MSD}(t)}{6t} \tag{2.27}$$

$$\lambda = \lim_{t \to \infty} \frac{e^2}{6tVk_\text{B}TN} \sum_{i,j}^N q_i q_j \langle [\mathbf{r}_i(t) - \mathbf{r}_i(0)][\mathbf{r}_j(t) - \mathbf{r}_j(0)] \rangle \tag{2.28}$$

The MSD of the ions obtained from MD simulations with the many-body and two-body polarizable force fields are compared in Figure 2.15. Simulations with both force fields predict the displacement of BF$_4^-$ anions to be ~40 percent higher than that of Li$^+$ cations, with ions moving much slower in the simulations with the two-body polarizable force field. The much slower Li$^+$ ion dynamics observed

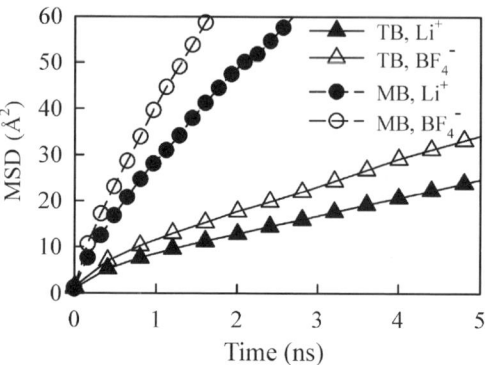

Figure 2.15. Li^+ and BF_4^- MSD from simulations.

in simulations with the two-body polarizable force field is due to slower Li^+ motion (hopping) along the PEO chain that requires breaking more EO–Li transient bonds due to the higher Li^+–O coordination numbers (Figure 2.14).

The Li^+ self-diffusion coefficients from MD simulations are compared with results of NMR pulsed-field gradient measurement on high molecular weight oxymethylene-linked PEO ($M_w = 10^5$)/LiPF$_6$ [86] as the ion self-diffusion coefficients are expected to be similar for PEO–LiBF$_4$ and PEO–LiPF$_6$ and no experimental data are available for PEO–LiBF$_4$. We used a factor of 1.3 based upon experimental investigation of the PEO molecular weight dependence of the Li^+ self-diffusion coefficient [87] to account for the difference between the Li^+ diffusion in PEO ($M_w = 10^5$ Da) and PEO ($M_w = 2380$ Da). The scaled Li^+ self-diffusion coefficients from NMR measurements are approximately two to three times lower than that obtained from MD simulations with the many-body polarizable force field and about three times higher than that obtained from MD simulations with the two-body polarizable force field. Similar quality agreement between simulation and experiment is seen for the anion [89] and overall system conductivity [88], as can be seen in Figure 2.16.

2.3.4.3. Conclusions about the Role of Polarization in Simulations of PEO–LiBF$_4$.

We have found that inclusion of many-body polarization effects is very important for accurate prediction of the structural and dynamic properties of polymer electrolytes. The PEO–LiBF$_4$ polymer electrolytes exhibit a few orders of magnitude slower dynamics or phase separation if many-body polarization interactions are directly included into the two-body dispersion terms or omitted, respectively. A scheme for including the polarization interactions approximately in the two-body force field was developed. MD simulations with the two-body force field yielded the Li^+ environment, ion aggregation and ion transport in reasonable agreement with the results of MD simulations with explicit inclusion of many-body polarization and experiments. However, a more detailed quantitative comparison of Li^+ environment and transport from MD simulations using the many-body and two-body polarizable force fields indicated some shortcomings of the two-body

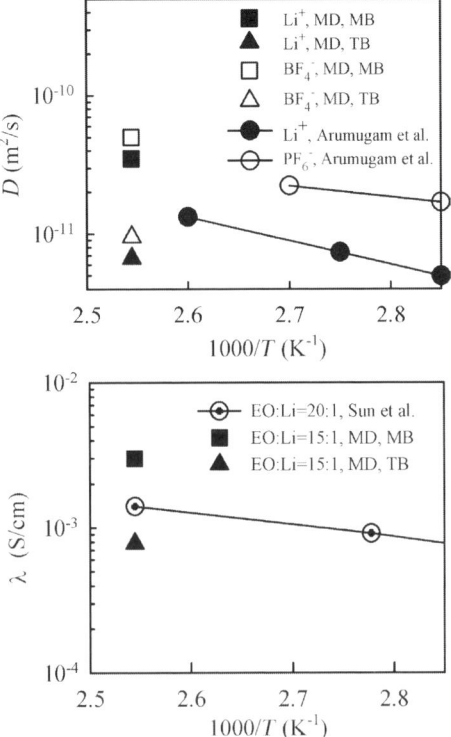

Figure 2.16. Ion self-diffusion coefficients and ionic conductivity of PEO-based polymer electrolytes from simulation and experiments of Arumugam et al. [86] and Sun et al. [88].

approximation, namely, a higher number of ether oxygen atoms in the first coordination shell of a Li^+ cation that hinders Li^+ transport, yielding five to six times slower ion dynamics in simulations using the two-body polarizable force field compared with those obtained with the many-body polarizable force field.

ACKNOWLEDGMENT

The authors are indebted to the American Chemical Society (Petroleum Research Fund), the National Science Foundation, NASA, and the Department of Energy (Lawrence Berkeley Laboratory) for financial support for the work comprising the majority of this contribution.

REFERENCES

[1] W.D. Cornell et al., *J. Am. Chem. Soc.* **117**, 5179–5197 (1995).
[2] J.W. Ponder, D.A. Case, *Adv. Protein Chem.* **66**, 27–85 (2003).

[3] W.L. Jorgensen, D.S. Maxwell, J. Tirado-Rives, *J. Am. Chem. Soc.* **118**, 11225–11236 (1996).
[4] D.W. Brenner, *Phys. Rev. B* **42**, 9458–9471 (1990).
[5] T.A. Halgren, *J. Comput. Chem.* **17**, 490–519 (1996).
[6] A. Toukmaji, C. Sagui, J. Board, T. Darden, *J. Chem. Phys.* **113**, 10912–10927 (2000).
[7] T.M. Nymand, P. Linse, *J. Chem. Phys.* **112**, 6152–6160 (2000).
[8] G. Ruocco, M. Sampoli, *Mol. Phys.* **82**, 875–886 (1994).
[9] J. Wang, P. Ceiplak, P. Kollman, *J. Comput. Chem.* **21**, 1049–1074 (2000).
[10] P. Cieplak, J. Caldwell, P. Kollman, *J. Comput. Chem.* **22**, 1048–1057 (2001).
[11] H. Sun, *J. Phys. Chem. B* **102**, 7338–7364 (1998).
[12] A.D. MacKerell et al., *J. Phys. Chem. B* **102**, 3586–3616 (1998).
[13] S.L. Mayo, B.D. Olafson, W.A. Goddard III, *J. Phys. Chem.* **94**, 8897–8909 (1990).
[14] A.K. Rappé, C.J. Casewit, K.S. Colwell, W.A. Goddard, W.M. Skiff, *J. Am. Chem. Soc.* **114**, 10024–10035 (1992).
[15] P. Ren, J.W. Ponder, *J. Phys. Chem. B* **107**, 5933–5947 (2003).
[16] W.L. Jorgensen, J. Chandrasekhar, J.D. Madura, R.W. Impey, M.L. Klein, *J. Chem. Phys.* **79**, 926–935 (1983).
[17] H.J.C. Berendsen, J.R. Grigera, T.P. Straatsma, *J. Phys. Chem.* **91**, 6269–6271 (1987).
[18] G.D. Smith, D. Bedrov, O. Byutner, O. Borodin, T.D. Sewell, *J. Phys. Chem. A* **107**, 7552–7560 (2003).
[19] O. Borodin, G.D. Smith, *J. Phys. Chem. B* **107**, 6801–6812 (2003).
[20] G.D. Smith, O. Borodin, D. Bedrov, *J. Comput. Chem.* **23**, 1480–1488 (2002).
[21] O. Borodin, G.D. Smith, R. Douglas, *J. Phys. Chem. B* **107**, 6824–6837 (2003).
[22] O. Borodin, G.D. Smith, D. Bedrov, *J. Phys. Chem. B* **106**, 9912–9922 (2002).
[23] F. Sato, S. Hojo, H. Sun, *J. Phys. Chem. A* **107**, 248–257 (2003).
[24] D.E. Woon, T.H. Dunning Jr, *J. Chem. Phys.* **98**, 1358–1371 (1993).
[25] R.A. Kendall, T.H. Dunning Jr, R.J. Harrison, *J. Chem. Phys.* **96**, 6796–6806 (1992).
[26] A.D. Becke, *J. Chem. Phys.* **98**, 5648–5652 (1993).
[27] C. Lee, W. Yang, R.G. Parr, *Phys. Rev. B* **37**, 785–789 (1988).
[28] C. Møller, M.S. Plesset, *Phys. Rev.* **46**, 618–622 (1934).
[29] M. Head-Gordon, T. Head-Gordon, *Chem. Phys. Lett.* **220**, 122–128 (1994).
[30] A.J. Stone, Oxford University Press, New York, 1996.
[31] Y.-P. Liu, K. Kim, B.J. Berne, R.A. Friesner, S.W. Rick, *J. Chem. Phys.* **108**, 4739–4755 (1998).
[32] J. Applequist, *Acc. Chem. Res.* **10**, 79–85 (1977).
[33] B. Ma, J.-H. Lii, N.L. Allinger, *J. Comput. Chem.* **21**, 813–825 (2000).
[34] J.M. Stout, C.E. Dykstra, *J. Am. Chem. Soc.* **117**, 5127–5132 (1995).
[35] J. Gasteiger, M. Marsili, *Tetrahedron* **36**, 3219–3228 (1980).
[36] F.B. van Duijneveldt, J.G.C.M. van Duijneveldt-van de Rijdt, J.H. van Lenthe, *Chem. Rev.* **94**, 1873–1885 (1994).
[37] R.A. Sorensen, W.B. Liau, L. Kesner, R.H. Boyd, *Macromolecules* **21**, 200–208 (1988).
[38] D.E. Woon, *J. Chem. Phys.* **100**, 2838–2850 (1994).

[39] G.D. Smith, W. Paul, *J. Phys. Chem. A* **102**, 1200–1205 (1998).
[40] G.D. Smith, W. Paul, M. Monkenbusch, L. Willner, D. Richter, D., X.H. Qiu, M.D. Ediger, *Macromolecules* **32**, 8857–8865 (1999).
[41] O. Byutner, G.D. Smith, *Macromolecules* **34**, 134–138 (2001).
[42] O. Byutner, O., G.D. Smith, *J. Polym. Sci.: Polym. Phys.* **39**, 3067–3071 (2001).
[43] G.D. Smith, O. Borodin, D. Bedrov, W. Paul, X.H. Qiu, M.D. Ediger, *Macromolecules* **34**, 5192–5199 (2001).
[44] O. Byutner, G.D. Smith, *Macromolecules* **35**, 3769–3771 (2002).
[45] G.D. Smith, O. Borodin, W. Paul, *J. Chem. Phys.* **117**, 10350–10359 (2002).
[46] R.H. Gee, R.H. Boyd, *J. Chem. Phys.* **101**, 8028–8038 (1994).
[47] E. Kim, S. Misra, W.L. Mattice, *Macromolecules* **26**, 3424–3431 (1993).
[48] W. Paul, D.Y. Yoon, G.D. Smith, *J. Chem. Phys.* **103**, 1702–1709 (1995).
[49] K.B. Wiberg, E. Martin, *J. Am. Chem. Soc.* **107**, 5035–5041 (1985).
[50] K.B. Wiberg, S.L. Schreiber, *J. Org. Chem.* **53**, 783–785 (1988).
[51] I. Petterson, K. Gundertofte, *J. Comp. Chem.* **12**, 839–843 (1991).
[52] J.L. Broeker, R.W. Hoffman, K.N. Houk, *J. Am. Chem. Soc.* **113**, 5006–5017 (1991).
[53] G.D. Smith, R.L. Jaffe, *J. Phys. Chem.* **100**, 18718–18724 (1996).
[54] C. Tosi, F. Ciampelli, *Eur. Polym. J.* **5**, 759–766 (1969).
[55] S. Kondo, E. Hirota, Y. Morino, *J. Mol. Spectrosc.* **28**, 471–489 (1968).
[56] J.R. Durig, D.A.C. Compton, *J. Phys. Chem.* **84**, 773–781 (1980).
[57] G.D. Smith, D.Y. Yoon, R.L. Jaffe, R.H. Colby, R. Krishnamoorti, L.J. Fetters, *Macromolecules* **29**, 3462–3469 (1996).
[58] D.R. Paul, A.T. DiBenedetto, *J. Polym. Sci. C* **10**, 17 (1965).
[59] B. Frick, D. Richter, C. Ritter, *Europhys. Lett.* **9**, 557–562 (1989).
[60] D.J. Gisser, S. Gowinkowski, M.D. Ediger, *Macromolecules* **24**, 4270–4277 (1991).
[61] G.W. Marini, N.R. Texler, B.M. Rode, *J. Phys. Chem.* **100**, 6808 (1996).
[62] L.X. Dang, *J. Phys. Chem. B* **106**, 10388–10394 (2002).
[63] F.M. Gray, *Polymer Electrolytes*. The Royal Society of Chemistry, Cambridge, 1997.
[64] H.F. Warriner, P. Davidson, N.L. Slack, M. Schellhorn, P. Eiselt, S.H.J. Idziak, H.W. Schmidt, S.R. Safinya, *J. Chem. Phys.* **107**, 3707–3722 (1997).
[65] S. Rex, M.J. Zuckermann, M. Lafleur, J.R. Silvius, *Biophys. J.* **75**, 2900–2914 (1998).
[66] E. Evans, W. Rawicz, *Phys. Rev. Lett.* **79**, 2379–2382 (1997).
[67] R.D. Rogers, M.A. Eiteman (eds), *Aqueous Biphasic Separations*, Plenum Press, London, 1995, pp. 1–17.
[68] S.I. Jeon, J.H. Lee, J.D. Andrade, P.G. de Gennes, *J. Colloid Interface Sci.* **142**, 129–158 (1991).
[69] G.D. Smith, O. Borodin, D. Bedrov, *J. Comput. Chem.* **23**, 1480–1488 (2002).
[70] F. Trouw, D. Bedrov, O. Borodin, G.D. Smith, *Chem. Phys.* **261**, 137–148 (2000).
[71] D. Bedrov, O. Borodin, G.D. Smith, *J. Phys. Chem. B* **102**, 5683–5690 (1998).
[72] D. Bedrov, O. Borodin, G.D. Smith, *J. Phys. Chem. B* **102**, 9565–9570 (1998).
[73] O. Borodin, D. Bedrov, G.D. Smith, *Macromolecules* **34**, 5687–5693 (2001).

[74] D. Bedrov, O. Borodin, G.D. Smith, F. Trouw, C. Mayne, *J. Phys. Chem. B* **104**, 5151–5154 (2000).

[75] G.D. Smith, D. Bedrov, O. Borodin, *J. Am. Chem. Soc.* **122**, 9548–9549 (2000).

[76] G.D. Smith, D. Bedrov, O. Borodin, *Phys. Rev. Lett.* **85**, 5583–5586 (2000).

[77] G.D. Smith, R.L. Jaffe, *J. Phys. Chem.* **97**, 12752–12758 (1993).

[78] O. Borodin, R. Douglas, G.D. Smith, F. Trouw, S. Petrucci, *J. Phys. Chem. B* **107**, 6813–6823 (2003).

[79] N. Goutev, K. Ohno, H. Matsuura, *J. Phys. Chem. A* **104**, 9226–9232 (2000).

[80] J.S. Higgins, H.C. Benoît, *Polymers and Neutron Scattering*, Clerendon Press, Oxford, 1996.

[81] O. Borodin, G.D. Smith, R. Douglas, *J. Phys. Chem. B* **107**, 6824–6837 (2003).

[82] G.D. Smith, R.L. Jaffe, H. Partridge, *J. Phys. Chem. A.* **101**, 1705–1715 (1997).

[83] O. Borodin, G.D. Smith, R.L. Jaffe, *J. Comput. Chem.* **22**, 641–654 (2001).

[84] J.D. Londono, B.K. Annis, A. Habenschuss, O. Borodin, G.D. Smith, J.Z. Tirner, A.K. Soper, *Macromolecules* **30**, 7151–7157 (1997).

[85] G. Mao, M.-L. Saboungi, D.L. Price, M. Armand, W.S. Howells, *Phys. Rev. Lett.* **84**, 5536–5539 (2000).

[86] S. Arumugam, J. Shi, D.P. Tunstall, C.A. Vincent, *J. Phys.: Condens. Matter* **5**, 153–160 (1993).

[87] J. Shi, C.A. Vincent, *Solid St. Ionics* **60**, 11–17 (1993).

[88] H.Y. Sun, Y. Takeda, N.Y. Imanishi, O. Yamamoto, H.-J. Sohn, *J. Electrochem. Soc.* **147**, 2462–2467 (2000).

3

MONTE CARLO SIMULATIONS OF BINARY POLYMER LIQUIDS

MARCUS MÜLLER

Institut für Physik, WA331, Johannes Gutenberg-Universität, D55099 Mainz, Germany
and
Department of Physics, University of Wisconsin, Madison, WI, USA

CONTENTS

3.1. Introduction	95
3.2. Background	99
3.2.1. Coarse-Grained Models	99
3.2.2. Predictions of the Mean-Field Theory	101
3.3. Models and Techniques	106
3.4. Results	111
3.4.1. Phase Behavior: Fluid Structure and Composition Fluctuations	111
3.4.2. Entropic Contributions to the χ Parameter	124
3.4.3. Films and Two-Dimensional Systems	134
3.4.4. Interfaces	136
3.5. Conclusion and Outlook	143

Molecular Simulation Methods for Predicting Polymer Properties, edited by Vassilios Galiatsatos.
ISBN 0-471-46481-3 Copyright © 2005 John Wiley & Sons, Inc.

3.1. INTRODUCTION

Melt blending of polymers is a promising route for tailoring materials to specific application properties: Polymeric materials in daily life are generally multicomponent systems. Chemically different polymers are 'alloyed' to design a material which combines the favorable characteristics of the individual components [1]. Clearly the miscibility behavior of the blend is crucial for understanding and tailoring properties relevant for practical applications. Miscibility on a microscopic length scale is desirable for a high tensile strength of the material. Unlike metallic alloys, however, chemically different polymers often do not mix at microscopic length scales. Rather a complicated morphology of droplets of one component dispersed into the other component forms on a mesoscopic length scale, and the blend can be conceived as an assembly of interfaces. While the detailed structure on this mesoscopic length scale depends strongly on the way the material is processed, the local properties of interfaces are certainly crucial for understanding the material properties. For instance, the interfacial width sets the length scale on which entanglement between polymers of the different components forms. Experiments [2] suggest that the mechanical strength increases if the interfacial width exceeds the entanglement length. Alternatively, the interfacial tension is important for the break-up of droplets under shear [3, 4]: The lower the interfacial tension is, the finer the two components are dispersed.

The behavior of complex macromolecular systems can be analyzed at different time and length scales. The choice of the level of detail, or abstraction, depends on the question one wants to address. For instance, one may be interested in the structure and dynamics at the segment level in a binary blend [5] or, at the other extreme, the morphology of a phase-separated state on the length scale of several micrometers. Just as different experimental techniques are suited to studying different phenomena, so theoretical approaches vary in their ability to describe different aspects of polymer liquids.

In the following we consider coarse-grained models that do not capture the structure at the atomistic scale but lump together a small number of chemical repeat units into a monomer of the coarse-grained model. These monomers interact via coarse-grained, simplified interactions. Electrostatic and torsional potentials are typically neglected in these models. The reduced number of degrees of freedom and the softer interactions on a coarse scale lead to a significant computational speed-up. Hence, large system sizes and long time scales can be studied that are inaccessible in atomistic simulations [6].

The use of coarse-grained models to describe polymeric systems has a long-standing tradition [7–10]. In polymer solutions and melts, the elimination of the degrees of freedom is justified by the self-similar structure on a large range of length scales from the statistical segment length, b, to the polymer's end-to-end distance, R_e. The coarse-graining procedure in such systems can formally be performed exactly within the framework of the renormalization group [11]. The behavior of polymers regarding the limit of long chain length, $N \to \infty$, corresponds to the critical behavior of the n-component vector model in the limit $n \to 0$. Similar to the universality at a second-order phase transition, the behavior of long chains (i.e. at the critical point) does not depend on the details of the microscopic interactions, as long as the relevant interactions—connectivity along the backbone of the polymer and excluded volume of the segments—are incorporated into the coarse-grained model.

To predict properties of polymer liquids within the framework of coarse-grained models, one has to identify which interactions on the coarse-grained scale are relevant for the phenomena under investigation and relate the properties of the coarse-grained model to experimental systems. Roughly, two (not mutually exclusive) strategies can be distinguished.

Minimal coarse-grained models incorporate only the relevant interactions to bring about the phenomena one is interested in. Those models are characterized by a small number of parameters, for instance, the Flory–Huggins parameter χ that parameterizes the repulsion between the segments in a blend, or the end-to-end distance, R_e, that describes the chain conformations on large distances. These parameters can be identified by comparing the coarse-grained model to experiments, and then one can use the coarse-grained model to make further predictions.

The main advantage of minimal coarse-grained models is that their predictions apply to a whole class of systems (e.g. binary blends with negligible volume change upon mixing). However, the uncertainty as to which are the relevant interactions and which interactions can be omitted for the sake of computational efficiency, limits their potential to make quantitative predictions.

Therefore, much effort has been directed towards coarse-grained polymer models which not only capture the generic features of polymers on the coarse-grained scale, but also retain information about the underlying chemical structure. These systematic coarse-graining approaches aim to design models that bridge the length and time scales from atomistic to macroscopic [8, 9, 12]. To this end, one chooses a set of structural and thermodynamic quantities of the underlying atomistic systems (e.g. extracted from an atomistic simulation or measured in experiments), and constructs interactions between the coarse-grained degrees of freedom so as to reproduce those quantities. Typical choices [7, 8, 13] include geometrical characteristics of the molecules, the distribution of distances between entities, and thermodynamic properties.

An example of this type of coarse-graining is the modeling of polycarbonate at a nickel surface [14]. Car–Parinello type quantum mechanical density functional calculations reveal a strong and orientation-dependent adsorption of the chain ends onto the metal surface. Unfortunately, these *ab initio* calculations are only feasible for a fragment of a chemical repeat unit of polycarbonate. Nevertheless, those calculations identified a relevant interaction for this material and this interaction has been accounted for in a coarse-grained model. Then, this coarse-grained model can be employed to investigate a melt of short polycarbonate polymers confined between two nickel surfaces. This strategy allows one to investigate the consequences of this specific interaction of the chain ends with the nickel surface at the atomistic scale for the molecular conformations on the length scale of the whole polymer and the segmental density profiles of a melt in contact with the surface.

In principle, those models allow the re-introduction of atomic degrees of freedom and the smaller length scales they entail once the coarse-grained model has equilibrated on a large length scale. Although these coarse-grained models hold the promise of quantitatively predicting properties of polymer liquids, there are caveats: As the interactions on the coarse-grained scale differ qualitatively from the atomistic interactions, they are in general not transferable [8]; that is, the systematic coarse-graining procedure is specific to a particular state point described by temperature, pressure, and so on. Moreover, a tiny inaccuracy in the free energy on the atomistic scale can give rise to dramatic changes on mesoscopic or macroscopic length scales. Note that the interactions on the atomistic scale (on the order 1 eV $\approx$ 40 k_BT for bonded and about k_BT for non-bonded interactions) are much stronger than at the coarse-grained scale. For instance, the Flory–Huggins parameter that measures the repulsion of monomers in units of k_BT typically takes values of 10^{-1}–10^{-4}. Recent large-scale simulation of a united-atom model of isotactic polypropylene–polyethylene blends [15] investigated the different contributions to the energy of mixing, arising from bonded and nonbonded interactions. Each contribution is only on the order of $10^{-3} k_BT$ per united atom site, and the interatomic potentials were considered a serious limitation of the simulation approach [15] to predict the miscibility behavior. This problem becomes particularly pronounced in the vicinity of phase transitions where one encounters a singular dependence on system parameters. Thus, much of the quality of the coarse-graining depends on a careful choice of the set of quantities used for the mapping and the type of interactions in the coarse-grained model.

Notwithstanding the limitations of coarse-grained models, they offer important qualitative insights into phenomena on the length scale from a few nanometers to a micrometer and they assess the validity and provide input parameters for phenomenological approaches (e.g. Landau–Ginzburg theories or effective interface Hamiltonians).

Of course, these two approaches—minimal coarse-grained models and systematic coarse-graining—are only the two extreme cases of a wide spectrum of coarse-grained models. This spectrum spans from united-atom models, where one lumps the hydrogen atoms bonded to a carbon atom together into a coarse-grained segments, to the Edwards Hamiltonian, which represents polymers as Gaussian strings.

Coarse-grained polymer models have been employed to address a large variety of questions: liquid–vapor [16–20] and liquid–liquid phase equilibria [21–27], polymers at surfaces and interfaces [28–32], single chain dynamics [33, 34] and kinetics of phase separation [35, 36], and the glass transition of macromolecules in the bulk and thin films [32, 37]. In the following we restrict ourselves to a deceptively simple system: a polymer mixture of two components A and B. We do not attempt to be comprehensive and draw examples mainly from our own research. On the one hand, we shall discuss how to identify coarse-grained parameters for a specific simulation model and how they are related to experiments. On the other hand, we shall briefly review some simulation techniques used to extract properties from coarse grained models that can subsequently be used in more phenomenological approaches.

This contribution is organized as follows: First, some brief background on the phase separation in binary polymer liquids is provided. Then, model and computational techniques are described. Subsequently, we discuss the phase behavior in the bulk, single chain conformations and interface properties. The article closes with an outlook on future problems.

3.2. BACKGROUND

3.2.1. Coarse-Grained Models

In the following we consider a mixture of two polymers—denoted A and B. Upon increasing the incompatibility between the two species, one encounters a critical point. At larger incompatibilities, the mixture will phase-separate into domains of an A-rich and a B-rich liquid. Of course, in compressible mixtures more complex phase behavior can exist; that is, in addition to liquid–liquid, a liquid–vapor phase coexistents between a dense polymer liquid and its vapor, or even three-phase coexistent lines can occur [38]. In fact, six qualitatively different types of phase diagrams can be distinguished for compressible binary mixtures according to a scheme by Konynenburg and Scott [39]. In the following we restrict our attention to liquid–liquid immiscibility, which is characterized by a single-order parameter, ϕ, which is the composition of the mixture.

Which coarse-grained parameters describe the liquid–liquid phase separation in binary polymer blends? This depends on the question one is interested in: If we were interested in the factors on the molecular scale that determine the miscibility behavior of a specific pair of substances, we would aim at describing the detailed fluid structure of the polymeric mixture. If we were interested in the qualitative behavior of dense multicomponent polymer systems and phenomena on large length scales, much insight could be gained from a highly coarse-grained description. The large size of the chain molecules imparts a rather universal behavior onto dense polymer mixtures, which can be characterized by only a small number of parameters: the end-to-end distance R_e of the molecules, and the incompatibility per chain, χN. The Flory–Huggins parameter χ describes the repulsion between unlike segments [40] and N is the number of segments per molecule. R_e sets the characteristic length scale of spatial inhomogeneities, for example, the width of interfaces between coexisting phases. Since this length scale is much larger than the size of a chemical repeat unit along the backbone of the polymer chain, one can use the Gaussian chain model, which captures the long-wavelength behavior of the polymer conformations in a melt. Note that R_e and χN denote properties on the scale of a whole molecule. They identify the length and energy scale of the model. In the same spirit, one can measure the polymer number density Φ in units of the chain volume R_e^3, and define a third dimensionless parameter $\mathcal{N} \equiv (\Phi R_e^3)^2 = (\rho R_e^3/N)^2$, where ρ denotes the number density of segments. In the dense melt, the chain conformations are Gaussian on large length scales, $R_e^2 = b^2 N$, and hence, $\mathcal{N} \approx N$. Therefore, we refer to this quantity as the invariant degree of polymerization. Phenomenologically it describes the number of other chains inside the volume of a reference chain.

Of course, one can also describe the properties of the blend by the characteristics on the scale of monomeric units: the number of segments N, the statistical segment length $b = R_e^2/N$, the Flory–Huggins parameter χ, and the monomer density $\rho = \Phi/N$. However, the results must not change upon representing the chains by a different number of effective segments N'. Such a reparameterization leaves the combinations χN, R_e, and $\mathcal{N}$ invariant.

Even in simple cases, the Flory–Huggins parameter χ typically results from subtle differences of dispersion forces between the different chemical constituents. The dispersion forces between all segments are strongly attractive when the liquid coexists with its vapor at vanishingly small pressure. The differences between the attractive interactions make up the Flory–Huggins parameter and these differences can be orders of magnitude smaller. Likewise, the extension of a polymer in a homogeneous melt results from a delicate screening of excluded volume interactions along the chain by surrounding molecules, and R_e depends on density and temperature. Accurately deriving those coarse-grained parameters, R_e and χN, from a microscopic model requires extremely accurate atomistic force fields. When these coarse-grained parameters are determined independently (e.g. by comparison to experiments) and used as an input, however, coarse-grained models are successful in making quantitative predictions. One role of coarse-grained models is to investigate the qualitative dependence of the coarse-grained parameters (e.g. like the Flory–Huggins parameter) on the gross features of the molecular architecture or structure of the polymeric liquid.

The parameters R_e, χN, and $\mathcal{N}$ encode the chemical structure of the polymers on microscopic length scales. The assumption of the coarse-grained modeling is that the properties on the mesoscopic length scale depend on the microscopic structure only via the parameters R_e, χN, and $\mathcal{N}$. In order to predict properties on large length scales, one has to identify these three parameters in the experimental system and the computational model, and one expects the behavior of the computational model to resemble the experimental system on mesoscopic length scales at the same values of the parameters, R_e, χN, and $\mathcal{N}$.

One can test this strategy and assess which properties on the mesoscopic length scale exhibit universal behavior by mutually comparing different experiments and computational models. The comparison between different computer simulation models is advantageous because a wealth of structural and thermodynamic properties on different time and length scales are simultaneously accessible.

Typically, computer simulations use particle-based models presenting one effective monomer by a particle on a lattice or in continuum space. Monomers that belong to a polymer are bonded together via a short-range attraction. To model liquid–liquid phase separation in a mixture, monomers of different species repel each other. The simulations yield the exact structural and thermodynamic properties of the model system without any approximation to the statistical mechanics. Statistical errors and finite size effects can be accounted for via rather sophisticated techniques [41, 42].

For the case of binary polymer blends, however, the coarse-grained parameter χN is often not extracted by a direct comparison to computer simulations but by the comparison with a mixture of Gaussian chains in continuum space interacting via zero-ranged segmental repulsions. Even though the particle-based models used in simulations and this field theoretical model [43, 44] differ substantially in their local structure, both models are expected to yield identical descriptions for properties on mesoscopic length scales by virtue of universality. This observation inspires reasonable confidence that coarse-grained models indeed capture the appropriate interactions to describe the universal aspects of liquid–liquid phase separation in experimental systems.

We emphasize, however, that the agreement between particle-based simulations and field-theoretical models can be significantly upset by the mean field approximation that is often invoked to extract predictions for the field-theoretic models. The mean field approximation [40] (or even stronger approximations) is necessary to obtain a tractable analytical description. In special limits, notably in the vicinity of the critical point (weak segregation) and at large incompatibilities (strong segregation), analytical expressions for various quantities as a function of χN and R_e are obtained and these simple expressions are routinely used in the analysis of experiments. Identification of the parameters χN and R_e by comparing experimental results with the mean field approximation might result in significant errors and different quantities might yield different, mutually inconsistent estimates. Of course, one should only use the mean field predictions to identify coarse-grained parameters by comparison with experiment and simulation in the regime of validity of the mean field approximation. Therefore, an additional role of coarse-grained computer simulations is to explore the validity of the mean field approximation for coarse-grained models.

We conclude this section with a summary of the mean field results, to provide some insight into the phenomena to be discussed and introduce our notation. We shall also point out some discrepancies between this mean field approach and experimental observations.

3.2.2. Predictions of the Mean-Field Theory

3.2.2.1. Spatially Homogenious Systems. We consider a dense mixture of polymers in a volume, V. For simplicity of notation, we restrict ourselves to a mixture where each molecule comprises the same number of segments, N, and the segmental volumes $1/\rho$ are identical. The considerations can be straightforwardly formulated for the more general case. Unlike segments repel each other with strength χ. The free energy of mixing per molecule takes the form [40]:

$$\frac{F_{\text{FH}}(\phi)}{T\Phi V} = \phi \ln \phi + (1-\phi)\ln(1-\phi) + \chi N \phi(1-\phi) \qquad (3.1)$$

where Φ denotes the number density of polymers and ϕ the composition of the incompressible mixture. T is the temperature and we set Boltzmann's constant at $k_B = 1$. The first two terms describe the entropy of mixing; it stems entirely from the translational entropy. The last term describes the energy of mixing per molecule. Note that this Flory–Huggins free energy [40] does not include any contribution from the conformational entropy of the extended macromolecules. Implicitly, one assumes that the conformations of a single chain in a homogeneous system are independent from the environment (i.e. local composition and temperature).

In the original lattice model used by Flory and Huggins [40], polymers are represented by self-avoiding walks on a simple cubic lattice. Interactions between monomers are extended to the nearest lattice sites, and are of strengths ε_{AA}, ε_{BB}, and ε_{AB}, respectively. Within this model the segmental interactions are related to the Flory–Huggins parameter explicitly via:

$$\chi = z_c \left(\varepsilon_{AB} - \frac{\varepsilon_{AA} + \varepsilon_{BB}}{2} \right) \quad (3.2)$$

When χ is used as an adjustable parameter or extracted from experiments, this mean field theory [40] of Flory and Huggins is quite successful in describing many experimental observations. Most notably the theory rationalizes the fact that long macromolecules tend to demix at very high temperatures. The critical point of the blend is located at:

$$\frac{N}{T_c} \sim \chi_c^{MF} N = 2 \quad \text{and} \quad \phi_c^{MF} = \frac{1}{2} \quad (3.3)$$

The Flory–Huggins theory provides simple analytical expressions for the free energy of mixing. In a symmetric blend, the binodals are given by

$$\chi N = \frac{1}{2\phi - 1} \ln \frac{\phi}{1 - \phi} \quad (3.4)$$

Since free energies are not easily accessible in Monte Carlo simulations, one often considers the relation between the exchange potential $\Delta\mu \equiv \mu_A - \mu_B$ and the composition of the mixture, where μ_A and μ_B denote the chemical potentials per molecule.

$$\frac{\Delta\mu}{T} = \ln \phi - \ln(1 - \phi) + \chi N(1 - 2\phi) \quad (3.5)$$

The simple form of the bulk free energy lays at the basis of self-consistent field (SCF) calculations [43–47] for spatially inhomogeneous polymer systems in the framework of the Gaussian chain model. Nevertheless, the Flory–Huggins theory cannot rationalize the following observations:

(1) The temperature dependence of the measured values on the χ parameter often takes the form $\chi = a + b/T$. Following a common convention, b is denoted as the enthalpic contribution, whereas a is referred to as an additional entropic contribution. Entropic contributions might arise, for instance, from the dependence of the packing of the segments (fluid structure) on the local environment. In experiments, however, the temperature dependence of χ is often only accessible over a rather small temperature range (limited by the glass transition temperature and/or thermal degradation) and there is a strong interplay between entropic and enthalpic contributions, which makes the standard decomposition of χ into an entropic and an enthalpic part difficult; for example, the temperature dependence of the chain conformations or equation of state effects might yield a nonlinear relation between

χ and the inverse temperature. Moreover, the χ parameter often depends on composition and chain length as well.

(2) According to the original mean field theory, the demixing temperature is independent of the molecular architecture. Hence, blends of two homopolymers and two ring or star polymers with the same number of monomeric units are predicted to have the same miscibility behavior. Similarly, the theory does not capture the dependence of the miscibility behavior of the chain stiffness or the degree of branching [48, 49].

(3) Being a mean field theory, the Flory–Huggins theory neglects long-range critical fluctuations of the local composition and invokes a random mixing approximation. Thus, the behavior in the vicinity of the critical point is described by mean field exponents and the binodals have a parabolic shape. In the ultimate vicinity of the critical point, the correlation length grows very large and the polymeric properties become irrelevant. In this critical region the behavior is characterized by the 3-D Ising universality class [21], which applies to all binary mixtures with short-range interactions. The latter behavior manifests itself in a much flatter binodal at the critical point and a stronger divergence of composition fluctuations. This has been observed in neutron scattering experiments extremely close to the critical point [50, 51].

(4) The same intermolecular forces which determine the miscibility behavior alter the conformation of the extended flexible macromolecules. Monte Carlo simulations [15, 21, 22, 52–55] for rather short chain lengths reveal a contraction of the polymer coils in the minority phase. Experiments in highly incompatible poly(methyl methacrylate) (PMMA) and poly(vinyl acetate) (PVAc) blends of rather low molecular weight indicate a relative contraction of isolated PMMA chain extensions by 13–15 percent [56]. These findings indicate a possible coupling between the single chain conformations and the thermodynamic state (i.e. temperature and composition of the mixture).

These observations can be qualitatively divided into two classes: The first two issues refer to short-ranged correlations on the scale of a segment. The Flory–Huggins free energy of mixing does not predict how the molecular architecture alters the local structure of the polymeric fluid. As we shall discuss in Section 3.4.1.1, these deficiencies can be accounted for by an effective identification of the Flory–Huggins parameter. The last two deviations from the Flory–Huggins theory stem from composition fluctuations on the length scale of the whole molecule or larger. They often become less important if the invariant degree of polymerization, $\mathcal{N}$, increases. Specific examples are discussed in Sections 3.4.1, 3.4.2.3 and 3.4.3.

3.2.2.2. Self-consistent field theory. In spatially inhomogeneous systems, the structure of the polymer molecules becomes more apparent. The SCF theory [43–47] considers a mixture of Gaussian chains with end-to-end distance R_e. The statistical mechanics of the interacting multichain system is analytically intractable and therefore one invokes a mean field approximation: The interaction of an A-chain with its neighbors is approximated by an effective external field $W_A[\phi]$, which, in turn, depends on the local composition $\phi(\mathbf{r})$. Then, the local density of A-chains is calculated as the Boltzmann average of the single chain density in the external field, and likewise for B-polymers. This relation and $W_A[\phi]$ constitute a closed, self-consistent set of equations, which allows inhomogeneous systems to be described [43–47].

The single chain problem in the spatially varying external field can be solved only numerically. Using the Gaussian chain model, one has to solve a modified diffusion equation that describes the propagation of the Gaussian, random walk polymer in the external field [43, 44]. The equations can be solved efficiently in real space, using a Crank–Nicholson scheme or pseudospectral algorithms [57]. If one is interested in self-assembled phases with a long-range periodicity, it is advantageous to expand the spatial dependencies of the fields and densities in eigenfunctions of the Laplace operator which possess the symmetry of the phase under consideration [47]. Then, the modified diffusion equation becomes a linear matrix equation for the Fourier coefficients.

In some limiting cases simple analytic expressions can be obtained, which are often used to analyze experimental and simulational results.

2. Weak Segregation Limit.
In the vicinity of the critical point, the variation of the composition can be treated as a small parameter. Expanding the spatial variation of the composition in a Fourier basis, $\phi(\mathbf{q}) = \phi_A(\mathbf{q}) = -\phi_B(\mathbf{q})$ for $q \neq 0$, and only considering terms to quadratic order does one derive the random-phase approximation (RPA) for the free energy per molecule [58]:

$$\frac{F_{\text{RPA}}[\phi(\mathbf{q})]}{T\Phi V} = \frac{1}{2}\sum_{\mathbf{q}} \phi(\mathbf{q})\left\{\frac{1}{\phi g_A(\mathbf{q})} + \frac{1}{(1-\phi)g_B(\mathbf{q})}\right\}\phi(-\mathbf{q}) - \frac{1}{2}\sum_{\mathbf{q}} \phi(\mathbf{q})2\chi N\phi(-\mathbf{q})$$

(3.6)

where the Debye function g characterizes the single chain structure factor of a Gaussian chain,

$$g(\mathbf{q}) = \frac{12}{q^2 R_e^2}(\exp[-q^2 R_e^2/6] - 1 + q^2 R_e^2/6)$$
(3.7)

The first two terms in equation (3.6) represent the conformational entropies of the two polymer species; the last term is the energy of mixing.

The experimentally accessible collective structure factor $S(\mathbf{q})$ of composition fluctuations is given by:

$$\frac{N}{S(\mathbf{q})} = \frac{1}{\phi g_A(\mathbf{q})} + \frac{1}{(1-\phi)g_B(\mathbf{q})} - 2\chi N$$
(3.8)

From the wavevector dependence one determines the correlation length, ξ, of composition fluctuations around the homogeneous phase in the one-phase region:

$$\frac{\xi}{R_e} = \frac{1}{\sqrt{18[1 - 2\chi N\phi(1-\phi)]}} \qquad (3.9)$$

The divergence of the structure factor at $q = 0$ signals the macrophase separation. The maximum of the structure factor in the one-phase region is routinely used to estimate the Flory–Huggins parameter experimentally:

$$\frac{N}{S(q_{max}=0)} \equiv \frac{1}{V\Phi}1\langle\phi^2\rangle - \langle\phi\rangle^2 = \frac{1}{\phi} + \frac{1}{(1-\phi)} - 2\chi N \stackrel{\phi=\phi_c=\frac{1}{2}}{=} 2(\chi_c^{MF}N - \chi N) \qquad (3.10)$$

where the last equation hold for critical blends only and $\chi_c^{MF} N = 2$.

For larger incompatibilities $\chi N > \chi_c N = 2$, the blend phase separates into an A-rich and a B-rich bulk. The two phases are separated by an interface, which is described by a tanh-profile

$$\phi(z) = \frac{1}{2}(1 + \tanh[z/w_{WSL}]) \quad \text{with} \quad w_{WSL} = 2\xi \qquad (3.11)$$

In the weak segregation limit (WSL) a polymer mixture behaves similarly to a mixture of small molecules, because the characteristic size of composition variations is on the order of the molecular extension R_e, and therefore the Gaussian statistics of the chain conformations is not important; only the overall size of the molecule matters.

3. Strong Segregation Limit. Far below the critical temperature, $\chi N \gg 2$, the SCF calculations take a particularly simple form (ground state dominance). In this case, the entropy loss the chain suffers at a spatial inhomogeneity can be described by the Lifshitz formula [59]:

$$\frac{S[\phi]}{\Phi V} = -\frac{R_e^2}{24V}\int d^3\mathbf{r}\left\{\frac{(\nabla\phi)^2}{\phi} + \frac{[\nabla(1-\phi)]^2}{1-\phi}\right\} \qquad (3.12)$$

The free energy as a function of the local composition is given by:

$$\frac{F[\phi]}{T\Phi V} = \frac{\chi N}{V}\int d^3\mathbf{r}\phi(1-\phi) - \frac{S[\phi]}{\Phi V} \qquad (3.13)$$

Minimizing this free energy functional with appropriate boundary conditions, one obtains the profile $\phi(z)$ across an interface between an A-rich and a B-rich domain. It is also a tanh-profile, as in equation (3.11), but with a width

$$\frac{w_{SSL}}{R_e} = \frac{1}{\sqrt{6\chi N}} \qquad (3.14)$$

This result can be heuristically understood as follows: The properties of the interface are dominated by excursions of A-polymers into the B-rich phase, and vice versa. The energy cost of each loop is comparable to the thermal energy scale T. Each monomer of a loop contributes to the energy an amount of the order χT, and, consequentially, the number of monomers per loop scales like $1/\chi$. The spatial extension of the loops determines the interfacial width. In the Gaussian chain model the conformational distribution is Gaussian on all length scales, and therefore the spatial extension of loops is on the order of $R_e/\sqrt{\chi N}$.

Each monomer in the interfacial region contributes to the interface tension an amount χ, and the interface tension scales as $\chi \Phi N R_e/\sqrt{\chi N}$. The result of the SCF theory (including prefactor of order unity) is [60]:

$$\frac{\sigma_{SSL} R_e^2}{T} = \sqrt{N}\sqrt{\frac{\chi N}{6}} \qquad (3.15)$$

In the strong segregation limit (SSL), loops determine the interface properties, which depend sensitively on the conformational statistics of the molecule. The SCF theory describes both limits, (i) WSL and (ii) SSL, and the crossover between them.

3.3. MODELS AND TECHNIQUES

Simulational models of various degrees of coarse-graining have been employed to study the structure and thermodynamics of binary polymer blends. Models range from the representation of polymers as self-avoiding walks on a simple cubic lattice—as in the original treatment of Flory and Huggins [40]—to simulations of the effect of branching in hydrocarbon melts in the framework of a united atom model [15, 61]. The choice of the simulation model is a compromise between computational efficiency and a more faithful representation of the details of molecular architecture.

Biased by our own research, we present Monte Carlo (MC) simulations in the framework of the bond fluctuation model [62], which incorporates the relevant universal characteristics of polymer blends: connectivity of the monomers along a chain, excluded volume of the segments, and a thermal interaction between monomers. In the framework of this coarse-grained lattice model, a monomer occupies the eight corners of a unit cell from further occupancy. Monomers along a polymer are connected by one of 108 bond vectors of length $2, \sqrt{5}, \sqrt{6}, 3$ and $\sqrt{10}$. The bond vectors are chosen such that the excluded volume interactions prevent a crossing of bond vectors during local hopping motion. Therefore the algorithm captures the effect of entanglements. The large number of bond vectors allows for 87 different bond angles—an indication of the rather good approximation of continuous space properties by this complex lattice model. This property also allows for a rather realistic implementation of the bending rigidity.

Here and in the following, all length scales are measured in units of the lattice spacing. When atomistically detailed simulations are mapped onto the bond fluctuation model a lattice unit corresponds to roughly 2 Å and a monomer in the bond fluctuation model represents a small number—say three to five—of chemical repeat units [12]. If not noted explicitly, we work at a monomer number density of $\rho \equiv \Phi N = 1/16$; that is, due to the extended structure of the monomers half of the lattice sites are occupied. These parameters correspond to a concentrated solution or a melt. On the one hand, the presence of vacancies allows a reasonably fast equilibration of the chain conformations on the lattice. On the other hand the size disparity between vacancies and extended monomers gives rise to packing effects. Indeed, the monomer–monomer density pair correlation function $g(\mathbf{r})$ exhibits oscillations at small distances, indicating a fluid-like packing due to the local compressibility. Moreover, the relation between osmotic pressure and density can be well described via the Carnahan–Starling equation [63]—an approximation for the equation of state of hard spheres in continuum space. This shows that this lattice model shares many features with off-lattice models. It is a compromise between simple lattice models, which represent monomers by a single lattice site and have only very limited possibilities to include details of the molecular architecture, and off-lattice models, which are computationally considerably more exacting (cf. also Indrakanti et al. [64] for a comparison between 'fine-grained' lattice models and continuum models).

The conformations of the polymers on the lattice evolve via local random monomer hopping [62]—a randomly chosen monomer attempts to move one lattice constant in a random direction—or slithering snake-like moves [23, 65]—a segment of the chain is removed at one end of the chain and added at the opposite end. While the former allows for a dynamical interpretation of the MC simulation in terms of a purely diffusional dynamics [62], the latter relaxes the chain conformations a factor N faster [65].

The blends comprise two components—denoted A and B. Monomeric units of the same type attract each other, whereas different monomers repel each other via a square well potential

$$\varepsilon \equiv -\varepsilon_{AA} = -\varepsilon_{BB} = \varepsilon_{AB} \qquad (3.16)$$

The potential is extended over the first peak of the pair correlation function; that is, it incorporates the first 54 neighbors up to a distance $\sqrt{6}$. The form of the potential is chosen by computational convenience; we expect our results to be qualitatively independent from the specific potentials used. However, if we were to model the interactions as (strongly) attractive with (slightly) different strengths between unlike species—a more faithful modeling of interactions in view of the experimental situation—the presence of vacancies would allow for a liquid–vapor phase separation between a concentrated polymer melt and a dilute phase in our ternary system. This liquid–vapor phase separation is common to both lattice based models and models in continuous space [16–20]. The temperature scale of this liquid–vapor coexistence is set by the Θ temperature, which is chain length-independent. This contrasts with the temperature scale of the liquid–liquid phase separation into A-rich and B-rich phases with similar content of vacancies. The latter temperature scale increases linearly with the chain length. Therefore, the two phenomena are well separated in blends of high molecular weight. In this paper we focus on the liquid–liquid phase separation at high temperatures.

Being a lattice model, the bond fluctuation model is highly computationally efficient. It allows for the investigation of rather large chain lengths and large system sizes. The latter is necessary to accurately locate the critical temperature via finite size scaling analysis. As we shall illustrate, the large chain length is crucial for reaching the high molecular weight scaling limit and extrapolating some quantities to experimentally relevant chain lengths. For the present investigation, chains with up to 512 (2048 for athermal systems [66]) monomeric units have been employed.

Various simulation techniques have been used to explore the miscibility properties of polymer blends. The most direct one is the simulation of both the coexisting phases in the simulation cell. This method has been employed by Madden [67] and Cifra [68] for well-segregated blends. In addition to the bulk properties (e.g. composition of the coexisting phases), this setup also yields information about interfacial properties. However, it requires rather large simulation cells in order to extract 'bulk' properties. In particular, the scheme is not very well suited to cope with the growing of the correlation length and vanishing of the difference between the phases as the critical point is approached. The direct estimation of the chemical potential of each individual species as a function of temperature and composition seems to be more computationally efficient. At coexistence the chemical potentials of the species in both phases are equal and the coexistence curve can be mapped out. This technique has been applied successfully by Kumar [69] using the incremental chemical potential method [70]. Kumar explored the influence of pressure and compressibility on the miscibility behavior. If one point on the coexistence curve is known, a Gibbs–Duhem integration technique can be employed to obtain the coexistence under constant pressure conditions [71, 72]. Both methodologies are particularly useful for blends in which the constituents are characterized by very different chain architecture.

Sariban and Binder [21] employed simulations in the semigrandcanonical ensemble for investigating the phase behavior at constant volume. In this ensemble, the total monomer density is fixed, the composition of the blend fluctuates, and the chemical potential difference $\Delta\mu$ between the species is controlled. The MC scheme comprises two types of moves. Canonical updates relax the conformation of the macromolecules on the lattice, whereas semigrandcanonical ones transfer A polymers into B polymers and vice-versa. Sariban and Binder [21] investigated strictly symmetric chains, for which the semigrandcanonical moves consist of a mere exchange of labels. The algorithm can be extended to some degree of structural asymmetry (e.g. different chain lengths between the species [23]). Overall speaking, it is reasonably efficient for a modest degree of structural asymmetry between the different constituents, but the extension to pronounced structural asymmetries is a challenging task. Improvement might be achieved by gradually 'mutating' one species into another [73]. The major advantages of the methodology stem from the study of composition fluctuations the mixture.

The relaxation times are much smaller than in the canonical ensemble, where the composition is conserved and composition fluctuations decay via the slow diffusion of polymers in a melt. The semigrandcanonical ensemble allows the straightforward application of finite size scaling techniques known from simple mixtures [42]. Therefore we can measure the critical temperature in symmetric and asymmetric mixtures accurately from the MC simulations of modest system size. Moreover, the relation between the composition of the mixture and the difference in the chemical potentials $\Delta\mu$ of the species is directly accessible with high accuracy. The latter is particularly important to establish a direct contact to analytical approaches outside of the ultimate critical region where 3-D Ising critical behavior dominates, that is, in the region where the mean field theories are applicable. Additionally, it is possible to determine excess interfacial properties (e.g. interfacial tension [29], excess energy [74], and enrichment of a third component [74]).

An important quantity in the semigrandcanonical simulations is the probability distribution of the composition $P(\phi)$. At phase coexistence, it exhibits two peaks that correspond to the two coexisting phases. At coexistence, the exchange chemical potential $\Delta\mu = \mu_A - \mu_B$, conjugated to the composition, has to be chosen such that both peaks have equal weight [75]. (For a symmetric blend, $\Delta\mu_{coex} = 0$, of course.) Thus, the simulation provides information about the composition of the two coexisting phases. From the width of the peaks, one additionally estimates the strength of composition fluctuations, that is, the maximum of the collective structure factor [cf. equation (3.10)].

At the critical point, the probability distribution adopts a universal shape when scaled to unit norm and variance. The shape characterizes the universality class of the transition. In case of unmixing this is the 3-D Ising universality class. Adjusting the temperature (i.e. ε/T) so as to match the probability distribution obtained from simulations and the predetermined universal curve, we accurately determine the critical temperature from the simulation [76].

At lower temperatures, the system has to tunnel between the two coexisting states, which are separated by a large free energy barrier (cf. Figure 3.1a). Therefore the probability of finding the system at $\phi_{middle} = (\phi_{coex}^{(1)} + \phi_{coex}^{(2)})/2$ is very low. In the middle of the miscibility gap, the typical configuration consists of a slab of A-rich phase, which is separated by two interfaces of size L^2 from the B-rich phase. The probability of these configurations is suppressed by an amount $P_{middle}/P_{coex} = \exp(-2L^2\sigma/T)$ [77]. Thus, we not only obtain information about the coexisting phases (composition, fluctuations, and coexistence chemical potential), but also about interface properties. One can also monitor other quantities (e.g. the excess of a third species, see Figure 3.1b) as a function of the composition and obtain the difference (excess) between the bulk and the system containing two interfaces.

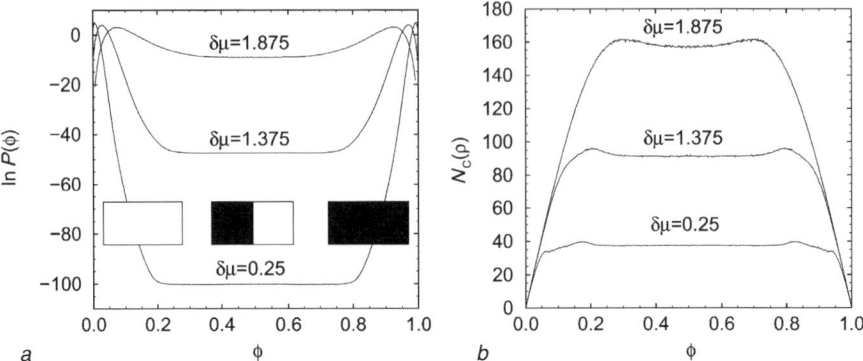

Figure 3.1. Ternary blend containing two homopolymers A and B and a symmetric AB diblock copolymer. All polymers have length $N = 32$ (see Section 3.4.3 for a detailed description of the system). (a) Probability distribution at $\varepsilon = 0.054T$ and system size $48 \times 48 \times 96$ (in units of the lattice spacing). Upon increasing the chemical potential $\delta\mu$ of the copolymers the 'valley' becomes shallower, indicating that the copolymers decrease the interface tension, σ. One clearly observes a plateau around $\phi = 1/2$. This insures that the system size is large enough to neglect interfacial interactions in the measurement of the interface tension. (b) Average number of copolymers as a function of the composition for the same parameters as in (a). The copolymer number is enhanced in the configuration containing two interfaces, and from the simulation data one estimates the adsorption ν of copolymers at the interface. Adapted from Müller and Schick [74].

In order to overcome the free energy barrier, one applies reweighting techniques [78]: To this end one adds to the original Hamiltonian $\mathcal{H}$ of the system a weight function $Tw(\phi)$ that depends on the composition, but not on the microscopic conformations of the polymers. The probability distribution of the reweighted system is given by:

$$P_{\text{rw}}(\phi) \sim P(\phi) \exp[-w(\phi)] \qquad (3.17)$$

Choosing $w(\phi) \approx \ln P(\phi)$, the distribution $P_{\text{rw}}(\phi)$ in the simulation is flat; that is, the system samples all compositions with roughly equal probability. The crux is that the probability distribution P is not known *a priori*; it is rather the result of the calculation. Several schemes to overcome this dilemma have been proposed.

First, histogram-reweighting techniques [79] alleviate this problem by performing a sequence of reweighted simulations and extrapolations starting at a point where barriers are small and the system explores a broad range of n, that is, close to the liquid–liquid critical point. Those results can then be successively extrapolated to lower temperatures, where barriers are larger. To do so, one accumulates a joint histogram $H(\phi, E|T, \Delta\mu)$ of the composition ϕ and the energy E of the system at a state characterized by temperature T/ε and exchange potential $\Delta\mu$. The probability distribution at a neighboring state described by T' and $\Delta\mu'$ is given (up to a normalization factor) by:

$$P(\phi|T',\Delta\mu') \sim H(\phi,E|T,\Delta\mu)\exp\left(-\frac{E}{T'}+\frac{E}{T}+\frac{\Delta\mu'\phi V\Phi}{T'}-\frac{\Delta\mu\phi V\Phi}{T}\right) \quad (3.18)$$

This extrapolation only works if the histograms of both states significantly overlap in the ϕ–E plane. The verification of the extrapolation via a control simulation is also a fine test of the simulation code, because one verifies that the configurations are sampled with the appropriate statistical weight. More sophisticated methods combine results of multiple histograms [80] to enlarge the parameter range over which one can extrapolate the results. Second, weight factors can be obtained from the transition probabilities between macrostates [81, 82]. Third, multicanonical recursion [83] conducts a series of short trial runs. After each run $w(\phi)$ is adjusted until the simulation can access all relevant states. Fourth, the weight function can also be self-adjusted during the simulation [84–87]. One starts with an initial guess of the weight function $w(n)$ and increments its value by Δw, each time a state with composition ϕ is visited in the course of the simulation. This procedure tends to 'push' the system out of states that it has frequently visited, and allows it to explore all pertinent states. If the histogram of visited states is approximately flat, that is, all states have been visited with nearly equal probability, the increment Δw is decreased and the histogram of visited states reset. This method works in many practical cases; it requires, however, some fine-tuning of the initial value of the increment Δw. Moreover, for any $\Delta w > 0$, detailed balance [41] is violated and separation of statistical and systematic errors becomes difficult. Finally, one could alternatively use successive umbrella sampling [88] to generate the probability distribution, $P(\phi)$. To this end, one divides the interval of composition into smaller windows that overlap at their boundaries. Each window is sampled consecutively and the results are linked together at the boundaries of the window. Ideally, the statistical error for a given total amount of computation time is independent from the window size. Hence, choosing a small windows size is beneficial, because the difference in $P(\phi)$ within a single window becomes small and one can simulate without weight function or use a rather crude estimate for $w(\phi)$, for example, a polynomial extrapolation of the results from previous windows. Choosing the window size too small, however, might cause sampling difficulties.

3.4. RESULTS

3.4.1. Phase Behavior: Fluid Structure and Composition Fluctuations

3.4.1.1. Identification of the Flory–Huggins Parameter χ for Symmetric Blends. A wealth of properties of binary polymer mixtures can be extracted from computer simulations as well as from experiments. As mentioned above, comparing these results to the approximate mean field expressions might give rise to quite distinct estimates for the Flory–Huggins parameter, partially due to the neglect of fluctuations in the mean field approach. Let us mention some possible identifications:

(1) The relation between exchange potential, $\Delta\mu$, and composition, ϕ, of the blend, cf. equation (3.5), or composition fluctuations/scattering, cf. equation (3.8), in the one-phase region far away from the critical point can be used to identify χ.
(2) Measuring the composition of the two coexisting phases at large incompatibility and comparing the result with the prediction of the Flory–Huggins theory, equation (3.4), one can estimate the χ parameter.
(3) The mean field theory also relates the location of the critical point of demixing [cf. equation (3.3)], its dependence on chain length, and scattering in the vicinity of the critical point to the Flory–Huggins parameter.
(4) The SCF theory makes explicit predictions for the width of the interface between coexisting phases, and simple analytical expressions are known in the weak and strong segregation limit [see equation (3.11) or (3.14)]. Comparing experimental measurements or simulation results with those predictions additionally yields an estimate for χ.

The first three identifications basically treat the Flory–Huggins parameter, χ, as a phenomenological quantity, determined by matching the thermodynamic properties of the blend to the predictions of mean field theory and thereby absorbing all unknown information concerning blend miscibility [89]. This is the spirit of coarse-grained models, where the structure on the microscopic length scale determines the behavior on the large length scale only via a small number of parameters. The identification of the set of these parameters (for the problem one wants to study) is an important issue on its own. Certainly, it is instructive to relate the Flory–Huggins parameter to the structure of the polymeric fluid similar to equation (3.2), because this contributes to correlating the molecular architecture and miscibility behavior and thereby greatly enlarges the predictive power of the theory. However, even if this was not possible, the identification of coarse-grained parameters (like the Flory–Huggins parameter) would be useful, for instance, to predict the properties of interfaces between coexisting phases [43], the kinetics of phase separation [90, 91] or the behavior at surfaces.

Therefore, let us first discuss these thermodynamic identifications of the Flory–Huggins parameter and then try to relate these observations to the structure of the polymeric fluid.

Measuring the dependence of the composition on the exchange potential is straightforward in the semigrandcanonical ensemble, where the exchange potential $\Delta\mu$, is a parameter and one observes the composition and its fluctuations in the simulation. The dependence $\phi(\Delta\mu)$ [cf. equation (3.5)] and the composition fluctuations $\langle\phi^2\rangle - \langle\phi\rangle^2$ [cf. equation (3.8)] correspond to the first and second derivative of the Flory–Huggins free energy of mixing with respect to the composition, and therefore constitute a direct test of the Flory–Huggins free energy [40].

RESULTS

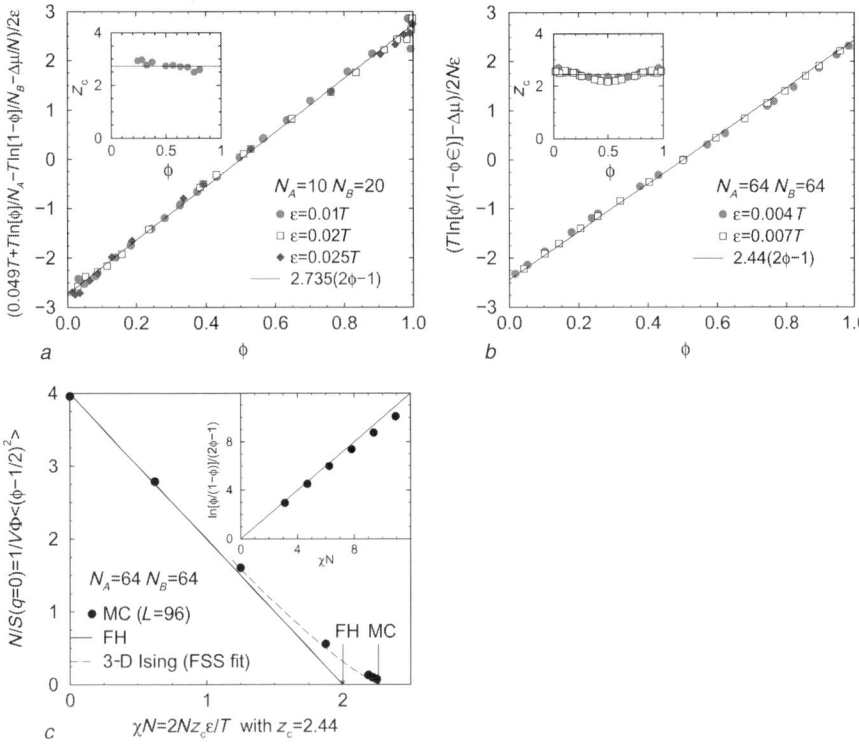

Figure 3.2. Estimation of the Flory–Huggins parameter: (*a*) first derivative of the Flory–Huggins free energy, equation (3.5), for a blend with chain length asymmetry, $N_A = 10$ and $N_B = 20$. The slope of the curve yields the effective coordination number $z_c = 2.735$. The inset presents the value of the coordination number $z_c = \chi T/2\varepsilon$ extracted from the collective structure factor, equation (3.10). The horizontal line marks the estimate from the main panel. From Müller and Binder [24]. (*b*) Same as (*a*) for a symmetric blend, $N \equiv N_A = N_B = 64$. (*c*) Inverse collective structure factor $1/S(q=0)$ vs Flory–Huggins parameter. Symbols denote simulation results for $N = 64$ and system size $L = 96$. The dashed line shows the finite size estimate using the Ising critical behavior. Arrows mark the location of the critical temperature (MC) and the mean field estimate (FH). The straight solid line corresponds to the mean field prediction, equation (3.10), using the effective coordination number from (*b*). The inset shows the concentration of the coexisting phases for $\chi N > 2$. The solid line is the mean field prediction, equation (3.4), using $z_c = 2.44$.

The simulation data for the bond fluctuation model are presented in Figure 3.2. In Figure 3.2*a* and *b* we plot the difference between the exchange chemical potential and the ideal gas contribution, $(\Delta\mu - T\ln[\langle\phi\rangle/(1-\langle\phi\rangle)])/2N\varepsilon$, vs the composition $\langle\phi\rangle$. In accord with the Flory–Huggins estimate, equation (3.2), we observe a linear dependence on composition for all temperatures far above the unmixing transition. This demonstrates that $\chi \approx \varepsilon$ in our model and we can read off the effective coordination number z_c. The excess of the exchange potential has also been used by Buta and Freed [92] to compare MC simulations to the predictions of the lattice cluster theory (LCT).

In the inset of Figure 3.2a and b, we present the estimate for the effective coordination number via the inverse of the collective structure factor, equation (3.10). Within the statistical uncertainties, the simulations give consistent results far above the critical point. Even though composition fluctuations are directly related to the experimental procedure, we prefer to extract the Flory–Huggins parameter from the excess of the exchange potential, equation (3.5), because the combinatorial contribution diverges only logarithmically for $\phi \to 0$ or 1. Also in experiments, strongly asymmetric compositions might induce inaccuracies [93]. In Figure 3.2c, we use this identification of the Flory–Huggins parameter to plot the inverse structure factor as a function of χN. For $\chi N \ll \chi_c N = 2$ the data are consistently describable by the mean field theory. Likewise, for strong incompatibilities, $\chi N \gg 2$, the mean field prediction, equation (3.4), yields a good description of the concentration of the two coexisting phases (see inset of Figure 3.2c). Hence, the first two methods lead to a mutually consistent identification of the Flory–Huggins parameter both above and below the critical point.

In the ultimate vicinity of the critical point, however, deviations from the mean field become pronounced. The inverse structure factor extracted from the MC simulations does not vanish at $\chi N = 2$, but the mean field theory overestimates the true critical temperature by about 12 percent. (Note that the chain length $N = 64$ corresponds to an invariant degree of polymerization of $\mathcal{N} = 240$.) Similar to previous simulations [21] and experiments [50, 51, 94, 95] we observe in Figure 3.2c the crossover from the mean field dependence to the Ising behavior: If we approach the critical point at $\phi_c = 1/2$, we do not observe a linear dependence of the inverse scattering factor with ε/T all the way up to the critical point, but we rather find $1/S(q=0) \approx (\varepsilon/T - \varepsilon/T_c)^\gamma$ close to $\chi_c N$, where $\gamma \approx 1.24$ denotes the critical exponent of the order parameter fluctuations in the 3-D Ising model. By virtue of the universality at second-order phase transition, a polymer mixture exhibits the same critical behavior as other mixtures characterized by a single scalar order parameter. The Flory–Huggins theory, however, being a mean field theory that assumes random mixing, cannot describe these composition fluctuations. If we extracted the Flory–Huggins parameter from the composition dependence of the concentration fluctuations at the critical point, χ would acquire a strong composition dependence [22].

These observed deviations from mean field theory in the simulations are in agreement with experiment [50, 51, 94, 95]. Hence, one can identify useful coarse-grained parameters χN, R_e, and $\mathcal{N}$ by matching the location of the critical point in the simulations and the experiments, because both include fluctuations. This resembles the 'principle of corresponding states' which is often employed to compare equation of states and phase diagrams [96]. However, it is inappropriate to use a mean field approximation to extract the χ parameter in a parameter region (i.e. the critical region) where the mean field approximation breaks down. Hence, we cannot expect the third method, which identifies the Flory–Huggins parameter by a comparison to mean field theory in the vicinity of the critical point, to give consistent results.

In Section 3.4.4 we will demonstrate that the fourth method, which tries to identify the χ parameter by comparing the width of an interface between coexisting phases to the mean field prediction, is also unreliable, because the mean field theory neglects capillary waves [97] of the local interface positions, which results in a length scale-dependent broadening of the profiles measured in experiments and simulations.

Having identified the Flory–Huggins parameter for our model of a symmetric polymer blend via a direct comparison of the simulation data to the Flory–Huggins free energy of mixing, we can also try to relate these thermodynamic results to the structure of the polymeric fluid. This is important for predicting the miscibility behavior for a specific pair of polymers; however, it also involves additional assumptions. The advantage of computer simulations is that one can start with the simplest possible model—a symmetric polymer blend—and successively increase the degree of architectural complexity. Thereby, one helps to clarify which deviations from the Flory–Huggins theory stem from the mean field assumption for composition fluctuations and which can be traced back to the local structure of the polymeric fluid.

In this section, we regard only symmetric blends and defer the discussion of structural asymmetries to Section 3.4.2.2. In the Flory–Huggins theory [40], the χ-parameter parameterizes the free energy of mixing in excess of the combinatorial (ideal gas) contribution. In general, this excess free energy of mixing will comprise contributions due to the interactions between the different species as well as contributions which stem from the change of the chain conformations and the packing structure of the fluid upon mixing.

In the framework of the original lattice model of Flory and Huggins the χ-parameter results only from the interactions between different species, that is, the energy of mixing. Both the chain conformations as well as the fluid structure are assumed to be independent of composition. In fact, this also holds true to a good approximation for our highly idealized model of a symmetric polymer blend. Identifying the Flory–Huggins parameter χ via the energy of mixing, we obtain the estimate:

$$\chi N = \Phi N^2 \int d^3\mathbf{r} \left[g_{AB}^{\text{inter}}(\mathbf{r}) v_{AB}(\mathbf{r}) - \frac{g_{AA}^{\text{inter}}(\mathbf{r}) v_{AA}(\mathbf{r}) + g_{BB}^{\text{inter}}(\mathbf{r}) v_{BB}(\mathbf{r})}{2} \right] \quad (3.19)$$

$$= \frac{2N z_c \varepsilon}{T} \quad \text{with} \quad z_c = \Phi N \int_{r \leq \sqrt{6}} d^3\mathbf{r}\, g^{\text{inter}}(\mathbf{r}) \quad (3.20)$$

We emphasize, that this identification of χ via the energy of mixing assumes random mixing and only can be employed in the regime where the mean field approximation is valid. $v_{IJ}(I, J = A, B)$ denotes the interaction potentials between the different segments. $g_{IJ}^{\text{inter}}(\mathbf{r})$ denotes the intermolecular pair correlation function, which describes the probability of finding a monomer of type J that belongs to a different chain a distance $\mathbf{r}$ away from a monomer of type I. Note that only the intermolecular energy contributes to the energy change upon mixing; the intramolecular energy does not depend on the composition, because the chain conformations are assumed not to depend on composition. In analogy to the original treatment of Flory and Huggins [40], we refer to z_c as the effective coordination number.

Equation (3.19) expresses the χ parameter in terms of the local fluid structure of the polymer liquid and the quantities involved are directly accessible in the MC simulations. Often, we shall approximate the intermolecular pair correlation functions by their athermal values [98]. Even in this simple form, it already is evident that the chain architecture has pronounced effects on the intermolecular pair-correlation function and the miscibility behavior: The more open the chains are, the larger the number of intermolecular contacts is, and the smaller is the miscibility. This partially rationalized the failure of the simple Flory–Huggins theory.

In Figure 3.3a we present the intermolecular pair correlation function of strictly symmetric polymer blends in the athermal limit and at the critical point of the mixture. In the athermal case, $\varepsilon = 0$, the distinction between the two species becomes irrelevant. The intermolecular pair correlation function mirrors two effects [23]: Owing to the extended monomer structure the pair correlation function vanishes for distances $r < 2$. The presence of vacancies introduces local packing effects, which give rise to a highly structured function at short distances. One can identify several neighbor shells, which are characteristic of the monomeric fluid. These packing effects are, of course, absent in simple lattice models where a monomeric unit occupies a single lattice site, and are less pronounced in the bond fluctuation model than in continuum models. The length scale of these packing effects is set by the monomeric extension or the statistical segment length; the detailed shape depends strongly on the model and the degree of structure on local length scales.

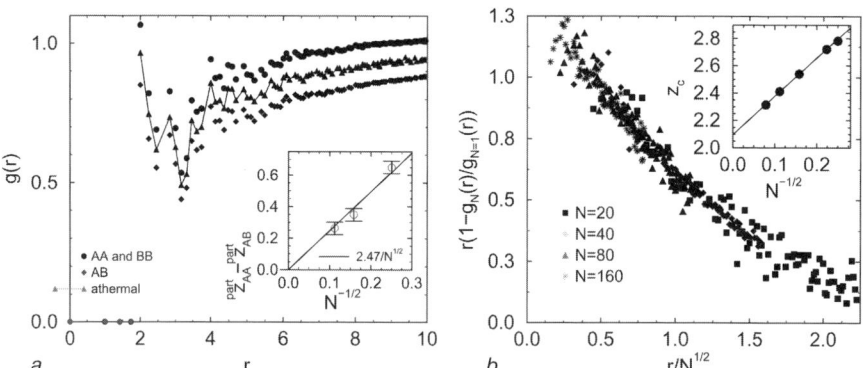

Figure 3.3. (a) Intermolecular pair correlation function for chain length $N = 80$ for the athermal system (triangles) and at criticality (circles and diamonds). The inset presents the scaling of the non-random mixing with increasing chain length. $z_{AA}^{\text{part}} = \langle \phi \rangle \int d^3 r g_{AA}^{\text{inter}}(\mathbf{r})$. From Müller [99]. (b) Correlation hole of linear chains: scaling behavior of the athermal intermolecular pair correlation function with chain length. The inset shows the chain length dependence of the effective coordination number. The line corresponds to $z_c = 2.1 + 2.8/\sqrt{N}$. Adapted from Müller and Binder [23].

RESULTS

Furthermore, the extended structure of the macromolecules manifests itself in a reduction of contacts with *other* chains on intermediate length scales [58]. The length of this polymeric correlation hole is set by the size of the molecules, R_e, and its shape is characteristic for the large scale conformations of the molecule.

To a first approximation, we assume that the fluid structure is determined by the packing of the hard cubes on the lattice; neither the connectivity of the monomers along a polymer nor the thermal interactions influences the total pair correlation function. Under this assumption, we can separate the fluid-like packing effects on the monomer scale and the polymeric correlation hole effect and approximate the intermolecular pair correlation function by [23]:

$$g^{\text{inter}}(r) = g_1(r)\left[1 - \frac{1}{\sqrt{\mathcal{N}}} f(r/R_e)\right] \quad (3.21)$$

where g_1 denotes the pair correlation function of the monomer fluid and the function f parameterizes the structure of the molecule on the scale R_e. The prefactor is determined by requirement that the correlation hole contains one polymer:

$$\Phi \int d^3\mathbf{r} \left[\frac{g^{\text{inter}}(r)}{g_1(r)} - 1\right] = 1 \quad (3.22)$$

Indeed, this factorization works excellently well for flexible molecules in the bond fluctuation model. The ratio $g^{\text{inter}}(r)/g_1(r)$ is largely independent of packing effects and permits a distinction between monomeric packing effects and polymeric correlation hole effects in the simulations—although the length scales are not clearly separated for short chains. Not surprisingly, the correlation hole becomes deeper and wider as we increase the molecular weight. This imparts a chain length dependence on the Flory–Huggins parameter or the effective coordination number, respectively. The effective coordination number is related to the short distance behavior of the intermolecular pair correlation function, $z_c \approx \Phi N \sigma_e^3 g^{\text{inter}}(\sigma_e)$, where σ_e describes the range of the thermal interactions. More quantitatively, we calculate:

$$z_c = \Phi N \int_{r \leq \sqrt{6}} d^3\mathbf{r}\, g^{\text{inter}}(r) = z_c^\infty \left(1 + \frac{\text{const}}{\sqrt{\mathcal{N}}}\right) \quad (3.23)$$

The scaling of the effective coordination numbers for athermal, flexible, linear chains is presented in the inset of Figure 3.3b. The effective coordination number approaches its limiting scaling behavior with a $1/\sqrt{\mathcal{N}}$ correction.

In Figure 3.3a we also present the intermolecular pair correlation functions g_{AA} and g_{AB} for chain length $N = 80$ close to criticality. In accord with intuition, AA contacts are more likely than AB ones and, hence, $g_{AA}^{\text{inter}} > g_{BB}^{\text{inter}}$. However, note that the sum of AA and AB correlations can be approximated well by the intermolecular pair correlation function $g_{\text{atherm}}^{\text{inter}}(\mathbf{r})$ in the athermal limit:

$$\frac{g_{AA}^{\text{inter}}(\mathbf{r}) + g_{AB}^{\text{inter}}(\mathbf{r})}{2} \approx g_{\text{atherm}}^{\text{inter}}(\mathbf{r}) \quad (3.24)$$

This relation shows that the weak difference in interactions between the monomers $\chi \approx 1/N$ does not alter the structure of the monomer fluid. The energy of the system is mainly determined by composition fluctuations. The approximative decoupling between density fluctuations/packing effects and composition fluctuations in our model makes it possible to use the athermal value of the intermolecular pair correlation functions in equation (3.19). This identification corresponds to the high temperature approximation in the framework of the P-RISM theory [98]. In this sense, the miscibility behavior in this model can be described to a good approximation by a purely enthalpic χ parameter.

The inset of Figure 3.3a presents the integral of the correlation functions over the range of the square well potential, that is, $z_{AA}^{\text{part}} = \langle \phi \rangle \int_{r \leq \sqrt{6}} d^3\mathbf{r} g_{AA}^{\text{inter}}(\mathbf{r})$ and similarly for z_{AB}^{part}. The MC simulations show that the difference between the AA and AB contacts decreases as $1/\sqrt{N}$, when χN is held constant. This exemplifies that the mean field approximation (or random mixing assumption) is justified in the limit $\mathcal{N} \to \infty$.

The vanishing of composition fluctuations for $\mathcal{N} \to \infty$ can be rationalized by estimating the scaling of nonrandom mixing effects with growing chain length. The strength of composition fluctuations in a volume V is of the order $1/\Phi V$. Expressing the composition fluctuations via the correlation functions, we obtain for a symmetric blend:

$$1 \approx \Phi V(\langle \phi^2 \rangle - \langle \phi \rangle^2) \sim \Phi \int d^3\mathbf{r} \left[\frac{g_{AA}^{\text{inter}}(\mathbf{r}) + g_{BB}^{\text{inter}}(\mathbf{r})}{2} - g_{AB}^{\text{inter}}(\mathbf{r}) \right]$$

$$\approx \sqrt{\mathcal{N}} \int d^3\mathbf{x} \left[\frac{g_{AA}^{\text{inter}}(\mathbf{x}) + g_{BB}^{\text{inter}}(\mathbf{x})}{2} - g_{AB}^{\text{inter}}(\mathbf{x}) \right] \quad \text{with} \quad \mathbf{x} = \mathbf{r}/R_e \quad (3.25)$$

Therefore, we expect the difference in the AA and AB correlations to vanish like

$$g_{AA}^{\text{inter}}(\mathbf{r}) - g_{AB}^{\text{inter}}(\mathbf{r}) \sim 1/\sqrt{\mathcal{N}} \quad (3.26)$$

upon increasing the chain length. These mean field arguments are in agreement with P-RISM calculations by Yethiraj and Schweizer [98].

The non-random mixing gives rise to a correction of the mean field critical temperature of the order $1/\sqrt{\mathcal{N}}$. This dependence also follows from the Ginzburg criterium [100], which estimates the region of validity of the mean field approximation. The neglect of fluctuations is justified when concentration fluctuations in one 'correlation volume' of size ξ^3 are small compared with the composition difference between the two coexisting phases. Using the Flory–Huggins free energy, we obtain for the binodals in the vicinity of the critical point: $\phi = 1/2\{1 + \sqrt{[3(\chi - \chi_c)/\chi_c]}\}$. The strength of composition fluctuations is determined by the second derivative of the free energy of mixing [cf. equation (3.10)], and the correlation length diverges as $\xi \approx R_e/\sqrt{(\chi N - 2)}$. Using the above expressions, one obtains:

$$|\chi N - \chi_c N| \ll \text{Gi} \sim \frac{1}{\mathcal{N}} \qquad (3.27)$$

for the mean field theory to be accurate. The Ginzburg criterion [100] asserts that composition fluctuations are important at the critical point, where simulations [21] and experiments [51, 95] find 3-D Ising critical behavior. However, unlike the situation in mixtures of small molecules, the region of Ising critical behavior shrinks as $1/\mathcal{N}$. General arguments [101] rationalize that fluctuations lead to an overestimation of the 'true' critical temperature T_c in the mean field theory T_c^{MF} by an amount $(T_c^{\text{MF}} - T_c)/T_c \approx 1/\sqrt{\mathcal{N}}$, in accordance with the nonrandom mixing behavior discussed above.

The scaling of the critical temperature with chain length was studied by Deutsch and Binder [22] in the framework of the bond fluctuation model. In accordance with the predictions of the Flory–Huggins theory [40], the simulations exhibited a linear scaling of the critical temperature with chain length, which was also observed in carefully designed experiments [95] and in integral equation theories [98]. The crossover between the mean field behavior away from the critical point and the ultimate Ising behavior at the critical point was unraveled via a sophisticated analysis of the MC data which simultaneously coped with finite size effects and the crossover from mean field to 3-D Ising critical behavior. This simulation study strongly supports the Ginzburg criterium and thereby clarifies in which regime the mean field assumption break down.

The linear scaling of the critical temperature also has been observed for mixtures comprising molecules of different length [23], and the behavior is also reproduced in off-lattice models [26, 27]. The overestimation of the critical temperature is investigated in more detail in Figure 3.4. Here, we plot the ratio between the MC results and our simple mean field estimate of the critical temperature. Upon increasing the chain length the difference between the MC result and the mean field estimate decreases. The figure also includes data for mixtures of different chain lengths and blends of ring polymers. Using the scaling variable $1/\sqrt{\mathcal{N}}$ we achieve a collapse of all data onto a common curve within the accuracy of the MC data which is of the order 1–5 percent. The figure also displays results for two choices of interactions ranges (for symmetric linear chains) [22]. Circles represent the results of a model where the square well potential is extended over the first 54 lattice sites (as in the remainder of this paper), while triangles denote the results of a model, where the interaction comprises only the six nearest lattice sites. For large chain lengths, the MC results are consistent with a linear dependence on the scaling variable. The collapse of the ratio T_c/T_c^{MF} with $\sqrt{\mathcal{N}}$ for all blends marks the regime of chain lengths where the universal polymeric behavior dominates. It is indicated by a straight line.

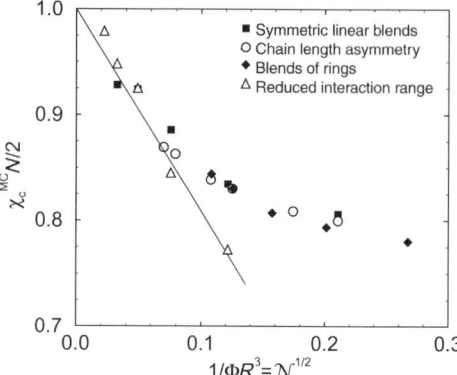

Figure 3.4. Ratio of the critical temperature (as determined in MC simulations) and the Flory–Huggins estimate for binary blends. Using the scaling variable $\sqrt{\mathcal{N}}$ the MC results for blends of linear chains and ring polymers collapse onto a common curve. Adapted from Müller [99].

For the smallest chain lengths used in the simulation, matching the mean field predictions to the observed critical behavior would impart an error onto the χ-parameter of more than 20 percent.

The bending towards a constant value of T_c/T_c^{MF} for small chains is due to the following effect: The deviations from the mean field behavior depend on the correlation length, ξ. The length scale is set by the amplitude of the square gradient term in the expansion of the free energy per chain with respect to long-range composition fluctuations. In a symmetric polymer mixture the prefactor is of the form $R_e^2/36\phi(1-\phi)$ [58]. Hence, the free energy cost of an inhomogeneous composition is due to the configurational entropy. However, for finite-range interactions there is an enthalpic contribution to the square gradient term with a prefactor of the order $\chi N \sigma_e^2$, where σ_e denotes the range of the monomeric interaction potentials. For long chain lengths the entropic contribution dominates ($R_e \gg \sigma_e$) while the enthalpic term becomes important when the range of the thermal interactions becomes comparable to the size of the chains. Consequently, the range of the interactions does not enter the Ginzburg criterion [100] to leading order. For very small polymers or rather long-ranged monomeric interactions, the interaction range σ_e might increase the correlation length and the shift between mean field and true critical temperature is smaller than estimated by $1/\sqrt{\mathcal{N}}$. When the range of interactions σ_e is decreased this effect should set in at smaller chain lengths. This is consistent with the simulation data: The data with the reduced interaction range show larger deviations at small chain lengths.

3.4.1.2. Single Chain Conformations. One of the basic assumptions in relating the Flory–Huggins parameter χ to the local fluid structure via the energy of mixing was that the chain conformations do not depend on composition. Sophisticated theoretical approaches have been developed to study the dependence of chain conformations on the environment: self-consistent P-RISM theory [102, 103], SCF theory for clusters of chains [55], and ellipsoid models [54, 104]. In the following we constrain ourselves to MC simulations and simple scaling arguments to predict the qualitative behavior.

MC simulations suggest that one possible mechanism of conformational changes in blends is associated with exchanging energetically unfavorable intermolecular contacts for attractive intramolecular contacts upon reducing the spatial extension of the molecule. Attributing this shrinking of the minority component to a balance between entropy loss due to deviations from the unperturbed conformations and energy gain upon shrinking, we can estimate the magnitude of conformational changes [55]: Within the Gaussian chain model a deviation from the unperturbed chain extension R_e gives rise to an entropic force of the form:

$$\frac{dS}{dR} \sim \frac{(R - R_e)}{R_e^2} \quad (3.28)$$

This is opposed to an enthalpic force dE/dR, where E denotes the single chain energy. E comprises energetically favorable interactions Nz^{intra} among monomers of the same chain and Nz^{inter} interactions with monomers of other polymers. The exchange of an intermolecular interaction with an intramolecular one lowers the single chain energy by an amount of the order χ. The number of intramolecular interactions per monomer z^{intra} is given by the density of monomers of the same chain inside of its volume $z^{intra} \approx 1/\Phi R_e^3 \approx 1/\sqrt{N}$. Under the assumption that the reduction of the chain extension does not affect the total number of interactions, but merely exchanges intermolecular interactions into energetically favorable intramolecular ones, we estimate the chain length dependence of the energy change as:

$$\frac{dE}{dR} \sim -\frac{\chi N}{\Phi R_e^4} \quad (3.29)$$

Balancing the entropic force against the enthalpic one, we obtain:

$$\frac{R_e - R}{R_e} \sim \frac{\chi N}{\Phi R_e^3} = \frac{\chi N}{\sqrt{N}} \quad (3.30)$$

These scaling arguments are similar to those for a chain in a marginal solvent and suggest that the perturbation of the chain conformations decreases upon increasing the chain length at $\chi N = $ constant. The conformations in high molecular weight blends are only very mildly perturbed in the minority phase.

The derivation of the scaling arguments relies on the number of intramolecular contacts and its dependence of the chain extension. Obviously, this estimate excludes contributions from the neighbors along the polymer, which give rise to $z^{intra} \approx N$. Although the neighbor contribution is important for the scaling of the number of intramolecular contacts with chain length, we assume them to be independent from the instantaneous shape/extension of the polymer.

Using the scaling estimate above, the dependence of the z^{intra} on the instantaneous extension R at fixed chain length is given by $dz^{\text{intra}}/dR \approx 1/\Phi R_e^4 \approx 1/N$. Clearly, a detailed verification of this scaling behavior by MC simulations is warranted [55]. Such a test is presented in Figure 3.5a. The inset presents the average number of intramolecular contacts at fixed end-to-end distance for an athermal melt of chain length $N = 256$. At the mean end-to-end distance $\sqrt{\langle R^2 \rangle_0}$ we determine the slope dz/dR as indicated by linear regression. The chain length dependence of the derivatives of the number of inter- and intramolecular contacts with respect to the chain extension decreases as $1/N$ for large chain lengths. This confirms the scaling predictions. Moreover, the sum of intermolecular and intramolecular contacts depends much more weakly on the spatial extension R_e and the dependence decreases faster than $1/N$. This indicates that the fluid structure of the monomers is mainly determined by packing and approximately decouples from the chain conformations. For long chains, the conformational changes merely result in an exchange of inter- and intramolecular contacts. From the MC data we estimate $dz/dR = 0.77(7)/N$ for the bond fluctuation model at density $\rho \equiv \Phi N = 1/16$.

The scaling predictions can be made more quantitatively in the framework of the Gaussian chain model. Let $P(\mathbf{R})$ denote the probability distribution of the end-to-end vector $\mathbf{R}$ which incorporates the dependence of the single chain energy E on the chain extension.

$$E(\mathbf{R}) \approx E\left(\sqrt{\langle R_e^2 \rangle_0}\right) + \frac{dE}{dR}\left[|\mathbf{R}| - \sqrt{\langle R_e^2 \rangle_0}\right] \quad (3.31)$$

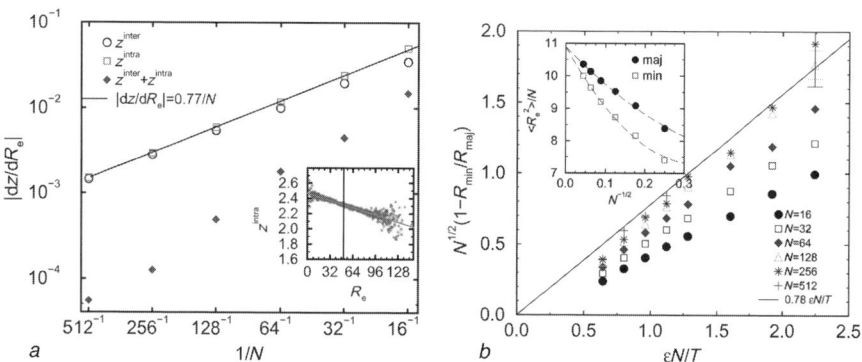

Figure 3.5. (a) Correlation between the chain extension and the number of intermolecular and intramolecular contacts. The straight line marks the prediction of the scaling considerations $dz/dR \approx 1/N$. (b) Shrinking of the end-to-end extension below the critical temperature. The solid line marks the expected behavior for long chain lengths. The inset in (b) shows the chain length dependence of end-to-end distance of the majority and minority component at constant $\varepsilon N/T = 0.64$. Lines in the inset are only guides to the eye. Adapted from Müller [55].

where $\langle R_e^2 \rangle_0 = b^2 N$ denotes the end-to-end distance in the athermal limit. Assuming Gaussian statistics for the unperturbed chain, we can write the probability distribution for a B polymer in an A-rich matrix in the form:

$$P(\mathbf{R}) \approx \exp\left(-\frac{3\mathbf{R}^2}{2\langle R_e^2 \rangle_0} - \frac{1}{T}\frac{dE}{dR}\left[|\mathbf{R}| - \sqrt{\langle R_e^2 \rangle_0}\right]\right) \quad (3.32)$$

The total energy change associated with the transfer of two intermolecular contacts of a B polymer $[\varepsilon(2\langle\phi\rangle - 1)]$ into an intermolecular contact $[-\varepsilon]$ and a contact between monomers not belonging to this B polymer $[-\varepsilon(2\langle\phi\rangle - 1)^2]$ amounts to $\Delta E = 4\varepsilon\langle\phi\rangle^2$. The number of intramolecular contacts per monomer increases by 2, and the number of intermolecular contacts decreases by the same amount. Therefore $dE/dR = 2\varepsilon\langle\phi\rangle^2 dz/dR$. Using this estimate and assuming that the conformational changes are small, we calculate the mean square end-to-end distance:

$$\langle R_e^2 \rangle \approx \langle R_e^2 \rangle_0 \left(1 - \sqrt{\frac{8}{27\pi}}\frac{1}{\sqrt{\mathcal{N}}}\left[\frac{\Phi R_e^4}{z_c}\frac{dz^{intra}}{dR}\right]\chi N\langle\phi\rangle^2\right) \quad (3.33)$$

We expect this asymptotic expression to hold only for very small values of $\chi N/\sqrt{\mathcal{N}}$, where the conformational changes can be treated perturbatively. This expression predicts a quadratic dependence of the chain extension on the composition of the mixture. The effect increases linearly with the χ parameter, but decreases at fixed χN as $1/\sqrt{\mathcal{N}}$. Hence, the conformations of long macromolecules are only very weakly dependent on composition. This is also in qualitative agreement with field theoretical calculations of Holyst and Vilgis [105], and Garas and Kosmas [106]. A decreasing dependence of the single chain conformations on the environment is also found in simulations and self-consistent P-RISM calculations for atomistic models [15].

In Figure 3.5b we explore the scaling of the shrinking of the chains in the minority phase at the binodal. The simple estimate (3.33) predicts that the chains in the majority phase are unperturbed, while the minority component reduces its size in a well segregated blend. The relative shrinking $1 - R_{min}/R_{max}$ increases linearly with εN and is at fixed εN of the order $1/\sqrt{\mathcal{N}}$. The straight line represents the prediction of equation (3.33). Surprisingly, the simulation data approach the asymptotic behavior very slowly. Only for chain length $N \geq 128$ do the simulations reach the scaling limit; for smaller chain length the estimate (3.33) overpredicts the shrinking.

There are at least two possible reasons for the pronounced corrections to the scaling behavior for small chain lengths: First, the considerations hold only in the regime $2 \ll \chi N \ll \sqrt{\mathcal{N}}$. The first limit is set by the condition that the blend is well segregated, that is, $\langle\phi\rangle_{min} \ll 1$. The second requirement corresponds to small conformational changes. This temperature regime is experimentally relevant because the concentration of the minority component is small but does not vanish for long chain lengths. However, for short chain lengths these conditions are rather restrictive. For very strong segregation the linear decrease of the chain dimensions with $\chi N/\sqrt{\mathcal{N}}$ will certainly break down and the changes cannot be treated perturbatively.

A second source of corrections to asymptotic scaling behavior might be deviations from the Gaussian chain statistics upon shrinking. The ratio between the end-to-end distance and the radius of a completely collapsed coil $(3/4\pi b^3 \rho \sqrt{N})^{1/3} = (3/4\pi\sqrt{\mathcal{N}})^{1/3}$ decreases only very weakly with chain length. Even for the chain length $N = 256$ the end-to-end distance exceeds the radius of the densely packed coil only by a factor of 5. If the extension of the shrunken chain becomes comparable to the size of the completely collapsed coil, it cannot reduce its size much further. In this case the data would not scale as a function of $\chi N/\sqrt{\mathcal{N}}$, but they would crossover to a temperature-independent end-to-end distance the earlier the smaller the chain length.

3.4.2. Entropic Contributions to the χ Parameter

3.4.2.1. Monomer Shape and Non-Additive Packing. The simple form of the χ parameter is based on the decoupling of density fluctuations or packing effects from molecular architecture, composition of the blend, and temperature. To illustrate the consequences of molecular architecture on the miscibility behavior, let us discuss a blend of polymers with indented monomer shapes. In Figure 3.6a we sketch such a symmetric mixture [24]. As one can observe, monomers of different types are separated by a spatial distance of at least $\sqrt{3}$, whereas monomers of the same species can approach each other up to a distance of 2. The properties of the pure phases are not altered, because the packing constraints only restrict the minimal distance between unlike species. Of course, this extremely simple monomer shape is not a faithful

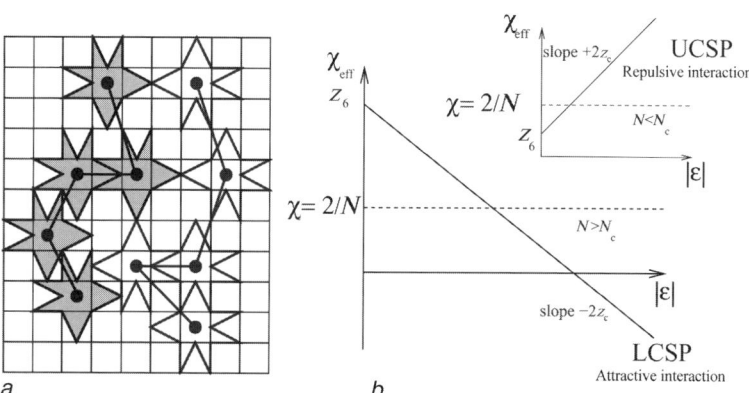

Figure 3.6. (*a*) Illustration of the intended monomer structure. *A* monomers are shaded, *B* monomers open. The effect of the indentation can be described by a nonadditive hard core repulsion between monomers. As mentioned, there is a strong entropic packing advantage which leads to phase separation. (*b*) Qualitative behavior of the χ-parameter for short chains (UCSP) and long chains (LCSP). Adapted from Müller [24].

representation of realistic monomer packing effects on a microscopic length scale. However, in the spirit that a monomer in the bond fluctuation model corresponds to a small number of chemical repeat units, we expect the model to capture some universal, long-wavelength properties on the coarse-grained length scale of a Kuhnian segment. The shape of the monomers leads to a non-additive packing between monomers of different species. In an A-rich phase, an A polymer possesses more conformational freedom than in a B-rich environment. Hence the free energy of mixing also acquires an entropic contribution.

We cannot resort to the simple estimate of the χ parameter [equation (3.19)], but we have to go back a step and approximate the partition function. Let us consider the ratio of the canonical partition functions of an additive mixture $\mathcal{Z}_{add}$ and a blend with nonadditive monomer shape $\mathcal{Z}_{nonadd}$:

$$\ln \frac{\mathcal{Z}_{nonadd}}{\mathcal{Z}_{add}} = \ln \frac{\sum \exp(-\beta E) \Pi_{i,j}[1 - \delta(r_{ij} - 2)]}{\sum \exp(-\beta E)}$$
$$= \ln \langle \Pi_{i,j}[1 - \delta(r_{ij} - 2)] \rangle_{|add} \quad (3.34)$$

where the sum comprises all configurations of the additive mixture. The index $i(j)$ runs through all monomers of type A(B) and the factor $[1 - \delta(r_{ij} - 2)]$ excludes all configurations violating the nonadditivity constraint. Neglecting correlations among the monomer positions, one can factorize the average and get to a first approximation:

$$\ln \frac{\mathcal{Z}_{nonadd}}{\mathcal{Z}_{add}} \approx \sum_{i,j} \ln \langle [1 - \delta(r_{ij} - 2)] \rangle_{|add}$$
$$\approx -\sum_{i,j} \langle \delta(r_{ij} - 2) \rangle_{|add}$$
$$= -n_A n_B \frac{z_6}{V\Phi N} = -V\Phi N \phi(1-\phi) z_6 \quad (3.35)$$

where z_6 corresponds to the mean number of particles at a distance 2 in an additive mixture. This quantity is accessible in the simulation via the intermolecular paircorrelation function in an additive mixture:

$$z_6 = \Phi N \sum_{i=1}^{6} g_{AB}(\vec{x}_i)|_{add} \quad (3.36)$$

where the pair correlation function is normalized such that $g(r) \to 1$ for $r \to \infty$. Neglecting all local packing effects, one gets $z_6 = 0.2625$ for $\Phi N = 0.35$. Finally the packing-induced contribution to the free energy and the effective χ parameter takes the form:

$$\frac{\Delta F}{T\Phi V} \approx N z_6 \phi(1-\phi) \qquad \chi_{\text{eff}} = \chi_0 + z_6 = \frac{2\varepsilon z_c}{T} + z_6 \quad (3.37)$$

where z_c is the effective coordination number of the thermal interaction, that is, the mean number of monomers of *other* chains within the range of the square well potential [23]. As anticipated, the effective χ-parameter contains an enthalpic part and an entropic contribution. The excess entropy of mixing comprises two concurrent terms. The combinatorial entropy stabilized the mixture, whereas the packing contribution favors phase separation. Since the former is reduced by a factor $1/N$, the packing entropy dominates the behavior in the long chain limit.

The consequences for the miscibility behavior are discussed in Figure 3.6*b*: If the chain length is small enough, the translational entropy, which favors mixing, dominates over the positive entropic contribution to the χ parameter. For $N < N_c \equiv 2/z_6$, the athermal blend is completely miscible. Upon lowering the temperature the blend phase separates at an upper critical solution temperature (UCST). For longer chains ($N > N_c$) the athermal blend is only partially miscible and to bring about a phase transition we have to assume an attractive interaction between unlike species. Keeping with the notation of additive blends, the attraction corresponds to negative values of ε. Upon increasing the absolute magnitude of the interactions $|\varepsilon|/T$, we can make the blend become miscible at a lower critical solution temperature (LCST):

$$\frac{-\varepsilon}{T_c} = \frac{z_6}{2z_c}\left(1 - \frac{2}{z_6 N}\right) \qquad (3.38)$$

The scaling of the lower critical solution temperature is in marked contrast to the scaling at the upper critical solution point with temperature. In the latter case the transition temperature is determined by a competition between the translational entropy of a polymer vs the monomeric repulsion; the critical temperature (UCST) increases linearly with chain length. In the former case, the conformational entropy per segment is balanced against the monomeric interactions and a chain length independent lower critical solution temperature (LCST) is approached from above. When expressed in terms of the χ parameter, the Ginzburg criterion [100] and the critical amplitudes of the magnetization take the same form than for the upper critical solution points of the symmetric, additive mixture. However, when these quantities are written in terms of temperature, the combination $(\chi_c - \chi)/\chi_c$ takes the form $Nz_c\varepsilon|T_c - T|/TT_c = Nz_6(2 - 1/z_6 N)|T - T_c|/T$. This gives rise to an additional chain length dependence of the critical amplitudes and the Ginzburg number, when the temperature instead of the χ parameter is used in the vicinity of a lower critical solution point. For instance, the binodals in the temperature–composition plane open wider the larger the chain length and the Ginzburg number, which measures the relative distance between the critical temperature and the temperature where mean field behavior sets become proportional to N^{-2}.

This mean field picture has been investigated via MC simulations [24] in the bond fluctuation model with the indented monomers sketched in Figure 3.6*a*. Owing to the additional excluded volume between unlike species, the simulations have been performed at a reduced monomer density $\rho = 0.35/8$ in order to allow for a reasonable acceptance ratio of the semigrandcanonical identity switches. Using a finite size scaling analysis we have determined the critical temperatures accurately. The

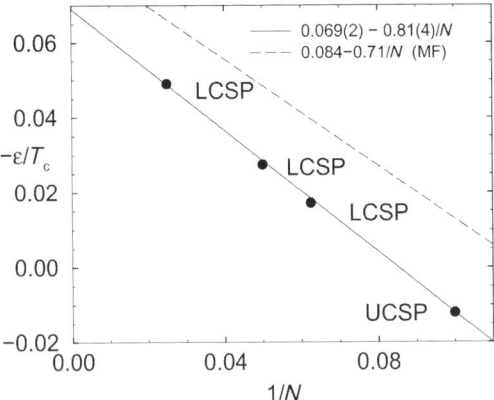

Figure 3.7. Chain length dependence of the critical temperature for mixtures of nonadditive monomers. Symbols denote the results of the MC simulations, while the dashed line depicts the mean field prediction. Adapted from Müller [24].

dependence of the critical temperature on the chain length is summarized in Figure 3.7. For chain length $N = 10$ we find an upper critical solution temperature while for chain length $N = 16$ and larger we observe lower critical solution points. We find Ising critical behavior at the lower critical solution points and the temperatures approach a limiting value upon increasing the chain length. The critical temperatures are describable by a dependence of the form: $-\varepsilon/T_c = 0.069(2) - 0.81(4)/N$. For chain length $N = 20$ we have measured $z_c = 1.41$ and $z_6 = 0.238$ in the MC simulations. This yields $\varepsilon/T_c = 0.084 - 0.71/N$ as mean field estimate for the critical temperatures which is also displayed in Figure 3.7. It describes the simulation data only qualitatively. Deviations are partially due to the chain length dependence of the parameters z_c and z_6 via the correlation hole effect and composition fluctuations. Moreover, pronounced deviations stem from the crude approximation of the composition dependence of the the configurational entropy which lead to the simple expression (3.37). Our approximation treats the (rather strong) non-additivity only perturbatively and neglects the composition dependence of the chain conformations (cf. Section 3.4.1.2).

If monomers of the same species pack less efficiently than monomers of the same species, a negative entropic contribution to the χ parameter results. An entropic contribution to the χ parameter occurs not only for nonadditive monomer shapes but also for large enough disparities in the segment size. Mixtures of small and large spheres demix [107] when the size difference between the species is large enough. We expect the consequences to be even more pronounced for polymers due to the small entropy on mixing. These effects have been explored in the framework of the lattice cluster theory [49].

3.4.2.2. Entropic Contribution to χ Parameter Due to Stiffness Asymmetry. Generally, the constituents of a blend are not symmetric, and asymmetry might have pronounced effects on the miscibility behavior. Since the properties of a blend deviate from the linear superposition of the individual properties of its components, the blend has new, possibly favorable characteristics.

A common asymmetry in polymer blends are differences in the statistical segment length [50]. This effect has attracted much attention recently because of synthesis techniques for saturated hydrocarbon with a controlled degree of branching and their practical applications [108]. Bates et al. [50] suggested that the degree of branching can be represented on a coarse-grained scale by a difference in statistical segment lengths. Graessley and co-workers [108] have systematically studied the miscibility behavior of this class of polymers. Many—although not all—blends were describable in terms of Hildebrand and Scott's solubility [109] parameters. This suggests that the incompatibility is chiefly determined by enthalpic effects.

In the framework of the bond fluctuation model, the effect of stiffness can be incorporated via an intramolecular bond angle potential of the form [24, 110]:

$$E(\theta) = f \cos(\theta) \tag{3.39}$$

where θ denotes the complementary angle to two successive bonds. Increasing the stiffness parameter f, we energetically favor straight bond angles and increase the spatial extension of the molecule. The more open the molecule is, the larger the number of intermolecular contacts. Upon increasing the stiffness parameter f from 0 (flexible chains) to 2 (semi-flexible), the chain extension for a polymer of $N = 32$ segments increases about a factor 1.5 and the number of intermolecular contacts, z_c, increases from 2.65 to 3.29 at $\varepsilon = 0.05T$ because the bond stiffness makes a folding back of the chain less probable [24, 110]. Unfortunately, the behavior of the intermolecular pair correlation function cannot be decomposed into packing effects of the monomeric units and polymeric correlation hole: On the one hand the chain structure is not Gaussian on all length scales—the rod-like behavior on short distances becomes more important upon increasing stiffness. On the other hand, the bond angle potential between neighboring monomers influences the packing structure of the liquid, which for semiflexible chains differs from the packing of the monomer fluid. The interplay between the packing arrangement of the monomers and the local conformations favored by the bond angle potential gives rise to an entropic contribution to the Flory–Huggins parameter.

To explore the possibility of entropic contributions to the χ parameter [24], we accurately measure the dependence of the chemical exchange potential, $\Delta\mu$, per polymer on the composition of the mixture. If there were no entropic contributions to the χ parameter, only the (exactly known) translational entropy would determine the relation between the exchange chemical potential $\Delta\mu$ and the composition ϕ. In Figure 3.8a we present the deviations $\Delta\mu - T \ln[\phi/(1 - \phi)]$ from the ideal mixing behavior. An entropic contribution to χ is related to a composition dependence of the form $-\chi N(2\phi - 1)$. Indeed, the MC results do reveal a composition dependence of this form and we accurately extract a small positive entropic contribution $\Delta\chi$ to the Flory–Huggins parameter [24]. $\Delta\chi = 0.0017(2)$ for $N = 16$ and $f = 1$, $\Delta\chi = 0.0018(2)$ for $N = 32$ and $f = 1$, and $\Delta\chi = 0.0031(3)$ for $N = 16$ and $f = 1.5$. For the parameters investigated, the entropic contribution $\Delta\chi$ increases with stiffness

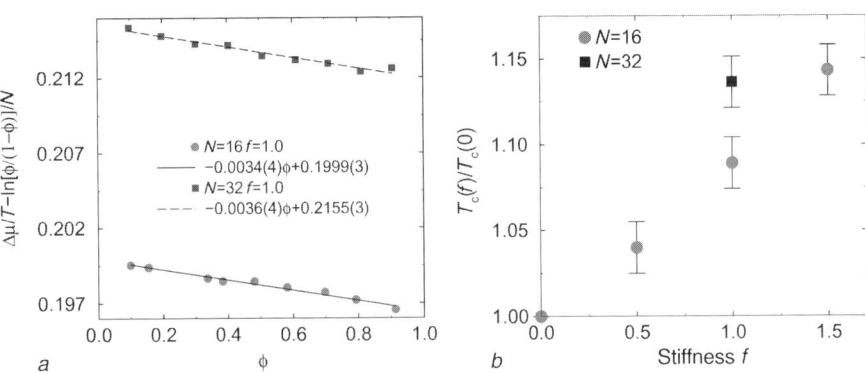

Figure 3.8. (*a*) Deviations from the semigrandcanonical equation of state for blends of polymers with different stiffness. The (negative) slope is proportional to the entropic contribution to the χ parameter. (*b*) Dependence of the critical temperature on the stiffness f for chain length $N = 16$ and 32. Adapted from Müller [24].

disparity in the blend and is roughly independent of chain length. For the chain lengths considered, it is only a few percent of the critical value $\chi_c = 2/N$ and we anticipate only a small increase in the critical temperature. If the chain length independence of the entropic contributions remains true in the long chain length limit, the data suggest that, for long macromolecules [$N \approx \mathcal{O}(1000)$], the stiffness disparity alone will be sufficient to cause phase separation or LCST behavior [24]. The positive entropic contribution to the χ parameter is consistent with field theoretical studies by Fredrickson et al. [111], P-RISM calculations by Singh and Schweizer [112] and lattice cluster theory by Foreman and Freed [113].

Unfortunately, the semigrandcanonical MC moves rapidly become less efficient as the chain length or the stiffness disparity is increased, because the typical conformations of the two species differ strongly. For the chain lengths accessible in the simulations the shift in the critical temperature is small and not chiefly determined by the entropic contribution to the χ parameter. The stiffness increases the size of the molecules and the number of intermolecular contacts. Hence, the enthalpic contribution to the χ parameter increases as well. The increase of the enthalpic and entropic contributions are both of the order of a few percent. We expect both effects to persist in the long chain length limit.

For the short chain length considered in the simulation the mean field theory overestimates the critical temperature by about 20 percent. According to the Ginzburg criterion the deviation between the critical temperature and the mean field estimate decreases with the chain extension [cf. equation (3.27)]. We expect composition fluctuations to decrease the critical temperature less the higher the stiffness. Even if the χ parameter were to remain unaltered, the critical temperature in the MC simulations (for short chains) would increase.

The measured shift of the critical temperature is presented in Figure 3.8*b*. Upon increasing the stiffness or the chain length, the ratio of critical temperatures between the strictly symmetric blend and the blend with stiffness disparity increases. For $N = 16$ and $f = 1$ we obtain a relative shift of 9 percent for T_c. The entropic contribution in the athermal system amounts to $\Delta \chi N/2 = 0.014$ while the relative increase of the effective coordination number is 6 percent as measured in the MC simulations. The remaining deviation is consistent with the dependence of the ratio T_c^{MF}/T_c on chain extension.

A similar study on the consequences of stiffness disparity was pursued in an off-lattice model by Weinhold et al. [114]. Using the increment chemical potential method [70], the authors explored the miscibility behavior. Upon blending, the chemical potential of the flexible chains increases, while the stiffer component lowers its free energy. This effect was rationalized via equation of state effects: At constant density the blend has a higher osmotic pressure than the pure flexible component and a lower osmotic pressure than the stiff component. The authors state that the behavior is in almost quantitative agreement [114] with the simulations in the bond fluctuation model. Again, this indicates that the lattice structure in the bond fluctuation model is a good approximation for the continuum space properties. The net excess free energy per monomer obtained from the simulations is essentially zero to within the $\pm 0.005T$ statistical error [114]. This is consistent with the small values, $\Delta \chi$, found in our simulations. They also illustrated the significance of the intermolecular pair correlation function for the χ parameter [115].

Structural asymmetries also can lead to a composition dependence of the Flory–Huggins parameter χ. Dudowicz et al. [89] studied a dependence of the form: $\chi_{SANS} = a + (b + c\phi)/T$. An important achievement of the lattice cluster theory consists of an explicit, but approximate, relation between these coefficients and the structure of the monomeric units [116].* Depending on the coefficients a, b, and c, different phase behaviors can be observed. Dudowicz et al. [89] use equation (3.10) to define the χ parameter (sometimes also denoted χ_{SANS} because it is directly accessible in small angle neutron scattering experiments) and not the Flory–Huggins free energy of mixing [equation (3.1)]. If the χ parameter does not depend on composition, that is, $c = 0$, those two definitions are equivalent, otherwise

$$\chi_{SANS} = -\frac{1}{2}\frac{d^2}{d\phi^2}[\chi\phi(1-\phi)] \qquad (3.40)$$

Let us briefly discuss the different miscibility behaviors for a symmetric blend (i.e. identical chain length $N_A = N_B = N$ and segmental volumes); explicit expressions for the general case can be found in Dudowicz et al. [89].

Some classes have already been considered: Class I is characterized by $a = 0$ and $b > 0$ and corresponds to the symmetric binary blend with an upper critical solution point and exhibits a miscibility behavior similar to the original Flory–Huggins theory (cf. Section 3.4.1.1). Class II is described by $a > 0$ and $b > 0$ and one

---

*For a blend with indented monomer shapes, we obtain from equation (3.37): $a = z_6$ and $b = 2z_c\varepsilon$.

finds an upper critical solution point only for short chains $N < N_c = 2/a$. A realization would be a blend with stiffness asymmetry or nonadditive packing (with an segmental repulsion, $\varepsilon > 0$). Class III comprises the case $a > 0, b < 0$ and $c = 0$ and yields lower critical solution points at a critical temperature that converges to a finite value $|b|/a$ for long chain lengths.

We again emphasize that for $c = 0$ the critical point is still given by $\chi_c N = aN + bN/T = 2$ and $\phi_c = 1/2$, and the binodals, the correlation length and the Ginzburg number obey the equations in Section 3.4.1.1 when expressed in χN rather than in terms of temperature. It is only via the temperature dependence of the Flory–Huggins parameter that the miscibility as a function of temperature adopts several distinct behaviors. This fact corroborates the Flory–Huggins parameter χ as a suitable coarse-grained parameter to parameterize the miscibility behavior.

In class IV of the classification scheme of Dudowicz et al. [89], $a > 0, b < 0$ and the concentration dependence $c \neq 0$ is essential. This class corresponds to a qualitatively different miscibility behavior. In the following, we assume $c < 0$ [similar expressions are obtained for $c > 0$ by exchanging ϕ and $(1 - \phi)$ and b and $b - c$] [89]. The location of the critical point (T_c, ϕ_c) is given by the two conditions:

$$\frac{d^2}{d\phi^2}\frac{F_{FH}(\phi)}{T\Phi V} = \frac{1}{\phi} + \frac{1}{1-\phi} - 2N\left(a + \frac{b+c\phi}{T}\right)$$
$$\frac{d^3}{d\phi^3}\frac{F_{FH}(\phi)}{T\Phi V} = -\frac{1}{\phi^2} + \frac{1}{(1-\phi)^2} - 2N\frac{c}{T}$$
(3.41)

The last equation shows that for large chain length N, the critical density decreases as $\phi_c = \sqrt{(T_c/2|c|N)}$. Inserting this in the first equation and keeping only leading order terms in N, one obtains $T_c = |b|/a + \mathcal{O}(1/\sqrt{N})$. The phase diagram is very asymmetric and the critical temperature converges towards a chain length-independent value. We note that this chain length-dependence of the critical point is similar to a compressible polymer solution or a very asymmetric binary blend, $N_A = N$ and $N_A \gg N_B$ = constant, of class I.

3.4.2.3. Effects of Pressure or Solvent Density. The effect of pressure on the miscibility behavior has attracted much interest. There is an interesting interplay between equation of state effects and the phase behavior. This has also a practical importance when a blend is mixed in an extruder or during injection molding of plastics [117]. If the two components differ strongly in their equation of state, more complex phase diagrams occur (cf. Konynenburg and Scott [39] for a classification of possible types of miscibility behavior in compressible blends). In addition to liquid–liquid phase separation, one encounters liquid–vapor phase coexistence and three-phase coexistence regions where two liquids and a vapor phase coexist.

In the following, we restrict ourselves to components that have a rather similar equation of state and do not consider liquid–vapor phase separation. In the framework of the bond fluctuation model the local fluid structure is mainly determined by athermal packing effects. In the athermal melt, the osmotic pressure largely is independent from chain length [118]. This chain length independence of the pressure at high densities is a universal property of polymer melts [58]. The excess free energy change upon mixing per monomer is of the order $\chi \approx 1/N$. The free energy cost for a density fluctuation (on the length scale of a monomer) is proportional to the compressibility and, hence, chain length independent. Therefore, in the one phase region, the interactions lead only to a small excess volume change upon mixing for high molecular weight (compatible) polymer blends. These kind of compressibility effects have been observed, for example, at polymer–polymer interfaces, where the energetic unfavorable interactions at the interface result in a slight decrease in density. However, even for rather strongly segregated blends [29] ($\chi N < 20$) in the bond fluctuation model, the effect yields only a density reduction by a few percent. This is also in agreement with the decoupling of composition and density fluctuations. The insignificance of compressibility effects in weakly interacting blends and the consequences for the analysis of neutron scattering data have been explored in Taylor-Maranas et al. [119, 120]. Gromov and dePablo [121] found in MC simulations of a Lennard-Jones chain model with length $N = 16$, a volume change of approximately 10 percent at constant pressure.

For our specific choice of interactions the total energy density per monomer is also of the order χ. Hence, the fluid structure corresponds to that of an athermal melt in the limit of long chain lengths and χN =constant. This observation is in accord with the temperature independence of the packing and the effective coordination number in the temperature range where the phase separation occurs. If we were to simulate at constant pressure, the density around the phase transition would be chiefly determined by the value of the athermal system in the limit of long chains. Therefore, we expect little change in the chain length dependence of the miscibility at constant volume or at constant pressure.

However, we would like to emphasize that the approximate decoupling between the fluid structure/density and the temperature (at constant pressure) is not a universal property and depends on the specific choice of interactions. Unlike the excess free energy change upon mixing, the total energy density per monomer needs not to be small for long polymers at χN =constant. In many experimental instances, concentrated polymer solutions and melts exhibit a temperature dependent equation of state. Therefore the density Φ is a function of temperature T at a given fixed pressure. If the Flory–Huggins parameter χ is mainly enthalpic, we still can use equation (3.19) to calculate the χ parameter. In a very crude approximation the intermolecular pair correlation function is independent of the density. Then, the effective coordination number is proportional to the monomer density $\rho = \Phi N$ and the critical temperature of the blend at constant pressure scales like $T_c \approx \rho(T_c)N$. For a blend with an UCSP the critical temperature increases with chain length and the monomer density $\rho(T_c)$ at the critical temperature decreases. This leads to a weaker dependence of the critical temperature with chain length. Such an effect has been observed in various experiments. Escobedo and dePablo [27] have investigated the scaling of the critical temperature in a symmetric blend under constant pressure. They found that the critical temperature increases effectively as $\sqrt{N}$ for the range of chain length investigated. Moreover, recent experiments on polyolefin blends by Lohse and co-workers found evidence for a temperature–pressure superposition [117]: Far from the UCST the interaction energies depend on the pressure P only via $\rho(P)$. However, deviations from this scaling are found for blends which demix upon heating. This might indicate an additional dependence of the local packing arrangements on density.

If the density gets lower, one obtains a semidilute solution instead of a melt: In a 3-D melt the chain statistics is Gaussian up to microscopic length scales and the correlation hole in the intermolecular pair correlation function has only a finite depth. In a semidilute solution, the chain statistics are Gaussian for distances larger than the excluded volume screening length, ξ_{ev}, but self-avoiding walk-like for smaller distances, that is, inside the excluded volume blob. The size of the excluded volume blob, ξ_{ev}, can be determined by requiring that the monomer density inside the blob is mainly created by monomers of the same chain [58]. Let g denote the number of monomeric units inside the blob, then $\rho = g/b_{ev}^3 g^{3\nu_{ev}}$, where $\nu_{ev} = 0.588$ characterizes the chain extension of a self-avoiding walk $R \approx b_{ev} N^{\nu_{ev}}$. Therefore the size of the blob decreases with density like $\xi \approx b_{ev}(\rho b_{ev}^3)^{-\nu_{ev}/3\nu_{ev}-1}$. Thus, the number of monomers of other chains inside the excluded volume blob is small and chains do not interdigitate on the length scale ξ_{ev} or smaller. The intermolecular pair correlation function exhibits a deep correlation hole and shows a power-law behavior $g^{\text{inter}}(\mathbf{r}) \approx (r/\xi_{ev})^\alpha$ for $r < \xi_{ev}$ (cf. Figure 3.9). The exponent adopts

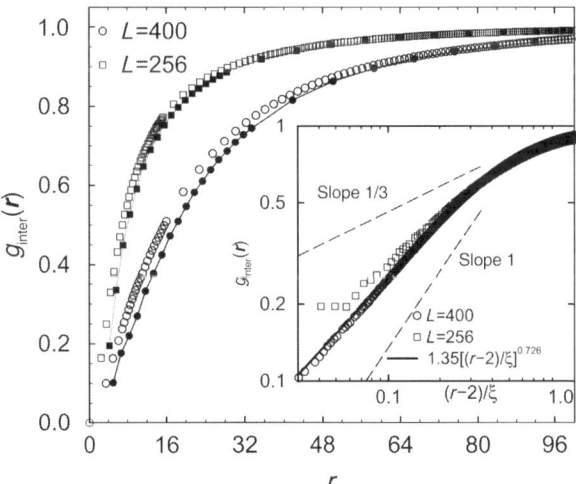

Figure 3.9. Real space dependence of the intermolecular correlations: (*a*) Intermolecular pair correlation function $g^{\text{inter}}(r)$ for chain length $N = 2048$ and monomer number densities $\rho = 0.0032$ ($L = 400$) and $\rho = 0.0122$ ($L = 256$). Open symbols correspond to the MC results, whereas solid ones represent renormalization group calculations. Inset: power-law dependence of $g^{\text{inter}}(\mathbf{r})$ at small distance. The data are compatible with a dependence of the form $g^{\text{inter}}(\mathbf{r}) \approx r^\alpha$ with an exponent 0.7(3), close to the predicted value $\alpha = 0.8$. Adapted from Müller et al. [66].

the value $\alpha = 3 - 2/\nu_{ev} - \omega_{12} \approx 0.8$ [66, 122], where $\omega_{12} \approx 0.4$ is the correction to scaling exponent that characterizes the contacts ($\sim N^{\nu\omega_{12}}$) of two mutually interdigitating self-avoiding walks. The number of intermolecular contacts scales like $\chi \approx z_c \approx \rho g^{\text{inter}}(\sigma_e) \approx (\rho b_{ev}^3)^{(1+\nu\omega_{12})/(3\nu_{ev}-1)} \approx \rho^{0.616}$. This effect—the Joanny renormalization of the Flory–Huggins parameter—has been worked out in detail by Leibler and coworkers [123] and Schäfer and Kappeler [124]. This weak density dependence leads to a smaller incompatibility of semidilute mixtures than expected from the Flory–Huggins theory [40].

3.4.3. Films and Two-Dimensional Systems

Much of the success of the mean field description is related to the strong interdigitation of the polymers in the bulk. This quantity is measured by $\mathcal{N} = (\Phi R_e^d)^2$. It increases like the chain length $\mathcal{N}$ in $d = 3$ spatial dimensions. If the polymers are confined into quasi two-dimensional configurations, the chain dimensions remain Gaussian ($R_e^2 \sim N$) [126] in a dense melt. In this case, however, the quantity $\mathcal{N}$ is, to the leading order, independent of the chain length; that is, a given molecule interacts only with a finite number of neighbors.

Let us consider a simple scaling argument for the behavior of the chain conformations upon confining a polymer into a thin film. According to Silverberg's argument [127], the chain conformations can be simply conceived as random walks reflected at the surface. If a finite stiffness (or bending rigidity) along the chain is considered, parallel and perpendicular chain dimensions are no longer independent, but this short-ranged correlation along the chain is not expected to affect the properties on long length scales.

At some film thicknesses, D, however, the description of the polymer conformations as mutually noninteracting Gaussian chains will fail. When the film thickness becomes very small, the chain folds back many times into its own volume and the density inside of the Gaussian coil increases. The fractal structure of the segments of a single chain gradually becomes compact (i.e. space-filling). When the density inside of the coil finally becomes comparable to the average density of the melt, the parallel chain extension begins to grow such that the density of the film remains laterally homogeneous. In this limit the chains adopt disk-like compact conformations of polymers in two dimensions. The stretching parallel to the surface is only negligible when $\rho D R_e^2 \ll N$ or

$$\frac{D}{R_e} \gg \frac{1}{\sqrt{\mathcal{N}}} \qquad (3.42)$$

where R_e denotes the unperturbed chain extension in the bulk. This reasoning suggests the following behavior of the average chain conformation in a thin film: (i) For $D \gg R_e$ the chain extensions parallel and perpendicular to the surface are unperturbed. Here, we expect also the intermolecular paircorrelation function not to deviate strongly from the bulk behavior. (ii) For $R_e \gg D \gg R_e/\sqrt{\mathcal{N}}$, the parallel chain extensions are unperturbed, but the chain folds back into the volume of its own coil. Other chains are gradually squeezed out of this volume; that is, the correlation hole in the intermolecular paircorrelation function deepens and the interdigitation of the chains decreases. In this regime, the number of intermolecular contacts, z_c, is reduced, but it is still a finite number. (iii) For $R_e/\sqrt{\bar{N}} \gg D$, the chains do not overlap strongly and stretch parallel to the surface as to maintain a laterally uniform density. The lateral extension scales as $R_\parallel^2 \approx R_e^3/(D\sqrt{\mathcal{N}})$. Note that this effect occurs when the film thickness is of the order of the excluded volume screening length $\xi_{ev} \approx R_e/\sqrt{\mathcal{N}}$.[*] In this quasi two-dimensional limit, the correlation hole is describable by [125]:

$$g^{\text{inter}}(r) = 1 - c\, \exp\left(-\frac{\text{const} \cdot r}{R_e}\right) \quad \text{for} \quad r > \xi \quad (3.43)$$

where $c \leq 1$ is a constant. The exponential term corresponds to the correlation function of a Gaussian walk in two dimensions and the functional form suggests that the correlation hole in the intermolecular pair correlation function in a dense melt is exactly canceled by the density of the monomers of the reference chain, as it is the case in three dimensions. The MC data for the intermolecular pair correlation function are presented in Figure 3.10a. The data are compatible with the value $c = 1$. Therefore, the intermolecular pair correlation function behaves at small distances like $1 - \exp(-\text{const} \cdot r/R_e) \approx \text{constant} \cdot r/R_e$ and the number of intermolecular contacts per monomer is proportional to $\rho g^{\text{inter}}(b) \approx 1/\sqrt{N}$. This implies a scaling of the critical temperature like $T_c \approx N z_c \approx \sqrt{N}$. Indeed, the simulation results [125] in Figure 3.10b for T_c/N decrease as $N^{-1/2}$, as suggested by the scaling arguments. The incompatibility in quasi two-dimensional polymer films is therefore strongly reduced. Unlike the situation in three spatial dimensions, the mean field theory does not become quantitatively correct in the limit of long chain lengths. In two dimensions, even long chains interact only with a finite number of neighbors, and therefore $T_c/z_c N$ does not tend to unity for $N \to \infty$.[†] This is confirmed by the simulation data presented in the inset of Figure 3.10b.

Of course, these considerations highlight the influence of the confined chain conformations on the miscibility. If the surfaces prefer one component of the mixture, one encounters wetting transitions that might also lead to a pronounced alteration of the phase diagram in confined geometry [30, 128, 129].

---

[*] Since we focus on semidilute solutions (and melts), we express all quantities for $r > \xi_{ev}$ in terms of R_e and $\mathcal{N}$. Conceiving a chain as a random walk of blobs of size ξ_{ev}, we can write: $R^2 = b^2 N \approx \xi_{ev}^2 N/g$ where g is the number of monomeric units per blob. To a first approximation, blobs do not interdigitate and $g \approx \rho \xi_{ev}^3$. Then we obtain $\xi \approx 1/\rho b^2 = R_e/(\sqrt{\mathcal{N}})$. To make contact with the dilute limit, we use $R_0 \approx b_{ev} N^{\nu_{ev}}$ with $\nu_{ev} = 0.588$ for the chain extension in the limit $\rho \to 0$, and we define the overlap $s = \rho R_0^3/N$. The extension in semidilute solutions is related to the properties in the dilute limit via $R_e^2 \approx R_0^2 \mathcal{F}(s) \stackrel{s \to \infty}{\to} R_0^2 s^{(1-2\nu_{ev})/(3\nu_{ev}-1)} \approx b_0(\rho b_0^3)^{(1-2\nu_{ev})/(3\nu_{ev}-1)} N$. Using this limiting scaling behavior we find for the screening length $\xi_{ev} \approx R_e/\sqrt{\mathcal{N}} \approx b_0(\rho b_0^3)^{\nu_{ev}/(1-3\nu_{ev})}$ and for the overlap $s \approx \mathcal{N}^{3\nu_{ev}-1}$.
[†] For the same reason, $T_c/z_c N$ does not tend to unity for $N \to \infty$, even for finite, chain-length-independent film thickness, even though $T_c \sim N$ for $D \gg R_e/\sqrt{\mathcal{N}}$.

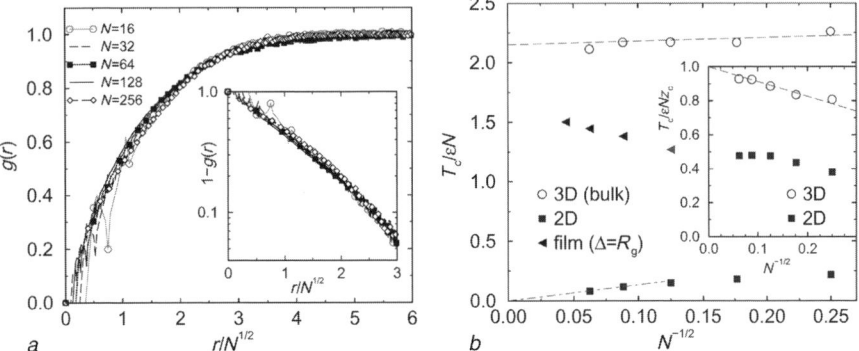

Figure 3.10. (a) Scaling plot of the intermolecular pair correlation function $g(r)$ vs $r/N^{1/2} \sim r/R$ in quasi two-dimensional polymer films. Data are always taken close to criticality, but refer to all chains independent of their species. Chain lengths $N = 16, 32, 64, 128,$ and 256 are included, as indicated in the key. For short chain lengths oscillations on the length scale of a few lattice units are visible and arise from packing effects. The inset shows $1 - g(r)$ on a logarithmic scale [cf. equation (3.43)]. (b) Scaling of T_c/N and T_c/Nz_c (inset) vs $N^{-1/2}$. For comparison, the results for the three-dimensional model are included. Adapted from Cavallo et al. [125].

3.4.4. Interfaces

Having identified the parameters of a coarse-grained model from the bulk behavior, we can use the same coarse-grained model to predict the behavior at interfaces and surfaces. Comparing the properties of interfaces as obtained from simulations or experiments with the predictions of mean field theory, one has to distinguish two types of properties: Excess quantities which do not make any explicit reference to the interface position—like, for example, the interface tension or the surface excess of a third component—can be directly compared. Profiles across the interfaces or observables that depend on the local position of the interfaces are affected by capillary waves [97, 130, 131].

In Figure 3.11 we compare the interface tension extracted from the simulations with the prediction of the SCF theory using the identification of the Flory–Huggins parameter, χ, as in the previous section. The simulation data confirm that σ/σ_{SSL} is only a function of χN and they agree quantitatively with the SCF theory [60, 133].

A comparison of profiles between simulations/experiments and SCF calculations has to take due account of the fluctuations of the local interface position. Interfaces are not perfectly flat, as assumed in the SCF calculations, but there are thermal fluctuations. A snapshot of the interface position in the MC simulations of a binary blend is shown in Figure 3.12a. To a first approximation the effect of these fluctuations is to increase the effective area of the interface. Let $u(\mathbf{r}_\parallel)$ denote the local interface position. Then, the free energy cost of deviations from perfectly flat configurations

RESULTS

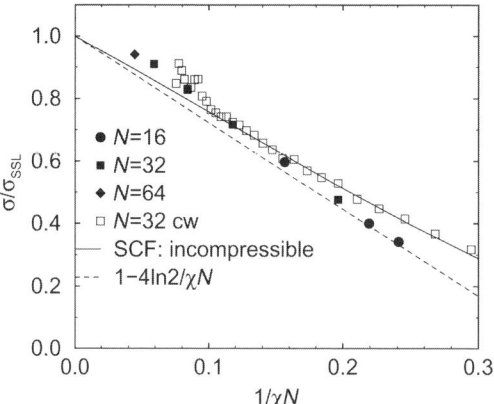

Figure 3.11. Interface tension in units of $\sigma_{SSL}R_e^2/T = \sqrt{\mathcal{N}}\sqrt{(\chi N/6)}$. The symbols correspond to MC simulations for three chain lengths, as indicated in the key. The interfacial tension has been extracted from the probability distribution of the composition in the semigrandcanonical ensemble, and the spectrum of capillary waves (for $N = 32$). Data are compared with SCF results and the exact asymptotic correction in the limit $\chi N \to \infty$ by Semenov [60]. Adapted from Schmid and Müller [132].

is described by the effective interface Hamiltonian, $\mathcal{H}$:

$$\mathcal{H}[u(\mathbf{r}_\parallel)] = \int d^2\mathbf{r}_\parallel \left\{ \frac{\sigma}{2}|\nabla u|^2 + \frac{\kappa}{2}|\Delta u|^2 \right\} \tag{3.44}$$

This is an expansion in terms of small u and its derivatives, σ denotes the interfacial tension and κ the bending rigidity of the interface. For interfaces between

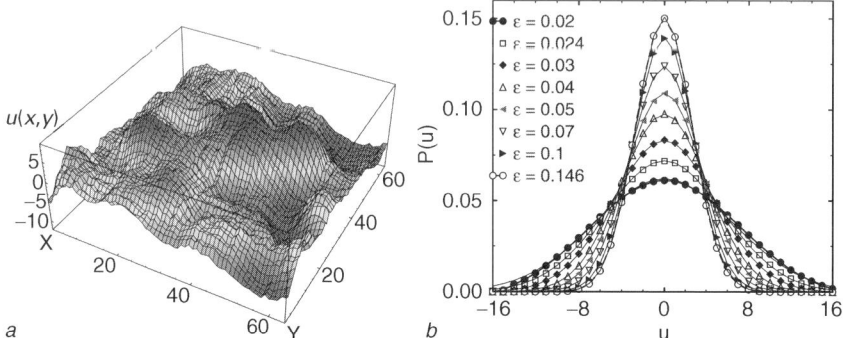

Figure 3.12. (*a*) Fluctuations of the local interface position in a binary polymer blend at $\varepsilon = 0.03T$ ($\chi N \approx 5.1$). The position has been averaged over a lateral size of $B = 8$. (*b*) Distribution of the local interface positions as a function of the incompatibility ε/T. From Werner et al. [130].

not-too-long homopolymers, κ is very small; for copolymer-laden interfaces, however, the second term becomes important, as we shall discuss below. This capillary wave Hamiltonian is diagonal and quadratic in terms of the Fourier components $u(q)$ and the equipartition theorem yields for the spectrum of fluctuations in thermal equilibrium:

$$\langle u^2(q) \rangle = \frac{T}{\sigma q^2 + \kappa q^4} \tag{3.45}$$

The local positions $u(\mathbf{r}_\parallel)$ are also Gaussian distributed $P(u)$ with variance s:

$$s^2 = \frac{1}{4\pi^2} \int d^2\mathbf{q}_\parallel \langle u^2(\mathbf{q}_\parallel) \rangle = \frac{T}{2\pi\sigma} \ln\left(\frac{q_{max}}{q_{min}}\right) \tag{3.46}$$

where a short and long length scale cut-off, q_{max} and q_{min}, have to be introduced to avoid the divergence at $q \to 0$ and $q \to \infty$. The bending rigidity κ has been neglected; it would make the cut-off at small distance obsolete. The MC result [130] for the distribution $P(u)$ of the local positions is presented in Figure 3.12b. Upon decreasing the incompatibility ε, the interface tension growth and we decrease the strength of fluctuations.

These capillary waves broaden the apparent profiles, p_{app}. Laterally averaged profiles, as obtained in experiments or MC simulation, are describable via the convolution of an intrinsic profiles p_{int} of ideally flat interfaces and the distribution of the local positions [130, 131, 133]

$$p_{app}(z) = \int du P(u) p_{int}(z - u) \tag{3.47}$$

where z denotes the coordinate perpendicular. When applied to a erfc-shape profile, one obtains [130, 131, 133]:

$$\frac{w_{app}^2}{R_e^2} = \frac{w_{int}^2}{R_e^2} + \frac{T}{4\sigma R_e^2} \ln\left(\frac{q_{max}}{q_{min}}\right) \tag{3.48}$$

$$\chi \stackrel{N \gg 2}{=} \frac{1}{\sqrt{6\chi N}} \left(1 + \frac{3}{2\sqrt{N}} \ln\frac{q_{max}}{q_{min}}\right) \tag{3.49}$$

where we have used the expression for the strong segregation limit [cf. equation (3.15)] in the last line [60]. The apparent width is broader than the intrinsic one and depends via the two cut-offs on the system geometry [130, 131]. For a free interface, the lower cut-off, q_{min}, is set by the lateral block size B, on which the interface is observed. This might be set by the size of the simulation cell or the coherence length of the neutron beam by which the interfacial structure is investigated. Gravitation or interactions with walls/surfaces also limit long-wavelength fluctuations. The upper cut-off q_{max} describes the crossover from capillary waves (on large distances) to 'intrinsic' fluctuations, which build up the smooth profile of the ideally flat interface. Measuring the width of an interface in experiments or simulations one cannot extract both the 'intrinsic' width, w_{int}, and the upper cut-off, q_{max}, individually. In principle, the 'intrinsic' profile (and its widths) depends on the definition of the upper cut-off, q_{max} [130, 134].

Consequentially, the experimentally accessible apparent width of the interface [cf. method (4) in Section 3.4.1] is not suitable for extracting the Flory–Huggins parameter, χ, by a direct comparison to the SCF prediction—the thus determined χ parameter would depend on the lateral size on which the interface is observed.

Polymer blends are well suited to examining the crossover from intrinsic fluctuations to capillary waves, because the strong interdigitation of the molecules makes SCF calculations describe the properties of interfaces accurately except for capillary wave fluctuations [130]. *Assuming* the SCF prediction as the 'intrinsic' width of an hypothetical flat interface, we use equation (3.48) to *define* the length scale $B_0 = 2\pi/q_{max}$ at which the crossover between 'intrinsic' fluctuations and capillary waves occurs. This procedure is illustrated in Figure 3.13a. There are three possible candidates for B_0: a microscopic length scale (e.g. the bond length) independent of temperature or chain length; the width of the interface, which depends on temperature but not on N; or the correlation length, which depends both on χN and the molecular size, R_e. The simulation data [130] in Figure 3.13b suggest a behavior of the form $B_0 = 3.8 w_{scf}(1 - 3.1/\chi N)$; that is, the intrinsic width of the interface sets the crossover length. This result is compatible with the calculations of Semenov [135].

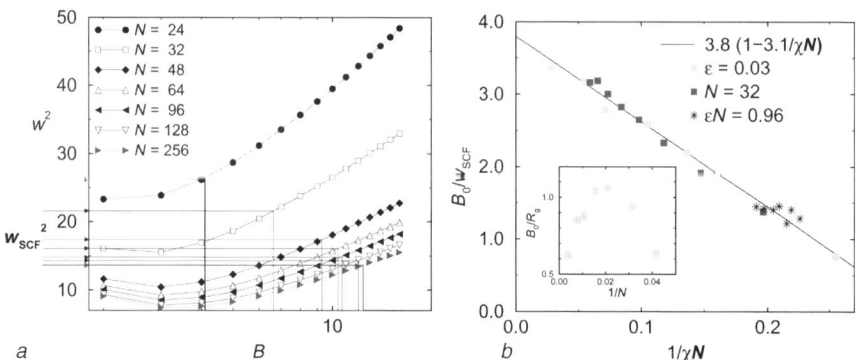

Figure 3.13. (*a*) Dependence of the apparent interfacial width (measured in units of the lattice spacing) on the lateral block size B at constant incompatibility $\varepsilon = 0.03T$ for different chain lengths. From the intersections of the simulation data with the SCF predictions (presented as horizontal lines), we determine the small length scale cut-off B_0. (*b*) The ratio B_0/w_{scf} approaches a constant value 3.8. Leading corrections are of the order $1/\chi N$. Note that data sets at constant N, χN and χ collapse in this representation. The inset shows the ratio B_0/R_e for fixed incompatibility $\varepsilon = 0.03T$. The ratio tends to zero for large N. From Werner et al. [130].

The spectrum of interfacial fluctuations is an alternative route for measuring the interfacial tension in MC simulations [74, 110]. This is illustrated for a blend with a stiffness asymmetry in Figure 3.14. When we increase the stiffness disparity, the semigrandcanonical identity switches become increasingly inefficient and σ cannot be obtained via preweighting techniques. Stiffness increases the incompatibility and this, in turn, results in a larger interfacial tension (cf. Figure 3.14a). This effect is quantitatively captured by the SCF calculations, which account for the detailed chain architecture by a partial enumeration technique [110, 136]. Qualitative agreement is also obtained within the Gaussian chain model. The deviations from the prediction of Helfand and Sapse [137] are mainly due to chain end corrections. The situation is qualitatively different for the intrinsic width of the interface (cf. Figure 3.14b). MC simulations and SCF calculations, which enumerate explicit single chain conformations extracted from the MC simulations [110], predict no increase or even a reduction of the width for larger stiffness, while the intrinsic width increases in the Gaussian chain model. The breakdown of the Gaussian chain model can be qualitatively rationalized as follows: The width of the interface is determined by loops of the polymers into the other phase. For large incompatibility the width of the interface becomes comparable with the persistence length and the conformation of a loop differs from the Gaussian statistic of the chain on large length scales. This is one example where the local structure of the molecular architecture is important.

The ability of the SCF calculations to provide a detailed description of the intrinsic interface structure is illustrated in Figure 3.15. We present the MC and SCF

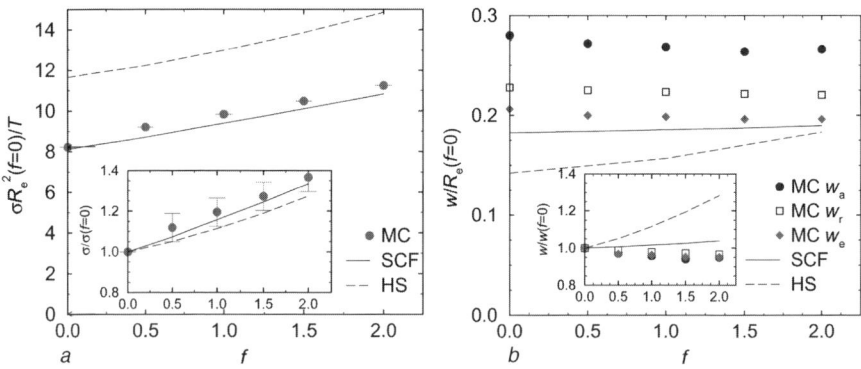

Figure 3.14. Interface tension (a) and interfacial width (b) in a blend of flexible ($f = 0$) and stiff (f as indicated) polymers at rather strong segregation, $\varepsilon = 0.05T$. Comparisons with detailed SCF calculations, which take account of the chain architecture on all length scales, and predictions of the Gaussian chain model by Helfand and Sapse (HS) are shown. In (b) w_a denotes the apparent width, which is averaged over the whole lateral system size $L = 64$, w_r represents the width on the block size $B = 16$, and $w_e = 2\Delta e/\chi T\rho$ denotes the width extracted from the excess energy Δe of the interface per unit area, respectively. From Müller and Werner [110].

RESULTS

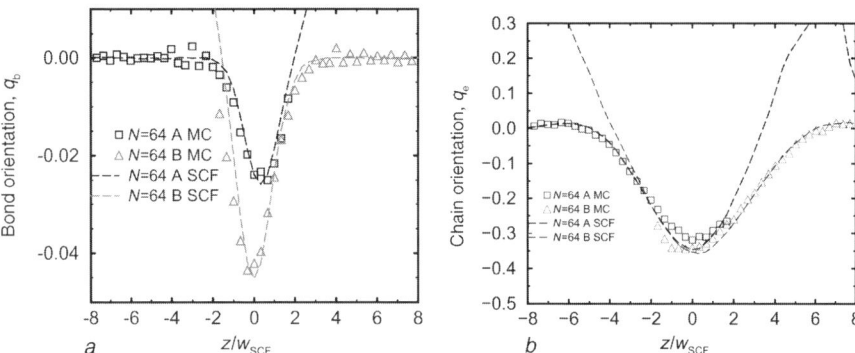

Figure 3.15. Orientations of the bond vector (*a*) and end-to-end $\mathbf{R}_e$ (*b*) in a blend of polymers with stiffness $f = 0$ (left-hand side) and $f = 1$ (right-hand side) at rather strong segregation $\varepsilon = 0.05T$ and chain length $N = 64$. Symbols denote the results of the MC simulation on a lateral length scale $B = 16$, while lines represent the SCF calculations. From Müller and Werner [110].

results for the orientations as measured by the second Legendre polynom of the angle between the bond vector or end-to-end distance $\mathbf{R}_e$ with respect to the interface. Both vectors align parallel to the interface, but the effect is much stronger and less stiffness dependent for $\mathbf{R}_e$ than for the bond vectors.

Diblock copolymers are model surfactants for the *AB* homopolymer blend. They adsorb at the interface as to extend both halves into the corresponding homopolymer phases. This decreases their enthalpy, but the localization at the interface reduces the translational entropy and the conformational entropy due to chain stretching at high copolymer excess at the interface [139]. Upon increasing the chemical potential $\delta\mu$ (or concentration) of the copolymers in the bulk, we observe the adsorption of copolymers at the interface and the concomitant reduction of the interfacial tension in Figure 3.16 (cf. also Figure 3.1). Both MC simulations and SCF calculations agree at high segregation. However, rather than forming a dense copolymer brush at the interface, a phase separation into a homopolymer-rich phase and a lamellar phase (swollen by homopolymers) is encountered [74].

For small ε/T the addition of copolymers drives the system to compatibility (cf. Figure 3.17*a*). At intermediate segregation we find a three-phase coexistence between two homopolymer-rich phases and a copolymer-rich disordered phase. The latter has a structure of a microemulsion (as revealed, e.g. by snapshots). SCF calculations by Janert and Schick [140] rather predict highly swollen lamellar phases in this region. Some insight into this discrepancy can be gained from the spectrum of interface fluctuations. Upon adding copolymers to the systems at $\varepsilon = 0.054T$, we decrease the interfacial tension and deviations from a simple q^2 dependence become apparent (see Figure 3.17*b*). Thus, we can obtain an estimate of the bending rigidity κ of the copolymer-loaded interface according to equation (3.45). The bending rigidity turns out to be much smaller than $T/2\pi$. It is this bending rigidity, however, which stabilizes the liquid-crystalline order of the

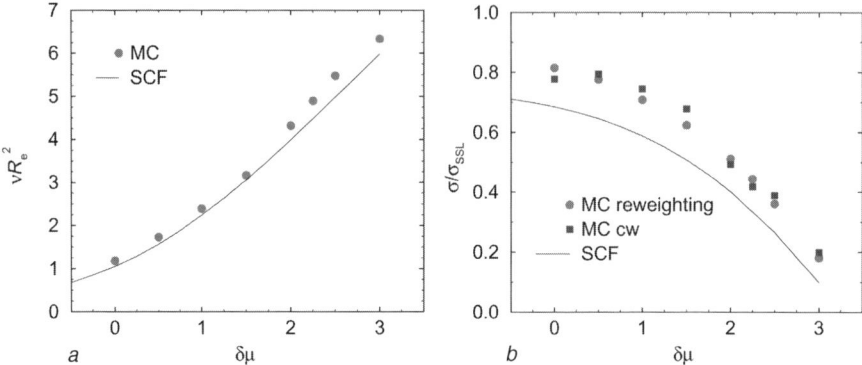

Figure 3.16. (a) Adsorption of diblock copolymers at a homopolymer–homopolymer interface as a function of the chemical potential of the copolymer at $\varepsilon = 0.1T$ and chain length $N = 32$. (b) Reduction of the interfacial tension upon adding copolymers. Symbols correspond to the MC results while lines denote the predictions of the SCF theory. Reproduced from Werner et al. [138] with permission.

lamellar phase. de Gennes and Taupin [141] argued that a small value of κ leads to the formation of a microemulsion. Indeed this is observed in the simulation [142] and experiments [143]. If one incorporates fluctuations into the SCF theory, one can successfully describe the formation of the polymeric microemulsion [144]. If we were to increase the chain length, we would increase the bending rigidity $\kappa \approx \sqrt{\mathcal{N}}$ [145] and stabilize the lamellar phases predicted by the SCF theory. Again, the deviations from the mean field theory are controlled by the parameter $\mathcal{N}$. This reasoning agrees with the Ginzburg criterium at the (tricritical) Lifshitz point [146]. The example shows that interface fluctuations can modify the phase behavior in the bulk and can result in strong deviations from mean field behavior.

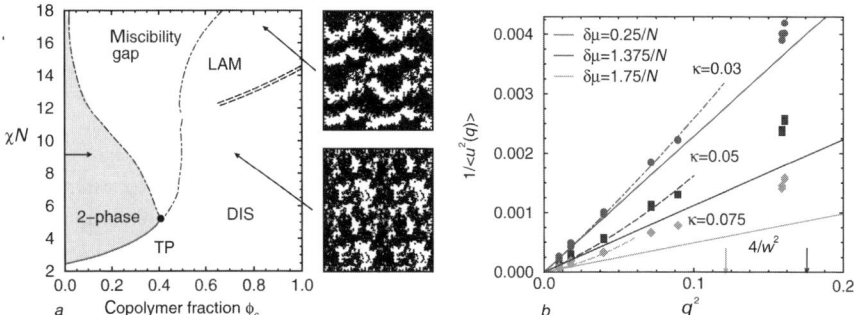

Figure 3.17. (a) Isopleth cut through the phase triangle of a ternary blend for $N = 32$. (b) Spectrum of interface fluctuations in a ternary blend of two homopolymers and a diblock copolymer at $\varepsilon = 0.054T$. Estimates for the bending rigidity of the interfaces are indicated. Adapted from Müller and Schick [74].

3.5. CONCLUSION AND OUTLOOK

By virtue of its simplicity, the Flory–Huggins theory provides a popular framework to parameterize miscibility in dense, multicomponent polymer systems. A large body of experimental data has been analyzed in terms of the mean field predictions in the bulk or self-consistent field theory (or approximations thereof) for spatially inhomogeneous systems.

Unfortunately, the prediction of blend miscibility for a specific multicomponent system has been an elusive goal [147, 148]; that is, no theory that could quantitatively and accurately predict the Flory–Huggins parameter on the basis of the atomistic structure has yet emerged. Nevertheless, much progress has been achieved in correlating qualitative features of molecular architecture and blend miscibility. Both the P-RISM approach of Schweizer and Curro [48] and the LCT of Freed and Dudowicz [49] provide much insight into the dependence of the Flory–Huggins parameter on the structure of the polymeric fluid and the molecular architecture.

In this contribution we have reviewed the results of coarse-grained MC simulations on the bulk phase behavior and interface properties of binary polymer blends. The advantage of the simulation is that one can simultaneously access a variety of thermodynamic and structural quantities and relate them to the Flory–Huggins theory [40] without invoking any approximations for the statistical mechanics of the underlaying model. Simulations often do not provide simple analytical expressions to which experimental data can directly be compared. One role of the simulations, however, is to explore the range of validity of these simple expressions of the mean field theories and thereby clarify whether deviations from theory are due to the idealizing assumptions of the theoretical model or the necessary approximations invoked to obtain analytically tractable expressions. The applicability of analytical theory can be enlarged by going beyond the mean field approximation, and much progress has recently been made [36, 91, 149].

Even in the simplest case of a completely symmetric polymer blend with no volume change upon mixing, several phenomena that contribute to the miscibility—local structure/packing, correlation hole effect, concentration fluctuations in the vicinity of the critical point, and changes of the chain conformations—could be delineated. Also different thermodynamic and structural identifications of the Flory–Huggins parameters have been compared in the framework of the same model without invoking any approximation in the statistical mechanics of the model. Then we proceeded to successively incorporate architectural details, for example, different monomer shapes or stiffness, to investigate their influence of the miscibility behavior. Although these simulations do not provide quantitative predictions for specific materials, they contribute to predicting the qualitative dependence of miscibility due to molecular architecture.

Even if we could not correlate the Flory–Huggins parameter to the molecular architecture, the Flory–Huggins parameter would still be a useful parameter to relate the thermodynamics of the bulk with the properties of spatially inhomogeneous systems that can be calculated within the framework of the SCF theory.

We have illustrated this by quantitatively comparing the properties of interfaces between the MC simulations and the SCF theory.

We have restricted ourselves to thermodynamic equilibrium properties in binary blends. A great deal of effort has been directed towards understanding the dynamics in (miscible) blends [5] and the kinetics of phase separation [150, 151]. An understanding of the miscibility behavior and interface properties is certainly a good starting point to study the kinetics of phase separation, which, in turn, determines the morphology of the phase-separated material on large length scales. We close by illustrating that the knowledge of the bulk phase diagram also allows conclusions to be drawn about the kinetics of phase separation. To this end we regard a quench from high temperatures to $\chi N = 5$ at the critical composition $\phi_c = 1/2$ and observe the spontaneous phase separation in our simple model of a symmetric blend. The early stage of spinodal decomposition has attracted much interest [151] and its theoretical description is based on two ingredients [152]: the free energy of a concentration fluctuation with wavevector q and an Onsager coefficient that relates the single chain dynamics to the dynamics of collective concentration fluctuations. The study of the equilibrium thermodynamic properties shows that the free energy of concentration fluctuations can be well described by equation (3.8) with no adjustable parameter. Hence the simulation allows a critical test of the predictions for the Onsager coefficient [152]. In Figure 3.18 we present the time dependence of the collective structure factor obtained from two versions of the dynamic SCF theory. In the first version, the Onsager coefficient is local [153], while the nonlocality of the Onsager coefficient in the second version, external potential dynamics [36, 154], captures effects of chain connectivity within the Rouse model [155].

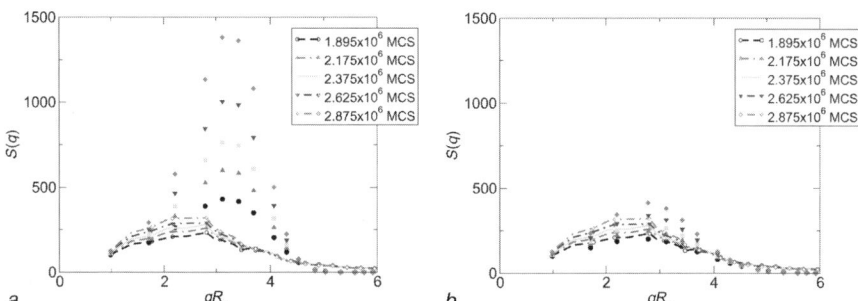

Figure 3.18. Global structure factor vs wave vector for different times for a quench from the mixed state to $\chi N = 5$. Time is measured in units of Monte Carlo steps (MCS), which correspond to one attempted local hopping move per monomer. The time it takes a chain to diffuse a distance that corresponds to its extension is $R_e^2/D = 1.5 \times 10^7$ MCS. Lines represent Monte Carlo results, solid symbols dynamic SCF theory results. (*a*) compares dynamic SCF theory using a local Onsager coefficient with the Monte Carlo simulations. Local dynamics obviously overestimates the growth rate and shifts the wavevector that corresponds to maximal growth rate to larger values. (*b*) compares dynamic SCF theory using a nonlocal Onsager coefficient that mimics Rouses dynamics with Monte Carlo results showing better agreement. From Reister et al. [36].

The quantitative comparison between the simulations and the dynamical SCF calculations shows that the nonlocal Onsager coefficient yields a much better description. Clearly this is only a modest, exploratory step towards using coarse-grained models to investigate the dynamics of phase separation in polymer blends, and many questions concerning nucleation [156], thermal fluctuations [91], and viscoelastic properties [157, 158] remain to be studied.

ACKNOWLEDGMENTS

Fruitful and enjoyable collaborations with K. Binder, A. Cavallo, M. Fuchs, E. Reister, L. Schäfer, M. Schick, F. Schmid, A. Werner, financial support from the DFG under grants Mu1674/1, Mu1674/3 and Bi314/17, and access to the CRAY T3Es at the NIC Jülich and the HLR Stuttgart are acknowledged.

REFERENCES

[1] R.W. Cahn, P. Haasen and E.J. Kramer, *Materials Science and Technology, a Comprehensive Treatment*, Vol. 12, VCH, Weinheim (1993). F. Garbassi, M. Morra and E. Occhiello, *Polymer Surface: From Physics to Technology*, Wiley, Chichester (2000).
[2] C. Creton, E.J. Kramer and G. Hadziioannou, *Macromolecules* **24**, 1846 (1991).
[3] G.I. Taylor, *Proc. R. Soc. Lond.* **A138**, 41 (1932).
[4] S.T. Milner, *MRS Bull.* **22**, 38 (1997).
[5] G.C. Chung, J.A. Kornfield and S.D. Smith, *Macromolecules* **27**, 964 (1994); K.-L. Nagai and D.J. Plazek, *Rubber Chem. and Technol.* **68**, 376 (1995); S.K. Kumar, R.H. Colby, S.H. Anastasiadis and G. Fytas, *J. Chem. Phys.* **105**, 3777 (1996). T.P. Lodge and T.C.B. McLeish, *Macromolecules* **33**, 5278 (2000); S. Kamath, R.H. Colby and S.K. Kumar, *Macromolecules* **36**, 8567 (2003).
[6] M. Müller, K. Katsov and M. Schick, *J. Polym. Sci. B* **41**, 1441 (2003).
[7] J. Baschnagel et al., *Adv. Chem. Phys.* **152**, 41 (2000).
[8] F. Müller-Plathe, *Chem. Phys. Chem.* **3**, 754 (2002).
[9] K. Kremer and F. Müller-Plathe, *MRS Bull.* **26**, 169 (2000).
[10] K. Binder (ed.), *Monte Carlo and Molecular Dynamics Simulations in Polymer Science*, Oxford University Press, New York (1995).
[11] K.F. Freed *Renormalization Group Theory of Macromolecules*, Wiley-Interscience, New York (1987); J. des Cloizeaux and G. Jannink, *Polymers in Solution: Their Modeling and Structure*, Oxford Science Publications, Oxford (1990); L. Schäfer, *Excluded Volume Effects in Polymer Solutions*, Springer, Berlin (1999).
[12] V. Tries, W. Paul, J. Baschnagel and K. Binder, *J. Chem. Phys.* **106**, 738 (1997).
[13] J.C. Shelly, M.Y. Shelly, R.C. Reeder, S. Bandyopadhyay and M.L. Klein, *J. Phys. Chem. B* **105**, 4464 (2001).
[14] L. Delle Site, C.F. Abrams, A. Alavi and K. Kremer, *Phys. Rev. Lett.* **89**, 156103 (2002). C.F. Abrams, L. Delle Site and K. Kremer, *Phys. Rev. E* **67**, 21807 (2003).
[15] D. Heine, D.T. Wu, J.G. Curro and G.S. Grest, *J. Chem. Phys.* **118**, 914 (2003).

[16] B. Smit, S. Karaborni and J.I. Siepmann, *J. Chem. Phys.* **102**, 2126 (1995).
[17] N.B. Wilding, M. Müller and K. Binder, *J. Chem. Phys.* **105**, 802 (1996).
[18] A.Z. Panagiotopoulos, V. Wong and M.A. Floriano, *Macromolecules* **31**, 912 (1998).
[19] F.A. Escobedo and J.J. dePablo, *Mol. Phys.* **87**, 347 (1996).
[20] A.D. Mackie, A.Z. Panagiotopoulos and S.K. Kumar, *J. Chem. Phys.* **102**, 1014 (1995).
[21] A. Sariban and K. Binder, *Macromolecules* **21**, 711 (1988); *J. Chem. Phys.* **86**, 5859 (1987); *Makromol. Chem.* **189**, 2357 (1988).
[22] H.-P. Deutsch and K. Binder, *Europhys. Lett.* **18**, 667 (1992); *Macromolecules* **25**, 6214 (1992); *J. Phys. II (France)* **3**, 1049 (1993).
[23] M. Müller and K. Binder, *Macromolecules* **28**, 1825 (1995); M. Müller and K. Binder, *Comput. Phys. Commun.* **84**, 173 (1994).
[24] M. Müller, *Macromolecules* **28**, 6556 (1995).
[25] S.K. Kumar, *Macromolecules* **27**, 260 (1994).
[26] G.S. Grest, M.-D. Lacasse, K. Kremer and A.M. Gupta, *J. Chem. Phys.* **105**, 10583 (1996).
[27] F.A. Escobedo and J.J. dePablo, *Macromolecules* **32**, 900 (1999).
[28] A. Yethiraj, *J. Chem. Phys.* **101**, 2489 (1994).
[29] M. Müller, K. Binder and W. Oed, *J. Chem. Soc. Faraday Trans.* **91**, 2369 (1995).
[30] M. Müller and K. Binder, *Macromolecules* **31**, 8323 (1998).
[31] M. Müller and L.G. MacDowell, *Macromolecules* **33**, 3902 (2000).
[32] J. Baschnagel and K. Binder, *Macromolecules* **28**, 6808 (1995).
[33] K. Kremer and G.S. Grest, *J. Chem. Phys.* **92**, 5057 (1990).
[34] W. Paul, K. Binder, D.W. Heermann and K. Kremer, *J. Chem. Phys.* **95**, 7726 (1991).
[35] A. Sariban and K. Binder, *Macromolecules* **24**, 578 (1991).
[36] E. Reister, M. Müller and K. Binder, *Phys. Rev. E* **64**, 41804 (2001).
[37] K. Binder, J. Baschnagel and W. Paul, *Prog. Polym. Sci.* **28**, 115 (2003).
[38] J.S. Rowlinson and F.L. Swinton, *Liquids and Liquid Mixtures*, 3rd edn, Butterworth, Oxford (1982).
[39] P. Van Konynenburg and R.L. Scott, *Phil. Trans. Soc. Lond. Ser.* **A298**, 495 (1980).
[40] M.L. Huggins, *J. Chem. Phys.* **9**, 440 (1941); P.J. Flory, *J. Chem. Phys.* **9**, 660 (1941).
[41] D. Frenkel and B. Smith, *Understanding Molecular Simulation*, Academic Press, Boston, MA (1996).
[42] D.P. Landau and K. Binder, *A Guide to Monte Carlo Simulations in Statistical Physics*, Cambridge University Press, Cambridge (2000).
[43] E. Helfand and Y. Tagami, *J. Chem. Phys.* **56**, 3592 (1972); E. Helfand, *J. Chem. Phys.* **62**, 999 (1975).
[44] J. Noolandi and K.M. Hong, *Macromolecules* **14**, 727 (1981); *Macromolecules* **15**, 483 (1982).
[45] K.R. Shull, *Macromolecules* **26**, 2346 (1993).
[46] J.M.H.M. Scheutjens and G.J. Fleer, *J. Phys. Chem.* **83**, 1619 (1979); *J. Phys. Chem.* **84**, 178 (1979); *Macromolecules* **18**, 1882 (1985).
[47] M.W. Matsen and M. Schick, *Phys. Rev. Lett.* **74**, 4225 (1995).

REFERENCES

[48] K.S. Schweizer and J.G. Curro, *Adv. Chem. Phys.* **XCVIII**, 1 (1997).
[49] K.W. Foreman and K.F. Freed, *Adv. Chem. Phys.* **103**, 335 (1998). K.F. Freed and J. Dudowicz, *Macromolecules* **31**, 6681 (1998).
[50] F.S. Bates, M.F. Schultz, J.H. Rosedale and K. Almdal, *Macromolecules* **25**, 5547 (1992); M.D. Gehlsen and F.S. Bates, *Macromolecules* **27**, 3611 (1994); F.S. Bates and G.H. Fredrickson, *Macromolecules* **27**, 1065 (1994).
[51] D. Schwahn, G. Meier, K. Mortensen and S. Janssen, *J. Phys. II (France)* **4**, 837 (1994); H. Frielinghaus, D. Schwahn, L. Willner and T. Springer, *Physica B* **241**, 1022 (1998).
[52] P. Cifra, E. Nies and J. Broersma, *Macromolecules* **29**, 6634 (1996); P. Cifra, F.E. Karasz and W.J. MacKnight, *Macromolecules* **25**, 192, 4895 (1992).
[53] M. Müller and K. Binder, *J. Phys. II (France)* **6**, 187 (1996).
[54] M. Murat and K. Kremer, *J. Chem. Phys.* **108**, 4340 (1998).
[55] M. Müller, *Macromolecules* **31**, 9044 (1998).
[56] K.A. Peterson, A.D. Stein and D.M. Fayer, *Macromolecules* **23**, 111 (1990).
[57] K.O. Rassmusen, G. Kalosakas, *J. Polym. Sci. B* **44**, 1777 (2002). S.W. Sides and G.H. Fredrickson, *Polymer* **44**, 5859 (2003).
[58] P.G. de Gennes, *Scaling Concepts in Polymer Physics*, Cornell University Press, Ithaca, NY (1979).
[59] I.M. Lifshitz, A.Y. Grosberg and A.R. Khokhlov, *Rev. Mod. Phys.* **50**, 683 (1978).
[60] A.N. Semenov, *J. Phys. II* **6**, 1759 (1996).
[61] J.K. Maranas, M. Momdello, G.S. Grest, S.K. Kumar, P.G. Debenedetti and W.W. Graessley, *Macromolecules* **31**, 6991, 6998 (1998); J.D. Londono, J.K. Maranas, M. Mondello, A. Habenschuss, G.S. Grest, P.G. Debenedetti, W.W. Graessley and S.K. Kumar, *J. Polym. Sci. B* **36**, 3001 (1998).
[62] I. Carmesin and K. Kremer, *Macromolecules* **21**, 2819 (1988); H.-P. Deutsch and K. Binder, *J. Chem. Phys.* **94**, 2294 (1991).
[63] N.F. Carnahan and K.E. Starling, *J. Chem. Phys.* **51**, 635 (1969). H.-P. Deutsch and R. Dickman, *J. Chem. Phys.* **93**, 8983 (1990).
[64] A. Indrakanti, J.K. Maranas, A.Z. Panagiotopoulos and S.K. Kumar, *Macromolecules* **34**, 8596 (2001); A.Z. Panagiotopoulos, *J. Chem. Phys.* **112**, 7132 (2000).
[65] K. Kremer and K. Binder, *Comp. Phys. Rep.* **7**, 261 (1988).
[66] M. Müller, K. Binder and L. Schäfer, *Macromolecules* **33**, 4568 (2000).
[67] W.G. Madden, A.I. Pesci and K.F. Freed, *Macromolecules* **23**, 1181 (1990).
[68] P. Cifra, K.F. Karasz and W.J. McKnight, *Macromolecules* **25**, 4895 (1992); E. Nies and P. Cifra, *Macromolecules* **27**, 6033 (1994).
[69] S.K. Kumar, *Macromolecules* **27**, 260 (1994); S.T. Cui, H.D. Cochran, P.T. Cummings and S.K. Kumar, *Macromolecules* **27**, 3375 (1997).
[70] S.K. Kumar, I. Szleifer and A.Z. Panagiotopoulos, *Phys. Rev. Lett.* **66**, 2935 (1991); *Phys. Rev. Lett.* **68**, 3456 (1992).
[71] F.A. Escobedo and J.J. dePablo, *Europhys. Lett.* **40**, 111 (1997); F.A. Escobedo and J.J. dePablo, *J. Chem. Phys.* **106**, 2911 (1997); F.A. Escobedo, *J. Chem. Phys.* **113**, 8444 (2000).
[72] D.A. Kofke, *J. Chem. Phys.* **98**, 4149 (1993).
[73] R.D. Kaminski, *J. Chem. Phys.* **101**, 4986 (1994).

[74] M. Müller and M. Schick, *J. Chem. Phys.* **105**, 8885 (1996).

[75] C. Borgs, R. Kotecky, *J. Stat. Phys.* **60**, 79 (1990); *Phys. Rev. Lett.* **68**, 1734 (1992).

[76] M. Müller and N.B. Wilding, *Phys. Rev. E* **51**, 2079 (1995).

[77] K. Binder, *Phys. Rev. A* **25**, 1699 (1982).

[78] B.A. Berg and T. Neuhaus, *Phys. Rev. Lett.* **68**, 9 (1992); B.A. Berg, U. Hansmann and T. Neuhaus, *Z. Phys. B* **90**, 229 (1993).

[79] A.M. Ferrenberg and R. Swendsen, *Phys. Rev. Lett.* **61**, 2635 (1988).

[80] A.M. Ferrenberg and R. Swendsen, *Phys. Rev. Lett.* **63**, 1195 (1989).

[81] J.S. Wang, T. Tay and R. Swendsen, *Phys. Rev. Lett.* **82**, 476 (1999).

[82] G. Smith and A. Bruce, *J. Phys. A* **28**, 6623 (1995).

[83] B. Berg, *J. Stat. Phys.* **82**, 323 (1996).

[84] F. Wang and D.P. Landau, *Phys. Rev. Lett.* **86**, 2050 (2001).

[85] Q. Yan and J.J. de Pablo, *Phys. Rev. Lett.* **90**, 035701 (2003).

[86] U. Hansmann and L. Wille, *Phys. Rev. Lett.* **88**, 068105 (2002).

[87] B. Schulz, K. Binder and M. Müller, *J. Mod. Phys. C* **13**, 477 (2002).

[88] P. Virnau and M. Müller, *J. Chem. Phys.* **120**, 10925 (2004).

[89] J. Dudowicz, K.F. Freed and J.F. Douglas, *J. Chem. Phys.* **116**, 9983 (2002); *Phys. Rev. Lett.* **88**, 095503 (2002).

[90] S.M. Wood and Z.G. Wang, *J. Chem. Phys.* **116**, 2289 (2002).

[91] Z.G. Wang, *J. Chem. Phys.* **117**, 481 (2002).

[92] D. Buta and K.F. Freed, *J. Chem. Phys.* **116**, 10959 (2002); D. Buta, K.F. Freed and I. Szleifer, *J. Chem. Phys.* **114**, 1425 (2001).

[93] B. Crist, *Macromolecules* **31**, 5853 (1998).

[94] F.S. Bates, J.H. Rosedale, P. Stepanek, T.P. Lodge, P. Wiltzius, G.H. Fredrickson, P.P. Helm, *Phys. Rev. Lett.* **65**, 1893 (1990).

[95] M.D. Gehlen, J.H. Rosedale, F.S. Bates, G.D. Wignall and K. Almdal, *Phys. Rev. Lett.* **68**, 2452 (1992).

[96] E.A. Guggenheim, *J. Chem. Phys.* **13**, 253 (1945).

[97] F.P. Buff, R.A. Lovett and F.H. Stillinger, *Phys. Rev. Lett.* **15**, 621 (1965).

[98] A. Yethiraj and K.S. Schweizer, *J. Chem. Phys.* **97**, 5927 (1992); *J. Chem. Phys.* **98**, 9080 (1993); K.S. Schweizer and A. Yethiraj, *J. Chem. Phys.* **98**, 9053 (1993); C. Singh, K.S. Schweizer and A. Yethiraj, *J. Chem. Phys.* **102**, 2187 (1995).

[99] M. Müller, *Macromol. Theory Simul.* **8**, 343 (1999).

[100] V.L. Ginzburg, *Sov. Phys. Solid State* **1**, 1824 (1960); P.G. de Gennes, *J. Phys. Lett. (Paris)* **38**, L-441 (1977); J.F. Joanny, *J. Phys. A* **11**, L-117 (1978); K. Binder, *Phys. Rev. A* **29**, 341 (1984).

[101] M.E. Fisher, *Rev. Mod. Phys.* **46**, 587 (1974).

[102] C.J. Grayce, A. Yethiraj and K.S. Schweizer, *J. Chem. Phys.* **100**, 6857 (1994). C.J. Grayce and K.S. Schweizer, *J. Chem. Phys.* **100**, 6846 (1994).

[103] J.D. Weinhold, J.G. Curro, A. Habenschuss and J.D. Londono, *Macromolecules* **32**, 7276 (1999).

[104] F. Eurich and P. Maass, *J. Chem. Phys.* **114**, 7655 (2001).

REFERENCES

[105] R. Holyst and T.A. Vilgis, *Macromol. Theory Simul.* **5**, 573 (1996); R. Holyst and T.A. Vilgis, *J. Chem. Phys.* **99**, 4835 (1993); *Phys. Rev. E* **50**, 2087 (1994); M.G. Brereton and T.A. Vilgis, *J. Phys. (France)* **50**, 245 (1989); A. Aksimentiev and R. Holyst, *Macromol. Theory Simul.* **7**, 447 (1998).

[106] G.E. Garas and M.K. Kosmas, *J. Chem. Phys.* **103**, 10790 (1995); *J. Chem. Phys.* **105**, 4789 (1996); *J. Chem. Phys.* **108**, 376 (1998); M.K. Kosmas, *Macromolecules* **22**, 720 (1989).

[107] M. Dijkstra and D. Frenkel, *Phys. Rev. Lett.* **72**, 292 (1994); M. Dijkstra, D. Frenkel and J.P. Hansen, *J. Chem. Phys.* **101**, 3179 (1996).

[108] W.W. Graessley, R. Krishnamoorti, N.P. Balsara, L.J. Fetters, D.J. Lohse, D.N. Schulz and J.A. Sissano, *Macromolecules* **27**, 2574, 3073, 3896 (1994); R. Krishnamoorti, W.W. Graessley, G.T. Dee, D.J. Walsh, L.J. Fetters and D.J. Lohse, *Macromolecules* **29**, 367 (1996); G.C. Reichart, W.W. Graessley, R.A. Register, R. Krishnamoorti and J.D. Lohse, *Macromolecules* **30**, 3036 (1996); W.W. Graessley, R. Krishnamoorti, G.C. Reichart, N.P.Balsara, L.J. Fetters and D.J. Lohse, *Macromolecules* **28**, 1260 (1995); R. Krishnamoorti, W.W. Graessley, L.J. Fetters, R.G. Garner and D.J. Lohse, *Macromolecules* **28**, 1252 (1995).

[109] J.H. Hildebrand and R.L. Scott, *The Solubility of Nonelectrolytes*, Dover, New York (1964); J.H. Hildebrand, J.M. Prausnitz and R.L. Scott, *Regular and Related Solutions*, Van Nostrand Reinhold, New York (1970).

[110] M. Müller and A. Werner, *J. Chem. Phys.* **107**, 10764 (1997).

[111] G.H. Fredrickson, A.J. Liu and F.S. Bates, *Macromolecules* **27**, 2503 (1994); G.H. Fredrickson and A.J. Liu, *J. Polym. Sci.* **B33**, 1203 (1995).

[112] C. Singh and K.S. Schweizer, *Macromolecules* **28**, 8692 (1995); *J. Chem. Phys.* **103**, 5814 (1995); *Macromolecules* **30**, 1490 (1997); K.S. Schweizer and C. Singh, *Macromolecules* **28**, 2063 (1995).

[113] K.W. Foreman and K.F. Freed, *Macromolecules* **30**, 7279 (1997); K.W. Foreman, K.F. Freed and I.M. Ngola, *J. Chem. Phys.* **107**, 4688 (1997); K.F. Freed and J. Dudowicz, *Macromolecules* **29**, 625 (1996).

[114] J.D. Weinhold, S.K. Kumar, C. Singh and K.S. Schweizer, *J. Chem. Phys.* **103**, 9460 (1995).

[115] S.K. Kumar and J. Weinhold, *Phys. Rev. Lett.* **77**, 1512 (1996).

[116] K.F. Freed and J. Dudowicz, *Macromolecules* **31**, 6681 (1998).

[117] M. Rabeony, D.J. Lohse, R.T. Garner, S.J. Han, W.W. Graessley and K.B. Migler, *Macromolecules* **31**, 6511 (1998).

[118] M. Müller and W. Paul, *J. Chem. Phys.* **100**, 719 (1994).

[119] J.K. Taylor-Maranas, P.G. Debenedetti, W.W. Graessley and S.K. Kumar, *Macromolecules* **30**, 6943 (1997).

[120] J.F. Joanny and H. Benoit, *Macromolecules* **30**, 3704 (1997); J.F. Joanny, H. Benoit and W.H. Stockmayer, *Macromol. Symp.* **121**, 95 (1997).

[121] D.G. Gromov and J.J. dePablo, *J. Chem. Phys.* **109**, 10042 (1998).

[122] M. Fuchs and M. Müller, *Phys. Rev. E* **60**, 1921 (1999).

[123] J.-F. Joanny, L. Leibler and R Ball, *J. Chem. Phys.* **81**, 4640 (1984).

[124] L. Schaefer and C. Kappeler, *J. Chem. Phys.* **99**, 6135 (1993); L. Schaefer and C. Kappeler, *J. Phys. (France)* **46**, 1853 (1985).

[125] A. Cavallo, M. Müller and K. Binder, *Europhys. Lett.* **61**, 214 (2003).

[126] B. Duplantier, *J. Stat. Phys.* **54**, 581 (1989).

[127] A.J. Silverberg, *Colloid Interface Sci.* **90**, 86 (1982).

[128] M. Müller, K. Binder and E.V. Albano, *Int. J. Mod. Phys.* **B15**, 1867 (2001).

[129] K. Binder, D.P. Landau and M. Müller, *J. Stat. Phys.* **110**, 1411 (2003).

[130] A. Werner, F. Schmid, M. Müller and K. Binder, *Phys. Rev. E* **59**, 728 (1999).

[131] A. Werner, F. Schmid, M. Müller and K. Binder, *J. Chem. Phys.* **107**, 8175 (1997).

[132] F. Schmid and M. Müller, *Macromolecules* **28**, 8639 (1996).

[133] M.D. Lacasse, G.S. Grest and A.J. Levine, *Phys. Rev. Lett.* **80**, 309 (1998).

[134] E. Chacon and P. Tarazona, *Phys. Rev. Lett.* **91**, 166103 (2003).

[135] A.N. Semenov, *Macromolecules* **27**, 2732 (1994).

[136] I. Szleifer, A. Ben-Shaul and W.M. Gelbhart, *J. Chem. Phys.* **85**, 5345 (1986); *J. Chem. Phys.* **86**, 7094 (1987); I. Szleifer and M.A. Carignano, *Adv. Chem. Phys.* **94**, 742 (1996).

[137] E. Helfand and A.M. Sapse, *J. Chem. Phys.* **62**, 1329 (1975).

[138] A. Werner, F. Schmid and M. Müller, *J. Chem. Phys.* **110**, 5370 (1999).

[139] J. Noolandi and K.M. Hong, *Macromolecules* **15**, 483 (1982); K.M. Hong and J. Noolandi, *Macromolecules* **16**, 1083 (1983); J. Noolandi and K.M. Hong, *Macromolecules* **17**, 1531 (1984); L. Leibler, *Makromol. Chem. Macromol. Symp.* **16**, 1 (1988); L. Leibler, *Physica A* **172**, 258 (1991); K.R. Shull and E.J. Kramer, *Macromolecules* **23**, 4769 (1990). K.R. Shull, E.J. Kramer, G. Hadziioanou and W. Tang, *Macromolecules* **23**, 4780 (1990); K.R. Shull and K.I. Winey, *Macromolecules* **25**, 2637 (1992).

[140] P.K. Janert and M. Schick, *Macromolecules* **30**, 3916 (1997).

[141] P.G. de Gennes and C. Taupin, *J. Phys. Chem.* **86**, 2294 (1982).

[142] M. Müller and M. Schick, *J. Chem. Phys.* **105**, 885 (1996).

[143] G.H. Fredrickson and F.S. Bates, *Polym. Sci. B* **35**, 2775 (1997). H.S. Jeon, J.H. Lee and N.P. Balsara, *Phys. Rev. Lett.* **79**, 3274 (1997).

[144] D. Düchs, V. Gansean, G.H. Fredrickson and F. Schmid, *Macromolecules* **36**, 9237 (2003).

[145] M.W. Matsen, *J. Chem. Phys.* **110**, 4658 (1999).

[146] M. Müller and G. Gompper, *Phys. Rev. E* **66**, 041805 (2002).

[147] K. Almdal, M.A. Hilmyer and F.S. Bates, *Macromolecules* **35**, 7685 (2002).

[148] A.V.G.Ruzette and A.M. Mayes, *Macromolecules* **34**, 1894 (2001).

[149] G.H. Fredrickson, V. Ganesan and F. Drolet, *Macromolecules* **35**, 16 (2002).

[150] C. Castellano and S.C. Glotzer, *J. Chem. Phys.* **103**, 9363 (1995).

[151] G. Müller, D. Schwahn, H. Eckerlebe, J. Rieger and T. Springer, *J. Chem. Phys.* **104**, 5326 (1996); D. Schwahn, S. Janssen and T. Springer, *J. Chem. Phys.* **97**, 8775 (1992); H. Jinnai, H. Hasegawa, T. Hashimoto and C.C. Han, *J. Chem. Phys.* **99**, 8154 (1993).

[152] P.G. de Gennes, *J. Chem. Phys.* **72**, 4756 (1980); P. Pincus, *J. Chem. Phys.* **75**, 1996 (1981); K. Binder, *J. Chem. Phys.* **79**, 6387 (1983).

[153] J.G.E.M. Fraaije et al., *J. Chem. Phys.* **106**, 4260 (1997).

[154] N.M. Maurits and J.G.E.M. Fraaije, *J. Chem. Phys.* **107**, 5879 (1997).

[155] P.E. Rouse, *J. Chem. Phys.* **21**, 1272 (1953); M. Doi and S.F. Edwards, *The Theory of Polymer Dynamics*, Oxford University Press, Oxford (1994).

[156] A.A. Lefebvre, J.H. Lee, N.P. Balsara and C. Vaidyanathan, *J. Chem. Phys.* **117**, 9063 (2002).

[157] H. Tanaka, *J. Phys.: Condens. Mater.* **12**, R207 (2000).

[158] G.H. Fredrickson, *J. Chem. Phys.* **117**, 6810 (2002).

4

MESOSCOPIC SIMULATIONS OF POLYMER MIXTURES

OLAF EVERS

BASF Aktiengesellschaft, Department of Polymer Physics,
D-67056 Ludwigshafen am Rhein, Germany

CONTENTS

4.1.	Introduction	154
4.2.	Homogeneous Polymer Mixtures	156
	4.2.1. Notation	156
	4.2.2. Flory–Huggins Theory	158
	4.2.3. Interaction Energy Density Parameter	159
	4.2.4. Chemical Potential	160
	4.2.5. Multi-component Phenomenological Approach	163
4.3.	Inhomogeneous Polymer Mixtures	172
	4.3.1. Notation	172
	4.3.2. What is a Functional?	173
	4.3.3. Grand Canonical Ensemble	176
	4.3.4. Basic Concepts of Density Functional Theory	178
	4.3.5. Kohn–Sham-Like Formalism	180
	4.3.6. Numerical Example: Excluded Volume Effects	184
	4.3.7. Ensemble of Quasi-ideal Systems	187
	4.3.8. Mean-Field Approximation	194
	4.3.9. Lagrange Multiplier Fields	196
	4.3.10. Numerical Example: The Scheutjens–Fleer Model	196
	4.3.11. Numerical Example: Continuous Field Model	199

Molecular Simulation Methods for Predicting Polymer Properties, Edited by Vassilios Galiatsatos.
ISBN 0-471-46481-3   Copyright © 2005 John Wiley & Sons, Inc.

4.4. Dynamic Density Functional Theory   205
    4.4.1. The DDFT Principle   205
    4.4.2. Excluded Volume Effects   208
    4.4.3. Numerical Example: Copolymer–Solvent Mixtures   209
    4.4.4. External Potential Dynamics   210

4.1. INTRODUCTION

Depending on the area of interest, the words microscopic, mesoscopic, and macroscopic are assigned different meanings. Even in material science what is understood by the term mesoscopic depends on the subject of research. We will use the term microscopic to refer to length scales of the order of a few nanometers. This means the microscopic world is the world of entities like atoms and electrons. To describe or simulate phenomena taking place at this scale (for instance diffusion of small molecules in polymer melts), one needs to apply quantum chemistry methods or molecular dynamics. In contrast to the microscopic scale we have the macroscopic scale, the world described in terms of continuum mechanics and thermodynamics. The macroscopic length scale is that of millimeters upwards. Processes taking place at this length scale (e.g. airflow around an airplane) are described using macroscopic parameters like viscosity, pressure and so on. The mesoscopic length scale describes the world in between and is typically of the order of 100 nm to micrometers. It typically links the macroscopic features to microscopic descriptors like the chemical structure of the molecules. Macroscopic parameters have no real meaning at this length scale, but a description of the system in terms of atoms and electrons makes no sense because the mesoscopic length scale is simply to large for it. In fact we should ask the question: What can be measured at this length scale? From a solution of polymer dispersion particles (latex), for instance, we obtain from the scattering of neutrons or electrons insight into the internal structure of the particles, but we do not get any information on the positions of individual molecules inside these particles. The best we can obtain experimentally is information on the so-called 'average density distributions' (or fields) of the various molecule types in the solution. Many polymer materials gain their (mechanical) properties from the mesoscopic morphology, like for instance the famous example of high-impact polystyrene (HIPS), just to mention one. In particular, polymer blends or melts of block copolymers show complex morphologies that include spheres, cylinders, gyroids, lamellae or even combinations of these. A simulation model for the mesoscale world should therefore be able to predict these morphologies at a resolution of about a few nanometers.

Writing a chapter on simulation techniques for polymer mixtures aiming at predicting mesoscopic morphologies implies making a decision on what to present and how. The major objective of this chapter must of course be the application of these techniques within an industrial environment. However, does this mean that we should present an endless series of examples of how people applied these simulation techniques

INTRODUCTION

to real industrial systems? The problem is that, when a simulation technique has been successfully adapted within an industrial research department, a model of the real application is not very likely to be published in great detail, if at all. This does not mean that examples cannot be found in literature; mostly they are the result of first investigations to answer the question of whether a certain simulation technique is of value. Most such first investigations are performed by academia and industry together, often in the framework of national or international (European Commission) goverment-funded projects, but fortunately also by direct industrial funding. In this chapter we will present a number of such examples but not put the emphasis on it. Why? Simply because there are enough review papers and books already available in which, among others, the original authors present their simulation technologies for a greater audience in the form of success stories. Should we then go the opposite way and present the simulation methods and the theories behind it in great detail? Probably not, as you will find this detail in the original publications.

In this chapter I will first of all show how the various self-consistent field methods (including dynamic density functional methods) are related; that is, they can all be derived from classical density functional theory. In fact, it is shown that they differ in the choice of the polymer chain model, the level of detail in the correlations taken into account and the numerical pathway to obtain self-consistent solutions. Special attention is paid to the level of coarse graining in the numerical approach taken. For example, although the well-known Scheutjens–Fleer theory is derived from lattice-based equations, you can derive it from density functional theory of polymer mixtures in the form of a very coarse-grained intergration scheme for the green functions. It might seems strange that somebody from industry puts effort into bringing various mesoscopic simulation methods under one umbrella, but it is not. If you are working in an industrial research department, the applications you are dealing with are complex, mostly too complex for theoretical investigations or simulation studies. The first step you take is of course to break down the complex system into a 'model' system. The next step is to find an adequate simulation method. This is where the content of this chapter comes into play. The choice of the simulation method appropriate for your application depends of course on the level of detail needed, the expert knowledge you might have gained already, and the issue of model parameterization. The latter is vital. Of course, a simulation model expressed in terms of 100 adjustable parameters will be able to describe quite a lot; however, you will have hardly any change to parameterize it for your application. Parameterization of a model is a major issue. In mesoscopic simulation methods there are two parameterization issues; the first is the parameterization of the enthalpy of mixing and the second is the parameterization of the polymer chain model. In addition to the parameterization, mesoscopic simulation techniques depend on the numerical device applied. The numerical mathematics and its level of coarse graining used by a simulation technique has a great impact on its applicability to the practical system of interest.

This chapter comprises three sections. The first deals with homogeneous polymer mixtures. It serves to formulate the basic thermodynamics. To put it more precisely, the phenomenological approach to the Flory–Huggins theory in its multicomponent formulation is derived and its parameterization is discussed in full detailed demonstrations of real practical systems. In the second section the basic concepts of classical density functional theory are introduced. In order to apply it to inhomogeneous polymer mixtures the 'quasi-ideal' system is introduced to the Kohn–Sham self-consistent scheme and the parameterization of the polymer chain model is discussed in detail. The dynamic density functional theory is described in the last section and special attention is paid to its so-called 'external potential dynamics' formulation.

4.2. HOMOGENEOUS POLYMER MIXTURES

Most mesoscopic density field simulation methods for inhomogeneous polymer mixtures are based on extending a free energy functional theory for homogeneous polymer mixtures. In other words, when applying these mesoscopic simulation methods to systems that do not microscopically phase separate, they must return the correct thermodynamics of the homogeneous mixture. We will concentrate in this section on the Flory–Huggins theory of homogeneous polymer mixtures (Section 4.2.2). In addition, we discuss the problem of applying the Flory–Huggins theory to practical systems and introduce the concept of expressing the Flory–Huggins theory in terms of the excess interaction energy density parameter (Section 4.2.3), which is customary for polymer blends. Special attention is paid to the phenomenological approach to the Flory–Huggins theory (Section 4.2.5) and its parameterization.

4.2.1. Notation

The state of a homogenous mixture is completely defined by its variables of state: S (entropy), V (volume), and n_i (amount of component i in moles). The total differential of the internal energy of a system, $U(S, V, n)$, is then formally given by:

$$dU = \left(\frac{\partial U}{\partial S}\right)_{V,N} dS + \left(\frac{\partial U}{\partial V}\right) dV + \sum_i \left(\frac{\partial U}{\partial n_i}\right)_{S,V,N_{j \neq i}} dn_i \quad (4.1)$$

The partial derivatives represent the intensive variables absolute temperature, T, pressure, p, and chemical potential, μ_i, respectively.

$$T = \left(\frac{\partial U}{\partial S}\right)_{V,N}$$
$$-p = \left(\frac{\partial U}{\partial V}\right)_{S,N} \quad (4.2)$$
$$\mu_i = \left(\frac{\partial U}{\partial n_i}\right)_{S,V,n_{j \neq i}}$$

HOMOGENEOUS POLYMER MIXTURES

Equation (1) can now be written into the more familiar form:

$$dU = T\,dS - p\,dV + \sum_i \mu_i dn_i \tag{4.3}$$

If we regard the intensive variables as generalized coordinates of the system, we can identify the internal energy, U, as a thermodynamic potential. The sets of pair state variables (T, S), (p, V), and (μ, n) are conjugated.

It is possible to define new thermodynamic potentials making use of the Legendre transformation:

$$\frac{\partial A}{\partial B} = C, \quad dA = C\,dB \tag{4.4}$$

which results in

$$d(BC - A) = B\,dC, \quad \frac{\partial(BC - A)}{\partial C} = B \tag{4.5}$$

Hence, if we want, instead of B, its conjugated variable C to be the independent variable, we should define a new potential $BC - A$. The following thermodynamic variables can thus be defined and are often used:

$$\begin{aligned} &U(S, V, n) \\ &H(S, p, n) \equiv U + pV \\ &F(T, V, n) \equiv U - TS \\ &G(T, p, n) \equiv H - TS = F + pV \end{aligned} \tag{4.6}$$

where H denotes the enthalpy, F the (Helmholtz) free energy and G the free enthalpy or Gibbs free energy. For most practical applications the independently controllable thermodynamic variables are T, p, and n. Therefore, the thermodynamic potential G, that is, the Gibbs free energy, is of most interest. Often the Gibbs free energy, G, is turned into an intensive state variable by dividing it either by the mass of the system (specific Gibbs free energy) or by the volume of the system (Gibbs free energy density, g_V), that is,

$$g_V \equiv \frac{G}{V} \tag{4.7}$$

For incompressible systems, g_V is an appropriate quantity to work with.

In this chapter we will always use the symbol n to refer to number of objects. For instance, n_i, is the number of molecules of type i, in units of number and not of mole. We will not use the unit mole, as is usually applied in homogeneous polymer mixture theories, as it normally not done in density field theories, which we will discuss in Sections 4.3 and 4.4.

4.2.2. Flory–Huggins Theory

The famous Flory–Huggins (FH) theory of polymer mixtures [29, 30, 50–52] has been reviewed in detail in many publications (e.g. Binder [8] and references therein). In this book the FH theory is introduced by Marcus Müller in Chapter 3. Therefore, in this sub-section we will only summarize some basic results of this theory needed in the derivation of density functional approaches to inhomogeneous mixtures of polymers.

By assuming incompressibility of the components (i.e. no volume changes upon mixing) Flory and Huggins formulated a lattice model to describe a homogeneous polymer mixture and derived by means of statistical thermodynamics [48, 65] an expression for the entropy of mixing, ΔS,

$$\Delta S = -k \sum_i n_i \ln \phi_i \qquad (4.8)$$

where n_i denotes the number of molecules of component i and the volume fraction, ϕ_i, of homopolymer component i is defined as:

$$\phi_i = \frac{n_i v_i}{\sum_j n_j v_j} \qquad (4.9)$$

where v_i is the volume of a homopolymer of type i. It is assumed that the polymer molecules can be segmented into N_i segments that exactly occupy (or fill) one lattice site. Figure 4.1 shows for instance a polymer chain in on a cubic lattice. The enthalpy of mixing, ΔH, in the FH approach is derived from a mean-field approach;

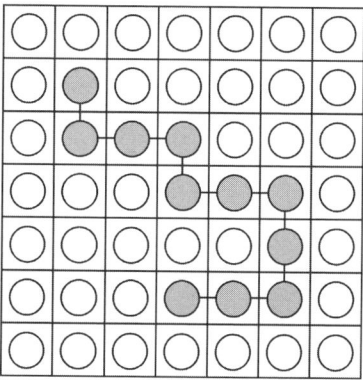

Figure 4.1. Lattice with sites of equal volume filled with solvent and polymer molecules.

HOMOGENEOUS POLYMER MIXTURES

that is, it is parameterized in terms of the so-called Flory–Huggins χ parameter (dimensionless),

$$\Delta H = kT \sum_{i,j>i} n_i N_i \phi_j \chi_{ij} \tag{4.10}$$

The sum is taken such that we never need χ_{ji}; only one FH parameter is needed: χ_{ij} with $j > i$. What does this mean? Suppose two chemically different species (1 and 2) have different volumes; you can be sure that $\chi_{12} \neq \chi_{21}$. That means, you can only apply the FH theory, in practice, if you have obtained, by whatever experimental or simulation methods, either χ_{12} or χ_{21}. This implies that, in order to compute ΔH you *must* apply the same χ parameter as determined from the experimental results. Quite often self-consistent field theories derived from some kind of generalized FH theory make the following approach:

$$\Delta H = kT \frac{1}{2} \sum_{i,j} n_i N_i \phi_j \chi_{ij} \tag{4.11}$$

This implies the assumption of $\chi_{ji} = \chi_{ij}$ and $\chi_{ii} = 0$ and only works nicely for molecules of equal size or, putting it more precisely, if the molecules are segmented into segments of equal size. Finally, the free enthalpy of mixing, ΔG, is given by:

$$\Delta G = kT \left\{ \sum n_i \ln \phi_i + \sum_{i,j>i} n_i N_i \phi_j \chi_{ij} \right\} \tag{4.12}$$

Although the FH lattice model is an extremely simplified model it has served for 50 years as the basis of most polymer mixing thermodynamics.

4.2.3. Interaction Energy Density Parameter

In polymer blend studies, it is most convenient to parameterize the enthalpy of mixing in terms of an excess interaction energy parameter per unit volume, B_{ij}; this interaction energy density parameter is defined as

$$B_{ij} \equiv kT \frac{N_i \chi_{ij}}{v_i}, \quad j > i \tag{4.13}$$

The enthalpy density of mixing according to the Flory–Huggins theory [equation (10)] becomes, in terms of this excess interaction energy parameter,

$$\Delta h_v = \sum_{i,j>i} \phi_i \phi_j B_{ij} \tag{4.14}$$

It seems tempting to parameterize the excess interaction energy or χ parameters from binary mixtures using microcalorimetry [11]; this is not a good idea. The theory of Flory and Huggins uses a very simple model to derive the entropic part of the free energy, which means that all entropic contributions not accounted for by this simple combinatorial entropic derivation must be compensated for by the interaction parameters. The enthalpy expressed by equation (4.14) is not a pure enthalpic term as it contains entropic contributions. Malcom and Rowlinson [71], already in the 1950s, showed the consequences of applying the FH enthalpy formula by a thermodynamic analysis of vapor pressure measurements and heats of mixing of polymer–solvent mixtures.

Inserting the FH expression (4.12) for ΔG into expression (4.7) and making use of equation (4.13), we obtain:

$$\Delta g_v = kT \sum_i \frac{\phi_i}{v_i} \ln \phi_i + \sum_{i,j>i} B_{ij} \phi_i \phi_j \qquad (4.15)$$

The Gibbs free energy density, given by expression (4.15), depends on the molar volumes of the various homopolymer components only and *not* on any intramolecular segmentation. The χ_{ij} parameter, on the other hand, does depend on the intramolecular segmentation. In practice, however, quite often χ parameters obtained from experiments are tabulated with respect to an arbitrary segmentation [100] using a reference volume, v_{ref}. This means the χ values refer to a segmentation: $N_i = v_i/v_{\text{ref}}$, so that from equation (4.13) we have:

$$B_{ij} = kT \frac{\chi_{ij}}{v_{\text{ref}}} \qquad (4.16)$$

4.2.4. Chemical Potential

The chemical potential plays a major role in the development of density functional theories, therefore its derivation for a homogeneous multicomponent mixture is discussed in more detail in this section. Another reason for this detailed derivation is simply that you will rarely find it in the literature.

The chemical potential, μ_i, of component i in a polymer mixture of volume V is by definition

$$\mu_i \equiv \left(\frac{\partial V \Delta g_v}{\partial n_i} \right)_{p,T,n_{j \neq i}} \qquad (4.17)$$

In deriving the chemical potential from the Gibbs free energy of mixing [equation (4.12)] within the framework of the Flory–Huggins model we encounter a problem. As the FH theory is based on complete incompressibility, it is not possible to add one chain of component i to the system without changing the volume of the system. In other words, the volume fractions of the various components, as was to be expected,

HOMOGENEOUS POLYMER MIXTURES

are by no means independent variables ($\sum_i \phi_i = 1$). In fact, we derive from equation (4.9):

$$\frac{\partial \phi_j}{\partial n_i} = \frac{-v_i v_j n_j}{(\sum_k v_k n_k)^2} = -\frac{v_i}{V}\phi_j, \quad j \neq i \tag{4.18}$$

and

$$\frac{\partial \phi_i}{\partial n_i} = \frac{v_i}{V}(1 - \phi_i) \tag{4.19}$$

By applying the differentiation chain rule to equation (4.17) we obtain:

$$\mu_i = v_i \Delta g_v + V \frac{\partial \Delta g_v}{\partial n_i} \tag{4.20}$$

In order to derive the differential in the right-hand side of equation (4.20), using the FH expression for g_v [equation (4.15)], we first concentrate on the entropic part of g_v, i.e. the first term in equation (4.15):

$$kT\left\{\frac{\phi_i}{v_i}\ln \phi_i + \sum_{k \neq i} \frac{\phi_k}{v_k} \ln \phi_k\right\}$$

Using the volume fraction differentials with respect to n_i [equations (4.18) and (4.19)] we have:

$$\frac{\partial kT \sum_j (\phi_j/v_j) \ln \phi_j}{\partial n_i} = \frac{kT}{V}\left\{(1-\phi_i)\ln \phi_i + (1-\phi_i)\right.$$

$$\left. \sum_{k \neq i}\frac{v_i}{v_k}\phi_k \ln \phi_k - \sum_{k \neq i}\frac{v_i}{v_k}\phi_k\right\} \tag{4.21}$$

Realizing that

$$\frac{kT}{V}\left\{-\phi_i \ln \phi_i - \sum_{k \neq i}\frac{v_i}{v_k}\phi_k \ln \phi_k\right\} = -\frac{kT}{V}\sum_k \frac{v_i}{v_k}\phi_k \ln \phi_k$$

and

$$\frac{kT}{V}\left\{(1-\phi_i) - \sum_{k \neq i}\frac{v_i}{v_k}\phi_k\right\} = \frac{kT}{V}\left\{1 - \sum_k \frac{v_i}{v_k}\phi_k\right\}$$

equation (4.21) can be reorganized into:

$$\frac{\partial kT \sum_j (\phi_j/v_j) \ln \phi_j}{\partial n_i} = \frac{kT}{V} \left\{ \ln \phi_i - \sum_k \frac{v_i}{v_k} \phi_k \ln \phi_k + 1 - \sum_k \frac{v_i}{v_k} \phi_k \right\}$$

Since $\sum_k \phi_k = 1$, the differential of the entropic part of g_v is thus given by:

$$\frac{\partial kT \sum_j (\phi_j/v_j) \ln \phi_j}{\partial n_i} = \frac{kT}{V} \left\{ \ln \phi_i - \sum_k \frac{v_i}{v_k} \phi_k \ln \phi_k + \sum_k \left(1 - \frac{v_i}{v_k}\right) \phi_k \right\} \quad (4.22)$$

For the enthalpic part of the differential of g_v with respect to n_i we derive, with the help of equations (4.18) and (4.19), the following expression:

$$\frac{\partial \sum_{k,j>k} B_{kj} \phi_k \phi_j}{\partial n_i} = \frac{1}{V} \sum_{j \neq i} v_i B_{ij} (1 - \phi_i) \phi_j - \frac{1}{V} \sum_{j \neq i} v_i B_{ij} \phi_i \phi_j$$

$$- \frac{1}{V} \sum_{k \neq i, j > k} v_i B_{kj} \phi_k \phi_j \quad (4.23)$$

By splitting the last term in this equation into:

$$-\frac{1}{V} \sum_{k \neq i, j > k} v_i B_{kj} \phi_k \phi_j = \frac{1}{V} \sum_{j \neq i} v_i B_{ij} \phi_i \phi_j - \frac{1}{V} \sum_{k, j > k} v_i B_{kj} \phi_k \phi_j$$

equation (4.23) becomes

$$\frac{\partial \sum_{k,j>k} B_{kj} \phi_k \phi_j}{\partial n_i} = \frac{1}{V} \sum_{j \neq i} v_i B_{ij} (1 - \phi_i) \phi_j - \frac{1}{V} \sum_{k, j > k} v_i B_{kj} \phi_k \phi_j \quad (4.24)$$

Putting it all together [equations (4.15), (4.20), (4.22) and (4.24)] the chemical potential, μ_i, is given by

$$\mu_i = kT \left\{ \ln \phi_i - \sum_k \left(1 - \frac{v_i}{v_k}\right) \phi_k \right\} + \sum_{j \neq i} v_i B_{ij} (1 - \phi_i) \phi_j \quad (4.25)$$

Let us validate this multicomponent result with the well known result for a binary mixture of a polymer and a solvent [48, 82]. Denoting the solvent as component 1 and the polymer as component 2 one obtains from equation (4.25) and realizing

HOMOGENEOUS POLYMER MIXTURES

that a completely incompressible system must satisfy the constraint $\phi_1 = 1 - \phi_2$,

$$\mu_1 = kT\left\{\ln(1 - \phi_2) + \left(1 - \frac{v_1}{v_2}\right)\phi_2\right\} + v_1 B_{12}\phi_2^2 \quad (4.26)$$

$$\mu_2 = kT\left\{\ln\phi_2 + \left(1 - \frac{v_2}{v_1}\right)(1 - \phi_2)\right\} + v_2 B_{12}(1 - \phi_2)^2 \quad (4.27)$$

From the definition of the excess interaction energy density [equation (4.13)], the chemical potentials in terms of the Flory–Huggins parameter, χ_{12}, for this binary mixture read:

$$\mu_1 = kT\left\{\ln(1 - \phi_2) + \left(1 - \frac{v_1}{v_2}\right)\phi_2 + N_1\chi_{12}\phi_2^2\right\} \quad (4.28)$$

$$\mu_2 = kT\left\{\ln\phi_2 + \left(1 - \frac{v_2}{v_1}\right)(1 - \phi_2) + \frac{v_2}{v_1}N_1\chi_{12}(1 - \phi_2)^2\right\} \quad (4.29)$$

The chemical potentials are in units of Joule per molecule and not in units of Joule per mole. These equations are the familiar Flory–Huggins expressions for the chemical potentials in a mixture of a polymer and a solvent. Equation (4.25) represents the generalized FH expression for the chemical potential in a homogeneous multi-component polymer mixture.

4.2.5. Multi-component Phenomenological Approach

How does one parameterize the Flory–Huggins model in order to make quantitative predictions on real applications? This is indeed an important question; from an industrial point of view it is the most important. If a model fails to be parameterized it is for practical use in an industrial research environment not really of value. To start with the bad news, it is hardly possible to parameterize the Flory–Huggins model for polymer mixtures in its original fashion. The good news is that, by taking a phenomenological approach to the Flory–Huggins interaction parameter, it is quite often possible to find for many systems a suitable parameterization.

4.2.5.1. Composition Dependent Interaction Parameter. Besides the famous Flory–Huggins theory, more detailed equations of state theories (e.g. the Sanchez–Lacombe theory [63, 93, 94]) have been developed in order to provide more quantitative predictions in terms of an increased number of parameters that have some physical meaning. However, in industry we should be pragmatic; that is, we might still apply the Flory–Huggins theory but obtain quantitative predictions by expressing its parameters in terms of more or less arbitrary mathematical relationships with adjustable parameters. In other words, we increase the number of parameters allowing the prediction of upper critical solution temperatures (UCST) as well as lower critical solution temperatures (LCST). Such an approach is called a phenomenological approach. There is much literature on composition- and temperature-dependent Flory–Huggins interaction parameters, especially the pioneering work of Kongingsveld and co-workers [58–60, 87] has received much attention.

Equation (4.15) can be transformed into a quantitative exact result by introducing an *effective* interaction energy parameter, B_{ij}^{eff} (ϕ_i, ϕ_j, p, T), which is a function of both volume fractions (ϕ_i, ϕ_j, p, T), the pressure and the temperature:

$$\Delta g_v = kT \sum_i \frac{\phi_i}{v_i} \ln \phi_i + \sum_{i,j>i} B_{ij}^{\text{eff}}(\phi_i, \phi_j, p, T)\phi_i \phi_j \qquad (4.30)$$

Actually, it is most appropriate to make B_{ij}^{eff} depend on the relative fraction of component j, that is, on $\phi_j/(\phi_i + \phi_j)$. In the following we will drop the explicit variable dependence of B_{ij}^{eff} ($\frac{\phi_i}{\phi_i+\phi_j}$, p, T). Within the phenomenological approach the enthalpic differential term of the chemical potential is derived as:

$$\frac{\partial \sum_{k,j>k} B_{kj}^{\text{eff}} \phi_k \phi_j}{\partial n_i} = \sum_{k,j>k} \phi_k \phi_j \frac{\partial B_{kj}^{\text{eff}}}{\partial n_i} + \frac{1}{V} \sum_{j \neq i} v_i B_{ij}^{\text{eff}}(1-\phi_i)\phi_j$$

$$- \frac{1}{V} \sum_{k,j>k} v_i B_{kj}^{\text{eff}} \phi_k \phi_j$$

where we have used equation (4.24). Applying the chain rule to the first term on the right-hand side gives:

$$\sum_{k,j>k} \phi_k \phi_j \frac{\partial B_{kj}^{\text{eff}}}{\partial n_i} = \sum_{k,j>k} \phi_k \phi_j \sum_{p=i,j} \frac{\partial B_{k,j}^{\text{eff}}}{\partial \phi_p} \frac{\partial \phi_p}{\partial n_i}$$

so that the chemical potential of component i is given by

$$\mu_i = kT \left\{ \ln \phi_i - \sum_k \left(1 - \frac{v_i}{v_k}\right) \phi_k \right\} + \sum_{j \neq i} v_i B_{ij}^{\text{eff}}(1-\phi_i)\phi_j$$

$$+ \sum_{k,j>k} \phi_k \phi_j \sum_{p=i,j} \frac{\partial B_{kj}^{\text{eff}}}{\partial \phi_p} \frac{\partial \phi_p}{\partial n_i} \qquad (4.31)$$

This general phenomenological expression for the chemical potential in a multi-component mixture can be validated by applying it to a two-component mixture where B^{eff} is parameterized such that it depends only on ϕ_2. We obtain the familiar

FH expression in the phenomenological approach:

$$\mu_1 = kT\left\{\ln(1-\phi_2) + \left(1 - \frac{v_1}{v_2}\right)\phi_2\right\} + v_1 B_{12}^{\text{eff}}\phi_2^2 - v_1\phi_1\phi_2^2 \frac{\partial B_{12}^{\text{eff}}}{\partial \phi_2}$$

or

$$\mu_1 = kT\left\{\ln(1-\phi_2) + \left(1 - \frac{v_1}{v_2}\right)\phi_2\right\} + v_1 \tilde{B}_{12}\phi_2^2 \quad (4.32)$$

with $\tilde{B}_{12}$ defined according to

$$\tilde{B}_{12}(\phi_2, T) \equiv B_{12}^{\text{eff}} - \phi_1 \frac{\partial B_{12}^{\text{eff}}}{\partial \phi_2} \quad (4.33)$$

Here we followed the notation introduced by Baulin and Halperin [5]. For the second component we have:

$$\mu_2 = kT\left\{\ln(1-\phi_2) + \left(1 - \frac{v_1}{v_2}\right)\phi_1\right\} + v_2 B_{12}^{\text{eff}}\phi_1^2 + v_2\phi_2\phi_1^2 \frac{\partial B_{12}^{\text{eff}}}{\partial \phi_2}$$

or

$$\mu_2 = kT\left\{\ln(\phi_2) + \left(1 - \frac{v_2}{v_1}\right)(1-\phi_2)\right\} + v_2 \tilde{B}_{21}(1-\phi_2)^2 \quad (4.34)$$

with $\tilde{B}_{21}$ defined according to

$$\tilde{B}_{21}(\phi_2, T) \equiv B_{12}^{\text{eff}} + \phi_2 \frac{\partial B_{12}^{\text{eff}}}{\partial \phi_2} \quad (4.35)$$

This result for a two-component mixture corresponds with usual Flory–Huggins two-component phenomenological approach.

In polymer literature the composition-dependent χ parameter corresponding to $\tilde{B}_{12}$ is often defined as

$$\chi = g + \phi_2 \frac{\partial g}{\partial \phi_2}$$

that is, χ_2^{eff} is mostly denoted by the symbol g. We have adopted the notation introduced by Baulin and Halperin [5] to prevent any confusion, denoting distributions by the symbol g.

4.2.5.2. Parameterization Issues.

For binary mixtures of polymers and solvent, the best route to parameterize a phenomenological model for $\tilde{B}_{12}$ is to use experimental vapor pressure measurements. The vapor pressure of a solvent, p_s, above a homogeneous polymer solution, as compared with that of the pure solvent, is directly related to the chemical potential of the solvent in the mixture (in units of Joule per molecule):

$$\frac{\mu_s(\phi_p) - \mu_w^0}{kT} = \ln \frac{p_s(\phi_p)}{p_s(0, T)} \quad (4.36)$$

It is assumed that the vapor phase behaves as an ideal gas. This relation allows the experimental determination of the chemical potential of the solvent in a polymer–solvent mixture, constrained to the requirement that the polymer–solvent mixture is indeed a one-phase system in equilibrium with the vapor phase. A famous series of such experiments was performed by Malcolm and Rowlinson [71] with polyethylene-glycol–water and polypropylene-glycol–water mixtures.

The solvent chemical potentials at different temperatures and composition determined from the measured vapor pressures with equation (4.36) allow the parameterization of the effective interaction density B_{12}^{eff} (ϕ_2, T) in terms of a model description. In the remaining of this subsection it is demonstrated how such a procedure is applied using a methodology published recently by Wolf and coworkers [7, 105, 112].

Wolf derives a model for the effective Flory–Huggins parameter, χ_{12}^{eff}, which requires four parameters to be extracted from experimental data. The model of Wolf [112] yields the following expression for the effective Flory–Huggins parameter:

$$\chi_{12}^{\text{eff}} = \frac{\alpha}{(1-\nu)(1-\nu\phi_2)} - \zeta[1 + (1-\lambda)\phi_2] \quad (4.37)$$

For most applications it is a good approximation to set λ equal to 0.5. The parameter α quantifies the enthalpic effect of opening a contact between polymer segments by insertion of a solvent molecule at inifinite dilution without modifying the chain conformation. Equilibrium is, however, only reached via conformational relaxation, quantified by the parameter ζ. Under theta conditions ζ becomes zero and α assumes the value of 0.5. Finally, ν accounts primarily for the entropic part in the interaction parameter (i.e. entropic contributions not accounted for by the ideal entropy of mixing expression of the FH model). For the differential Flory–Huggins parameter in Wolf's model one derives from equation (4.37) the expression:

$$\tilde{\chi}_{12}(\phi_2, T) = \frac{\alpha}{(1-\nu\phi_2)^2} - \zeta\lambda\left\{1 + 2\left(\frac{1}{\lambda} - 1\right)\phi_2\right\} \quad (4.38)$$

The four parameters α, ζ, λ, and v are not completely independent. Wolf points out that α and ζ cannot be independent, for vinyl polymers the following equation must hold [112]:

$$\alpha = 0.5 + D\zeta\lambda \tag{4.39}$$

where the constant D is less than unity. Hence, only the parameters ζ and v are temperature dependent. Stryuk and Wolf [105] have parameterized the phenomenological model for cyclohexane–polystyrene (CH–PS) mixtures at different temperatures by fitting the parameters against experimental UCST and LCST critical points. The molar volume of cyclohexane was used as reference volume, see equation (4.16). The critical point is computed from the Gibbs free energy of mixing by requiring that the second and third derivatives of the Gibbs free energy of mixing are zero. The parameter λ is approximated with the value 0.5 and from the critical point requirements it was found that $D = 0.59$. With the parameter values fitted in this manner, Struyk and Wolf predict the binodals and spinodals for this system (Figure 4.2).

Would it be possible to predict the phase diagram of CH–PS without any information on the experimental phase diagram. In other words, can we predict the phase diagram from vapor pressure measurements in the miscible region of the phase

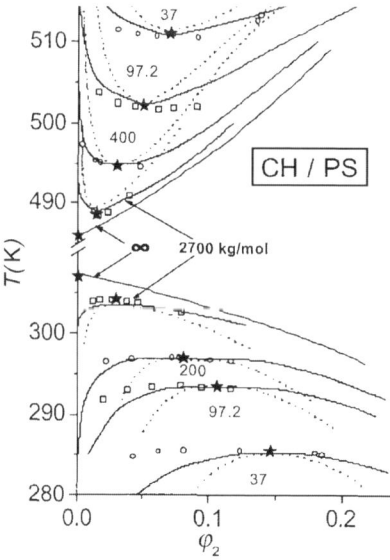

Figure 4.2. Phase diagram of the system cyclohexane–polystyrene for the indicated molar masses of the polymer (kg/mol). Cloud points are taken from Saeki et al. [92]; binodals (full lines) and spinodals (dotted lines) were computed using the model of Wolf using the critical point parameterization procedure by means of the temperature-dependent parameters ζ and v. The critical points are represented by full stars. Reproduced from Wolf [112] with permission.

diagram (i.e. between UCST and LCST). One can, of course, only obtain experimental vapor pressure near the UCST as the boiling point of the solvent is reached at around 335 K. Above this temperature the pressure rises constantly. Therefore, we concentrate on predicting the UCSTs from vapor pressures measured at temperatures just above the UCST. Figure 4.3 shows the differential Flory–Huggins parameter, $\tilde{\chi}$, extracted from experimental vapor pressure measurements of homogeneous CH–PS (233 kg/mol) solutions [85, 100].

The analysis of the temperature dependence of the parameters of Wolf's model for $\tilde{\chi}_{12}$ [equation (4.38)] is interesting, treating α, $\zeta\lambda$, and ν as independent parameters. As already argued before, the temperature dependence of α is marginal, and $\zeta = 0$ at theta temperature. Applying Wolf's model with the parameters obtained by extrapolating linearly to temperatures below 35°C (Figure 4.4), we obtain a UCST critical point at about 293 K. This is not bad at all. Actually, if we leave out the experimental values at 35°C for the linear fit of Wolf's parameters in dependence on temperature the result is excellent: a UCST critical point at 298 K. Leaving out experimental values near the critical point is even recommended. It is known that the mean-field approximation yields a critical behavior described by the Landau theory of phase transitions [65], which differs from the correct critical behavior [25, 104] in the 'universal' regime close to the critical point. See the review article by Binder [8] for a detailed analysis.

The model of Wolf seems to be a good choice for binary solvent–polymer mixtures that do not expose specific interactions. What is meant by specific interactions? Good examples are solutions of PEO or PVP and water (see Figure 4.5). The formation or break-up of hydrogen bonds definitely plays a major role in describing the composition-dependent behavior of the interaction parameters of these polymer solutions. Several theories have been formulated to describe the composition-dependent behavior of the interaction parameter of PEO and water [6, 17, 56]. It

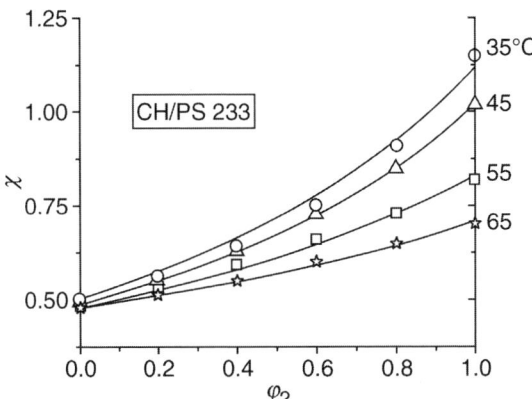

Figure 4.3. Measured [85, 100] and calculated [i.e. fitted with equation (4.38) for Wolf's model] concentration dependence of $\tilde{\chi}_{12}$ for the system CH–PS (233 kg/mol) for the indicated temperatures. Reproduced from Wolf [112] with permission.

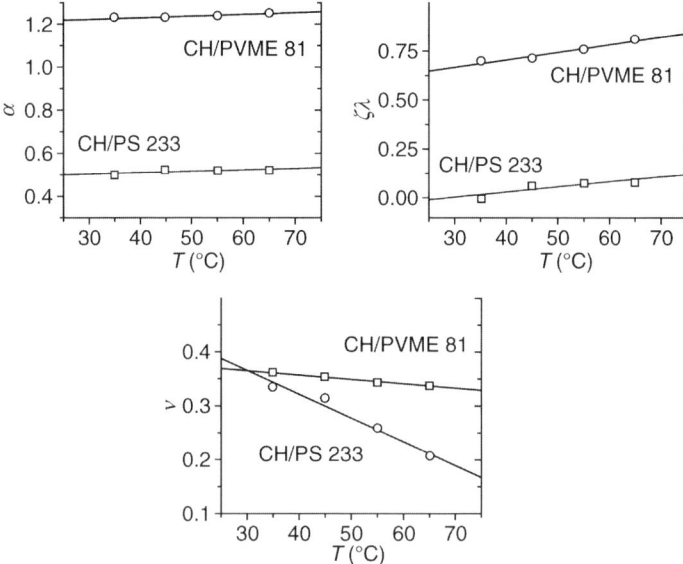

Figure 4.4. Temperature dependence of the parameters of Wolf's model for the system CH–PS 233 kg/mol: α, $\zeta\lambda$, and v. Reproduced from Wolf [112] with permission.

has been shown by Baulin and Halperin [5] that none of these models is capable of relating both the experimental vapor pressures and the experimentally determined interaction parameters with the phase diagrams. To put it more precisely, these models could be parameterized against phase diagrams and thus predict the phase behavior rather well, but the chemical potentials are way off the experimentally determined ones [18, 71] and, of course, vice versa.

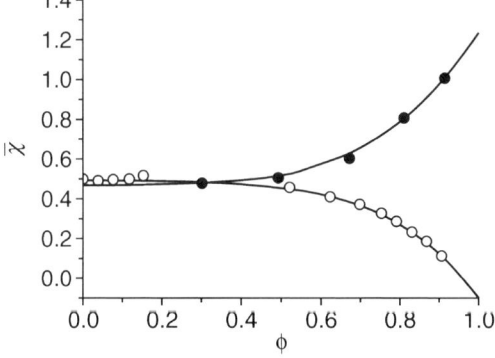

Figure 4.5. Measured $\tilde{\chi}_{12}$ as a function of ϕ_2 for aqueous solutions of the neutral water soluble polymers PEO $N = 300$, $T = 338$ K (solid circle) and PVP $T = 295$ K (open circle).

Determining the χ parameter for binary polymer–solvent mixtures from vapor pressure measurements is obvious. The question is, how to perform the parameterization of the interaction energy density parameter between highly incompatible polymers. An obvious way is to extract the interaction parameter from the interfacial thickness, λ, between films of both polymers. According to a theory of Broseta et al. [10], the equilibrium interfacial thickness is connected to the Flory–Huggins interaction parameter between two polymers A and B by

$$\lambda = \frac{2b}{\sqrt{6\chi_{AB}}}\left\{1 + \frac{\ln 2}{\chi_{AB}}\left(\frac{1}{N_A} + \frac{1}{N_B}\right)\right\} \qquad (4.40)$$

where b is the Kuhn segment length [61, 62]. Li et al. [68] studied the miscibility in polycarbonate–poly(styrene-co-acrylonitrile) blends by ellipsometry on bilayered specimens and determined the interaction parameter by applying equation (4.40) at different temperatures and AN content in the copolymer. The temperature dependence of χ parameter has a minimum at about 200°C. This behavior implies the coexistance of UCST and LCST for the mixture consisting of polymers with proper molecular weight. This prediction was confirmed by cloud point measurements of the mixtures having low molecular weight polycarbonate, although the location of the miscibility gap was not consistent with that predicted by the temperature dependence of χ. A more recent example of extracting the χ parameter from the interfacial thickness has been published by Zhang et al. [115]. The interfacial properties of a polybutadiene (PB) and a terpolymer [brominated poly(isobutylene-co-p-methylstyrene), BIMS] were studied by neutron reflectivity. Also, ternary mixtures were considered: PB, BIMS, and styrene-co-butadiene rubber (SBR). The results were complemented by scanning transmission X-ray microscopy and atomotic force microscopy in order to probe the morphology of the binary mixture. The density dependence of the χ parameter was fitted to a second-order polynomial

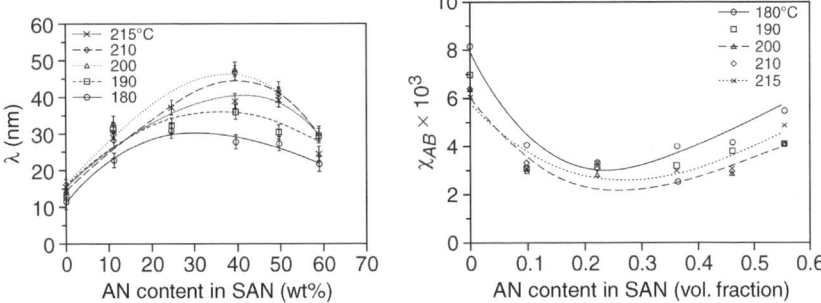

Figure 4.6. Interfacial thickness between bilayer of polycarbonate and poly(styrene-co-acrylonitrile) measured by ellipsometry (left) and the interaction parameter (right) extracted from these measurements using equation (4.40). These results were taken from Li et al. [68].

$\chi(\phi_2) = a + b\phi_2 + c\phi_2^2$. The number of authors of this paper [115] makes it clear that the approach needs a lot of resources and may exceed industrial research budgets. Still, the paper shows a way to deal with highly incompatible real industrial polymer blends.

Petri et al. [86] determined the interaction parameter between PS and polyisobutylene (PIB) from ternary mixtures of cyclohexane–PS–PIB. The interaction parameters in dependence of temperature and composition between both polymers and CH were determined from vapor pressure measurement of the corresponding binary mixtures as described earlier in this section. Cloud point curves and tie lines of the ternary mixtures were measured (Figure 4.7). Since the temperature- and composition-dependent $\chi_{CH,PS}$ and $\chi_{CH,PIB}$ parameters were determined independently from vapor pressure measurements of binary mixtures, the $\chi_{PS,PIB}$ parameter in dependence on temperature and composition can be evaluated from the experimental ternary phase diagram. The $\chi_{PS,PIB}$ parameter was found to increase at 35°C from 0.416 in the limit of pure PS to 0.449 in the limit of pure PIB. This result is in good agreement with information stemming from light scattering experiments [107] in a ternary system under 'optical theta conditions.' This experimental determination of polymer–polymer interaction parameters, that is, using a ternary mixtures of both polymers with a solvent, functions only if there are miscible regions for the temperatures of interest.

Finally, if there is no way to determine experimentally the interaction parameter because either the polymers of interest do not exist or one can not observe cloud points or interfaces for whatever reason, atomistic or coarse-grained Monte Carlo simulations might give an alternative route (see Chapter 3). You may be surprised by the argument 'the polymers of interest do not exists'—this is what industrial research is all about: 'innovations.'

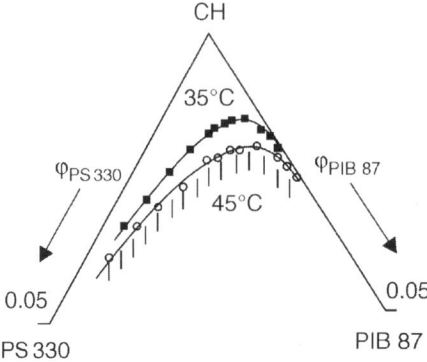

Figure 4.7. Visually determined cloud point curves of the ternary mixture CH–PS–PIB for the indicated temperature; the compositions are given in volume fractions and the two-phase region is indicated by small vertical dashes. The solid curves serve as a guide for the eye only. Reproduced from Petri et al. [86] with permission.

4.3. INHOMOGENEOUS POLYMER MIXTURES

Based on the pioneering work of Hohenberg and Kohn [84], who had already in the 1960s introduced a density functional description of quantum mechanical systems, a number of authors (see Evans [20] and references therein) derived a couple of years later the concept of a density functional description of ensembles of inhomogeneous classical systems. In both quantum mechanical systems and ensembles of classical systems, the basic objects of interest are the single particle density distributions of entities such as electrons, atoms or larger entities (e.g. beads). From a physical point of view the density distributions are sets of expectation values to observe the entities of interest at the various positions inside the system. The basic physics to obtain these density distributions involves (differential or integral) equations in terms of (intermediate multidimensional) functions such as the wave functions of quantum mechanics or probability functions in statistical mechanics (Ψ_0). The objective of density functional theory is to describe all basic physics in terms of functionals of the (one-particle or one-bead) density distributions. In this section we will briefly introduce the density functional theory (DFT) of classical systems following the treatment outlined by Evans [19, 20].

4.3.1. Notation

Consider a 'fluid like' mixture of polymers and simple molecules (both will be referred to as molecules). The type of molecule (chemical species) will be denoted by the indices i and j and the number of molecules of type i in the mixture is denoted by n_i.

At the mesoscale level, a molecule can be thought of as a (branched) chain of 'beads' (see Figure 4.8). A 'bead' is a rather abstract concept; it merely functions to effectively describe the intra- and intermolecular correlations as well as possible for a practical system on a course-grained level (1–2 nm) using an simplified

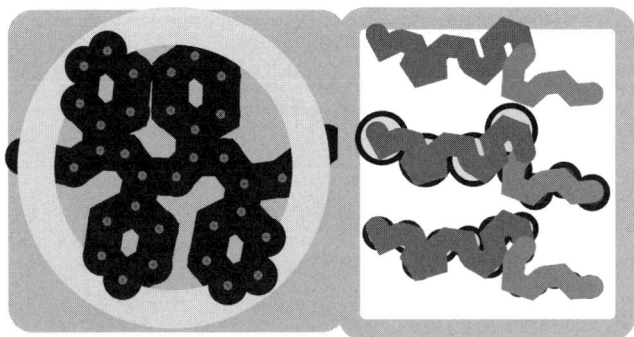

Figure 4.8. The concept of a 'bead' (left). At the mesoscopic level a polymer molecule can be thought of as a (branched) chain of 'beads' (right). A bead is a soft-core particle and allows for interpenetration as its excluded volume is lower than it is 'spatial volume.'

intramolecular correlation model such as the Gaussian chain model. The number of beads inside a molecule of type i is denoted by N_i. The total number of sth beads of molecules of type i is denoted by $n_{s,i}$ and their number density by $\rho(\mathbf{r}_{s,i})$. In the remainder of this chapter we will use coordinate indices to indicate the bead and molecule type for a function. Of course,

$$\int d\mathbf{r}_{s,i} \rho\,(\mathbf{r}_{s,i}) = n_i \tag{4.41}$$

where n_i is the total number of molecule of type i in the system. The index I will be used to refer to an individual bead in the system and thus

$$\int d\mathbf{r}_I \rho\,(\mathbf{r}_I) = 1 \tag{4.42}$$

The total number of beads in the system is denoted by N.

4.3.2. What is a Functional?

Briefly, a function is a mapping from one number or set of numbers to another; a functional is a generalized function that provides a mathematical rule to map a function space onto a number.

Consider first a uniform fluid of number density ρ. In this case a physical quantity of the fluid like the free energy F, which depends on the density, can be represented by an ordinary function $f(\rho)$. Such a function can be defined over a range of densities for which the fluid is thermodynamically stable. Next, we look at an inhomogeneous fluid, which is certainly a more interesting topic. Now the density varies through space and is not a constant. In fact the density itself is a function of the location, $\rho(\mathbf{r})$, where $\mathbf{r}$ is the coordinate vector specifying the position in space, $\mathbf{r} = (x, y, z)$. What is now the free energy? We denote this quantity by $F[\rho(\mathbf{r})]$ and call it the free energy functional of the density. Instead of acting on a variable (like a conventional function does), a functional acts on an entire function, in this case given by $\rho(\mathbf{r})$. The square brackets are used to indicate a functional operation and frequently the dependent variable inside is omitted, writing $F[\rho]$ rather than $F[\rho(\mathbf{r})]$.

Without some mathematics on functional analysis it might be hard to read the remainder of this chapter. If you are familiar with this topic please skip this section. We shortly introduce the Schwartz functional approach to generalized functions [37, 55, 103], following closely the introduction given by Farassat [24]. Schwartz [101] published his theory on distributions, which we call generalized functions, in the 1950s. Earlier, Dirac had introduced the famous delta function $\delta(x)$ by the shifting property:

$$\int_{-\infty}^{\infty} dx \phi(x) \delta(x) = \phi(0) \tag{4.43}$$

Equation (4.43) defines a mapping from function space to a number [i.e. $\phi(0)$], Dirac, of course, realized that no ordinary function could have this shifting property. As we agree, the Dirac-delta function is a useful function. In the theory of distributions the Dirac-delta function is a generalized function.

A conventional function $f(x)$ can be tabulated as a (infinite) series of ordered pairs, (x_i, f_i):

$$f_i = f(x_i) \qquad (4.44)$$

In generalized function theory the function $f(x)$ is, of course, still described in this way, but generated differently by the relation:

$$F[\phi] = \int_{-\infty}^{\infty} dx f(x) \phi(x) \qquad (4.45)$$

where $\phi(x)$ comes from a space of functions called the test function space. For a fixed function $f(x)$, expression (4.45) maps the test function space into real or complex numbers (i.e. F) and the function $f(x)$ is now described by a table of its functional values over a given space of test functions. We assume that the test functions belong to the space D of infinitely differentiable functions with bounded support. The latter means that the test functions are zero outside a specified domain for x. Now the question may be asked: are all continuous linear functionals generated by ordinary functions? The answer is no. Take the example:

$$\delta[\phi] = \phi(0) \qquad (4.46)$$

It is obvious that the functional $\delta[\phi]$ is linear and continuous. However, the functional has the shifting property the Dirac-delta function requires and there are no ordinary functions exposing this property. Therefore, the theory of generalized functions rigorously introduces Dirac's delta function into mathematics. Hence, generalized functions are defined as continuous linear functionals on the function space D. The formal definition for the Dirac-delta function, $\delta(x)$, is thus obtained by combining equations (4.43) and (4.46):

$$\delta[\phi] = \phi(0) = \int_{-\infty}^{\infty} dx \phi(x) \delta(x) \qquad (4.47)$$

The integral on the right-hand side of equation (4.47) does not stand for a conventional integration of a product of ordinary functions, rather it stands for $\phi(0)$. Farassat discusses in his NASA report [24] several approaches in mathematics to introduce and systematically develop generalized function theory.

In density functional theories, computing the derivatives of functionals is the key issue, or, better formulated, we will see in this section that the basic results of density functional theories are functional differential equations. This raises the issue: what is a derivative of a generalized function? Let $f(x)$ be an ordinary function with a continuous first derivative, that is, f is a C^1 function. If the functional $F[\phi]$ is a local functional, this means F depends only on the value of the test function $\phi(x)$ at a certain value of x, that is, $F[\phi] = f[\phi(x), x]$, we can express this local functional in terms of the Dirac-delta function:

$$F[\phi] \equiv f[\phi(x), x] = \int dx' f[\phi(x'), x'] \delta(x - x') \quad (4.48)$$

so that the functional derivative of $F[\phi]$ with respect to $\phi(x)$ is given by

$$\frac{\delta F[\phi]}{\delta \phi(x)} = \frac{df[\phi(x'), x']}{d\phi(x')}\bigg|_{x'=x} \quad (4.49)$$

Since the δ-functional is an odd functional, $\delta[-\phi] = -\delta[\phi]$, we must use $\delta(x - x')$ [24]. In case of non-local functionals like:

$$F[\phi] \equiv \int dx\, dx'\, g(x, x') \phi(x) \phi(x') \quad (4.50)$$

the functional derivative should tell us how much $F[\phi]$ changes for an infinitesimal change in ϕ at position x_0, which is accomplished by adding an infinitesimal amount of the Dirac-delta function, $v(x) \equiv \delta(x - x_0)$, to $\phi(x)$:

$$\begin{aligned}\frac{\delta F[\phi]}{\delta \phi(x_0)} &= \lim_{\epsilon \to 0} \frac{F[\phi + \epsilon v] - F[\phi]}{\epsilon} \\ &= \lim_{\epsilon \to 0} \int dx\, dx'\, g(x, x') \{\phi(x) v(x') + \phi(x') v(x) + \epsilon v(x) v(x')\} \\ &= \int dx \{g(x_0, x) + g(x, x_0)\} \phi(x) \end{aligned} \quad (4.51)$$

Once, again this functional depends on position x, but the dependence on the $\phi(x)$ is now local. The second functional derivative takes the simple form:

$$\frac{\delta^2 X[\phi]}{\delta \phi(x_0) \delta \phi(x_1)} = \{g(x_0, x_1) + g(x_1, x_0)\} \quad (4.52)$$

There is a lot more to tell about functional analysis, but it would be outside the scope of this chapter.

4.3.3. Grand Canonical Ensemble

For a classical ensemble the equivalent of the quantum mechanical ground state is the equilibrium state. Let $\mathbf{p}^N$ denote the set of momentum vectors $\{\mathbf{p}_I\}$ of the various beads I in the system and $\mathbf{r}^N$ the set of coordinate vectors $\{\mathbf{r}_I\}$. The Hamiltonian, $H(\mathbf{r}^N, \mathbf{p}^N)$, of such a classical system is given by:

$$H(\mathbf{r}^N, \mathbf{p}^N) = K(\mathbf{p}^N) + W(\mathbf{r}^N) + U(\mathbf{r}^N) \qquad (4.53)$$

where K denotes the kinetic energy and W the inter-particle potential energy for a system in the specified configuration $(\mathbf{r}^N, \mathbf{p}^N)$, respectively. The external potential energy U due to interaction of the beads with external potential fields $u(\mathbf{r}_I)$ is defined as:

$$U(\mathbf{r}^N) = \sum_I u(\mathbf{r}_I) \qquad (4.54)$$

From this definition of the external potential energy it follows that the Hamiltonian is a functional of the external potential fields; that is, $H[u]$ and u denote the set of external fields $\{u(\mathbf{r}_I)\}$.

In the thermodynamic limit, where the average numbers of particles and the volume go to infinity, keeping the average number densities of the beads constant, all ensembles in statistical mechanics (e.g. canonical or grand canonical) should give equivalent results. According to the nonuniformity of the system, here it is appropriate to work with a grand canonical ensemble of many particle systems, each with the same fixed volume V, temperature T and set of chemical potentials $\mu = \{\mu_i\}$. A grand canonical ensemble is an ensemble of open systems; that is, the number of molecules, is variable. In classical statistical mechanics [65, 77], the so-called 'grand canonical partition function' $\Xi(V, T, \{\mu_i\})$ is defined as:

$$\Xi(V, T, \{\mu_i\}) \equiv \mathrm{Tr}_{\mathrm{cl}} \exp\left[-\beta\left\{H[u] - \sum_i \mu_i n_i\right\}\right] \qquad (4.55)$$

with the classical trace $\mathrm{Tr}_{\mathrm{cl}}$ defined as

$$\mathrm{Tr}_{\mathrm{cl}}\{.\} = \sum_{\{n_i\}} \frac{1}{\prod_{i,s} n_{s,i}! h^{3n_i}} \left\{\prod_I \int d\mathbf{r}_I\right\} \left\{\prod_I \int d\mathbf{p}_I\right\} \{.\} \qquad (4.56)$$

where h denotes Planck's constant and $\beta \equiv 1/kT$. The factor $n_{s,i}!$ corrects for the fact that we can not distinguish between beads s of molecules of the same type i. The grand canonical partition function $\Xi(V, T, \{\mu_i\})$ is a natural functional of the external potentials fields u, that is, $\Xi = \Xi[u]$. The equilibrium probability distribution function $\Psi_0(\mathbf{r}^N, \mathbf{p}^N)$ is thus also a functional of the external fields:

$$\Psi_0[u] = \frac{1}{\Xi[u]} \exp\left[-\beta\left\{H[u] - \sum_i \mu_i n_i\right\}\right] \tag{4.57}$$

For reasons of readability we have omitted the variable dependence of the functionals, that is, $(\mathbf{r}^N, \mathbf{p}^N)$, in this equation. In the remaining of this chapter we will mostly omit the variable dependence of the functionals. The equilibrium number density of bead I at position $\mathbf{R}_I$ for a fixed set of external fields is denoted by $\rho_0[u](\mathbf{R}_I)$ and is found from the bead density operator, that is, the ensemble average of the Dirac-delta distribution $\delta(\mathbf{R}_I - \mathbf{r}_I)$,

$$\rho_0[u](\mathbf{R}_I) = \mathrm{Tr}_{\mathrm{cl}} \Psi_0[u] \delta(\mathbf{R}_I - \mathbf{r}_I) \tag{4.58}$$

The bead density distributions are in units of number of per m³ and not in mole per m³.

In equilibrium, all thermodynamic properties can be obtained from the grand canonical potential Ω (or 'grand potential' for short). The grand potential, $\Omega(V, T, \{\mu_i\})$, is defined as the Legendre transform of the Helmholtz free energy $A(V, T, \langle n \rangle)$, with respect to the set of average numbers of molecules, $\langle n \rangle = \{\langle n_i \rangle\}$,

$$\Omega(V, T, \{\mu_i\}) \equiv A(V, T, \langle n \rangle) - \sum_i \langle n_i \rangle \mu_i \tag{4.59}$$

According to statistical mechanics this grand potential can be expressed in terms of the grand canonical partition function $\Xi[u]$:

$$\Omega[u] = -\beta^{-1} \ln \Xi[u] \tag{4.60}$$

which makes it clear that the grand potential $\Omega[u]$ is a natural functional of the external potential fields. Equations (4.55) and (4.60) establish the link between (macroscopic) phenomenology and statistical mechanics. Taking the logarithm of both sides of equation (4.57) gives:

$$\ln \Psi_0[u] = -\beta\left\{H[u] - \sum_i \mu_i n_i\right\} - \ln \Xi[u] \tag{4.61}$$

or, after some rearrangements,

$$-\beta^{-1} \ln \Xi[u] = H[u] - \sum_i \mu_i n_i + \beta^{-1} \ln \Psi_0[u] \tag{4.62}$$

Multiplying both sides of equation (4.62) by Ψ_0, taking the classical trace afterwards and realizing that $-\beta^{-1} \ln \Xi[u]$ is equivalent to the grand potential $\Omega[u]$

[equation (60)], we arrive at the familiar relation:

$$\Omega[u] = \mathrm{Tr}_{\mathrm{cl}}\Psi_0[u]\left\{H[u] - \sum_i \mu_i n_i + \beta^{-1} \ln \Psi_0[u]\right\} \quad (4.63)$$

The above relation constitutes one of the central results of statistical mechanics.

4.3.4. Basic Concepts of Density Functional Theory

In analogy to equation (4.63) we introduce a functional $\Omega_u[\Psi]$ that, for a fixed set of external fields, $u = \{u_I\}$ is a functional of an *arbitrary* distribution function Ψ:

$$\Omega_u[\Psi] = \mathrm{Tr}_{\mathrm{cl}}\Psi\left\{H[u] - \sum_i \mu_i n_i + \beta^{-1} \ln \Psi\right\} \quad (4.64)$$

Clearly, where Ψ equals the equilibrium distributions function $\Psi_0[u]$, we have:

$$\Omega_u[\Psi_0[u]] = -\beta^{-1} \ln \Xi[u] = \Omega[u] \quad (4.65)$$

The functional $\Omega_u[\Psi]$ also satisfies the inequality (see Evans [20] for a derivation)

$$\Omega_u[\Psi] > \Omega[\Psi_0[u]] \quad \textit{for} \quad \Psi \neq \Psi_0[u] \quad (4.66)$$

From inequality (4.66) a variational principle can be formulated for the functional $\Omega_u[\Psi]$ in order to obtain, for a fixed set of external fields u, the corresponding equilibrium probability density distribution $\Psi_0[u]$:

$$\left.\frac{\delta \Omega_u[\Psi]}{\delta \Psi}\right|_{\Psi=\Psi_0[u]} = 0 \quad (4.67)$$

The problem in applying this variational principle of statistical mechanics to practical systems is the enormous multidimensionality of the distribution function Ψ, since it accounts for all correlations. As in the density functional approach to quantum mechanics, we would like to work around this problem by formulating all thermodynamics variables as functionals of single bead densities $\rho(\mathbf{r}_I)$.

Mermin [78] proved by *reductio ad absurdum* that the ensemble average densities ρ_0 are uniquely determined by a certain set of external potential fields, u_0, that is, $\Psi_0[u_0] \neq \Psi_0[u']$ if $u_0 \neq u'$. Consequently, for a given set of chemical potentials, $\{\mu_i\}$, there exists a unique set of external fields u and thus also a unique probability distribution $\Psi[u]$ that determines a given set of equilibrium density fields, ρ_0. In other words, there is a one-to-one (bijective) correspondance between the external potential fields and the equilibrium density fields:

$$u \rightleftharpoons \Psi_0 \rightleftharpoons \rho_0$$

Following Evans [20], we introduce a functional $F[\rho]$ that, if applied to the set of equilibrium density functions, is the *intrinsic* Helmholtz free energy of the system. This intrinsic free energy functional, $F[\rho]$, is defined as:

$$F[\rho] = \text{Tr}_{\text{cl}} \Psi[\rho] \{ K + W + \beta^{-1} \ln \Psi[\rho] \} \tag{4.68}$$

and does not contain any contribution of the external fields u. That it does reflect the Helmholtz free energy when applied to the equilibrium bead densities ρ_0 is easily proven. Evans shows how, based on the bijective relations between u, Ψ_0, and ρ_0, the set of equilibrium density functions must, in correspondance with the variational principle (4.67), satisfy the condition:

$$\left. \frac{\delta F[\rho]}{\delta \rho(\mathbf{r}_{s,i})} \right|_{\rho = \rho_0[u]} = \mu_{s,i} - u(\mathbf{r}_{s,i}) \tag{4.69}$$

where $\mu_{s,i}$ is the chemical potential of beads s of a molecule of type i,

$$\mu_{s,i} \equiv \frac{\mu_i}{N_i} \tag{4.70}$$

The functional derivative of the intrinsic free energy functional, $F[\rho]$, with respect to the density of bead s of molecules i is often called the intrinsic chemical potential, $\mu^{\text{intr}}[\rho](\mathbf{r}_{s,i})$, of those beads,

$$\mu^{\text{intr}}[\rho](\mathbf{r}_{s,i}) \equiv \frac{\delta F[\rho]}{\delta \rho(\mathbf{r}_{s,i})} \tag{4.71}$$

and is a functional of the bead density fields. Combining equations (4.69) and (4.71) gives for a system in a state of equilibrium the requirement:

$$\mu_{s,i} = u(\mathbf{r}_{s,i}) + \mu^{\text{intr}}[\rho_0](\mathbf{r}_{s,i}) \tag{4.72}$$

from which it follows that the r-dependence of the intrinsic chemical potentials $\mu^{\text{intr}}(\mathbf{r}_{s,i})$ must be exactly canceled by that of the external potential fields $u(\mathbf{r}_{s,i})$, since the chemical potential itself is a constant. Equation (4.72) is the fundamental equation in the theory of nonuniform fluids.

Basically, the functionals found until now constitute the formalism of classical density functional theory. The crucial point in applications of density functional theory is to determine an explicit form of the intrinsic free energy functional $F[\rho]$. It must be kept in mind that F contains all correlational information, and once $F[\rho]$ is known every physical quantity of the system under consideration can be computed. Because of the complexity of interparticle interactions in many-body systems, it will be clear that a definite formulation of the intrinsic free energy functional $F[\rho]$ for a specified system is only possible with some approximations.

4.3.5. Kohn–Sham-Like Formalism

In this section we derive a Kohn–Sham-like self-consistent field formalism [57] to obtain the equilibrium density distributions for ensembles of classical systems. The basic concept is to formulate the density functionals of a real system in terms of the single bead density fields of a corresponding ideal system. The advantage will become clear later in this section. In polymer density functional literature the term 'ideal system' is mostly used for a system of polymers in which the inter-molecular interactions are turned off but not the intramolecular interactions [76]. This is confusing when comparing the polymer density functional literature with that of quantum chemistry. In this chapter the prefix 'ideal' means no intermolecular and no intramolecular bead–bead interactions.

4.3.5.1. Ideal System. We start by deriving the expressions of an ideal system. The Hamiltonian of an ideal system, with no external field acting on it, is simply given by [77]:

$$H = \sum_I \frac{1}{2m_I} |\mathbf{p}_I|^2 \tag{4.73}$$

where m_I is the mass of bead I. Although, in the ideal system, there are no interactions between the various beads at all, we still treat the beads as if they belong to molecules or polymers. The grand potential Ω^{id} for such an ideal system is derived as [77]:

$$\tilde{\Omega}^{id} = \beta^{-1} \sum_{i,s} \langle n_{s,i} \rangle \left\{ \ln\left[\frac{\langle n_{s,i} \rangle}{Vol} \Lambda_{s,i}^3\right] - \beta\mu_{s,i}^{id} - 1 \right\} \tag{4.74}$$

where $\Lambda_{s,i}$ is the thermal or de Broglie wavelength of beads, s, belonging to molecules of type i, defined as:

$$\Lambda_{s,i} = \frac{h}{\sqrt{2\pi m_{s,i} kT}} \tag{4.75}$$

Suppose the ideal gas is not in a state of equilibrium. In this case we abstract a generalization [40] for a given set of nonequilibrium bead density distributions ρ from equation (4.74), by defining an intrinsic free energy functional,

$F^{id}[\rho]$, according to:

$$F^{id}[\rho] = \beta^{-1} \sum_{i,s} \int d\mathbf{r}_{s,i} \rho(\mathbf{r}_{s,i}) \{\ln[\Lambda_{s,i}^3 \rho(\mathbf{r}_{s,i})] - 1\} \quad (4.76)$$

For a given fixed set of bead density distributions, the corresponding set of external potential fields u^{id} at which the system is in a state of equilibrium is derived by applying equations (4.69) and (4.70) to $F^{id}[\rho]$:

$$u^{id}[\rho_0](\mathbf{r}_{s,i}) = \mu_{s,i} - \beta^{-1} \ln[\Lambda_{s,i}^3 \rho_0(\mathbf{r}_{s,i})] \quad (4.77)$$

where we have used expression (4.76) for the ideal intrinsic free energy functional. Equation (4.77) immediately leads to the result that, given a fixed set of external fields and chemical potentials, the equilibrium density distributions, ρ_0^{id}, for the ideal system must satisfy the relation:

$$\rho_0^{id}(\mathbf{r}_{s,i}) = \frac{1}{\Lambda_{s,i}^3} \exp[-\beta\{u^{id}(\mathbf{r}_{s,i}) - \mu_{s,i}\}] \quad (4.78)$$

This relation can be written into the familiar expression for an ideal gas:

$$\rho_0^{id}(\mathbf{r}_{s,i}) = z_{s,i} \exp[-\beta u^{id}(\mathbf{r}_{s,i})] \quad (4.79)$$

where $z_{s,i}$ is the so-called 'fugacity' of beads, s, of molecules, i:

$$z_{s,i} = \frac{\exp[\beta\mu_{s,i}]}{\Lambda_{s,i}^3} \quad (4.80)$$

If, instead of the chemical potentials, the set of ensemble average numbers of molecules, $\langle n_i \rangle$, for a system in the grand canonical ensemble is known, the constraining relations

$$\int d\mathbf{r}_{s,i} \rho_0^{id}(\mathbf{r}_{s,i}) = \langle n_i \rangle \quad (4.81)$$

must be satisfied. Substituting $\rho_0^{id}(\mathbf{r}_{s,i})$ in equation (4.81) by expression (4.79) results in the constraining relation:

$$\langle n_i \rangle = z_{s,i} \int d\mathbf{r}_{s,i} \exp[-\beta u^{id}(\mathbf{r}_{s,i})] \quad (4.82)$$

which fixes the mean level of the external potential fields.

Thus, in an ideal system there are no correlations between the beads and the equilibrium density distributions can be directly computed from the external potential fields $[u^{\text{id}}(\mathbf{r}_{s,i})]$.

4.3.5.2. Real System. We now turn our attention to a 'real' system and rewrite equation (4.68) in terms of a kinetic energy functional $K[\rho]$, an interaction energy functional $W[\rho]$ and the information entropy functional $I[\rho]$ [69]:

$$F[\rho] = K[\rho] + W[\rho] - \beta^{-1}I[\rho] \qquad (4.83)$$

with the functionals $K[\rho]$, $W[\rho]$, and $I[\rho]$ defined as

$$K[\rho] = \text{Tr}_{\text{cl}}\Psi[\rho]K(\mathbf{p}^N) \qquad (4.84)$$

$$W[\rho] = \text{Tr}_{\text{cl}}\Psi[\rho]W(\mathbf{r}^N) \qquad (4.85)$$

$$I[\rho] = -\text{Tr}_{\text{cl}}\Psi[\rho]\ln[\Psi[\rho]] \qquad (4.86)$$

respectively. We introduce an ideal system that equals the real system in all respects by construction, except there are no interparticle interactions at all. The intrinsic free energy functional $F[\rho]$ of the real system can be written in terms of the intrinsic free energy functional $F^{\text{id}}[\rho]$ of this ideal system but applied to the density distributions, ρ, of the real system:

$$F[\rho] = F^{\text{id}}[\rho] + W[\rho] + \{T[\rho] - T^{\text{id}}[\rho]\} - \beta^{-1}\{I[\rho] - I^{\text{id}}[\rho]\} \qquad (4.87)$$

If interbead interactions would have no affect on the correlations between the beads, then the intrinsic free energy is obtained from that of the corresponding ideal system ($F^{\text{id}}[\rho]$) and an uncorrelated pair interaction energy $J[\rho]$ defined as:

$$J[\rho] \equiv \text{Tr}_{\text{cl}} \sum_{I,J>I} \rho(\mathbf{r}_I)w(\mathbf{r}_I, \mathbf{r}_J)\rho(\mathbf{r}_J) \qquad (4.88)$$

However, the interbead interactions do introduce correlations. This needs some further discussion. The interaction functional $W[\rho]$ as defined in equation (4.85) is obtained by averaging the particle interactions, $W(\mathbf{r}^N)$. The uncorrelated pair interaction energy functional $J[\rho]$ constitutes a mean field approximation to the interaction functional $W[\rho]$. Exactly at this point the field theoretic methods differ from the classical density functional theories discussed here. Field theoretic methods aim at evaluating $W[\rho]$ directly. For an excellent introduction to field theoretic methods see the paper by Moreira and Netz [79].

Analogously to the exchange correlation functional in quantum mechanical DFT, we define a correlation functional F_{cor} according to:

$$F_{\text{cor}}[\rho] = \{W[\rho] - J[\rho]\} + \{T[\rho] - T^{\text{id}}[\rho]\} - \beta^{-1}\{I[\rho] - I^{\text{id}}[\rho]\} \qquad (4.89)$$

so that the intrinsic free energy functional $F[\rho]$ can formally be expressed as

$$F[\rho] = F^{id}[\rho] + J[\rho] + F_{cor}[\rho] \qquad (4.90)$$

The functionals $J[\rho]$ and $F_{cor}[\rho]$ are often taken together as the excess free energy functional: $F_{ex}[\rho] = J[\rho] + F_{cor}[\rho]$. The intrinsic chemical potential of beads s of molecules of type i, $\mu^{intr}[\rho](\mathbf{r}_{s,i})$, as defined in equation (4.71) is now given by:

$$\beta\mu^{intr}[\rho](\mathbf{r}_{s,i}) = \ln[\Lambda^3 \rho(\mathbf{r}_{s,i})] - c^{(1)}[\rho](\mathbf{r}_{s,i}) \qquad (4.91)$$

where we have used equation (4.76) to derive the functional derivative of $F^{id}[\rho]$ with respect to ρ. The functional $c^{(1)}[\rho](\mathbf{r}_{s,i})$ is the well-known one-particle direct correlation function [19, 20, 46], it is a functional of the bead densities and a function of $\mathbf{r}_{s,i}$:

$$c^{(1)}[\rho](\mathbf{r}_{s,i}) \equiv -\beta \left\{ \frac{\delta J[\rho]}{\delta\rho(\mathbf{r}_{s,i})} + \frac{\delta F_{cor}[\rho]}{\delta\rho(\mathbf{r}_{s,i})} \right\} \qquad (4.92)$$

and incorporates the interparticle interactions. From equations (4.69), (4.71), (4.80) and (4.91) it directly follows that the equilibrium bead density field of beads s of molecules i, $\rho_0(\mathbf{r}_{s,i})$, in terms of the direct correlation function and the external field, $u(\mathbf{r}_{s,i})$, is obtained from:

$$\rho_0(\mathbf{r}_{s,i}) = z_{s,i} \exp[-\beta u(\mathbf{r}_{s,i}) + c^{(1)}[\rho_0](\mathbf{r}_{s,i})] \qquad (4.93)$$

4.3.5.3. Kohn–Sham Scheme. At this point we propose a kind of Kohn–Sham self-consistent field scheme by introducing an effective potential field u^{eff}, which is a functional of the single field densities and a function of the bead coordinate vector, according to:

$$u^{eff}[\rho_0](\mathbf{r}_{s,i}) = u(\mathbf{r}_{s,i}) - \beta^{-1} c^{(1)}[\rho_0](\mathbf{r}_{s,i}) \qquad (4.94)$$

that, when applied to an ideal system,

$$\rho_0(\mathbf{r}_{s,i}) = z_{s,i} \exp[-\beta u^{eff}(\mathbf{r}_{s,i})] \qquad (4.95)$$

with the equilibrium density functions of the real system (by construction), ρ_0, would give the latter in return. In other words, the effective potential fields defined in equation (4.94) determine, in a self-consistent fashion, the equilibrium density distributions. It is analogous to the effective potential introduced in the one-electron Schrödinger equation by the Kohn–Sham [57] theory of the inhomogeneous gas.

The functional relation (4.94) constitutes a set of nonlinear equations, one equation for every bead type and position in the system, which has to be solved simultaneously for the unknown equilibrium densities [15]. In practice, it is not a good idea to take the equilibrium density functions as the iteration variables because they are constraints: $\rho_0 \geq 0$. Instead, because of the one-to-one relation between the effective potential field, u^{eff}, and the equilibrium densities, ρ_0, it is a better idea to take the set u^{eff} as the iteration variables.

The whole issue in applying density functional theory is to formulate approximations for the correlation functional F_{cor} that will still capture those correlations that govern the physics of the system under consideration. See the book *Fundamentals of Inhomogeneous Fluids*, edited by Douglas Henderson [46], for further information.

4.3.6. Numerical Example: Excluded Volume Effects

At this stage, after digesting so many rather abstract equations, it is probably appropriate to demonstrate the density functional formalism by applying it to a very simple system: soft-core particles between two flat hard walls. We must expect that the flat hard walls will introduce inhomogeneities within the mixture [46].

We will take a crude numerical approach to the problem as outlined by Sok and Evers [102]. We only consider inhomogeneities perpendicular to the walls, that is, in the z-direction. First, we completely neglect all correlations within the mixture; that is, we assume $F_{\text{cor}}[\rho] = 0$. Actually this approximation is well known in density functional literature as the van der Waals approach [19]. The noncorrelated pair interaction energy functional $J[\rho]$ in the van der Waals approximation is expressed by:

$$J[\rho] = \int\int d\mathbf{r}' d\mathbf{r} w(\mathbf{r}, \mathbf{r}')\rho(\mathbf{r})\rho(\mathbf{r}') \tag{4.96}$$

We apply for the pair interaction potential, $w(\mathbf{r}, \mathbf{r}')$, a purely repulsive Gaussian soft-core bead–bead interaction according to:

$$w(\mathbf{r}, \mathbf{r}') = \varepsilon \left(\frac{2}{\pi \sigma^2}\right)^{3/2} \exp\left[-\frac{2}{\sigma^2}|\mathbf{r} - \mathbf{r}'|^2\right] \tag{4.97}$$

where the parameter σ is a measure for the particle diameter, and the parameter ε gives the strength of the interaction in units of kT. The one-particle direct correlation function $c^{(1)}(\mathbf{r})$ for this system is easily derived from its definition (4.92) by substituting expression (4.96) for $J[\rho]$ and setting $F_{\text{cor}}[\rho] = 0$. The integrals are computed numerically by subdividing the space between the two flat walls into L planar layers with constant particle density and evaluating the kernel $w(\mathbf{r}, \mathbf{r}')$ using spherical shells, as shown in Figure 4.9. Both the planar field layers and spherical kernel shells are equally spaced by a distance d. For simplicity, we treat the hard walls by introducing a Stern layer adjacent to the wall, i.e. $\rho(z) = 0$ for $z < \sigma/2$ and $z > Ld - \sigma/2$. The integrals are calculated using a trapezoidal scheme [89] using the appropriate overlap volumina.

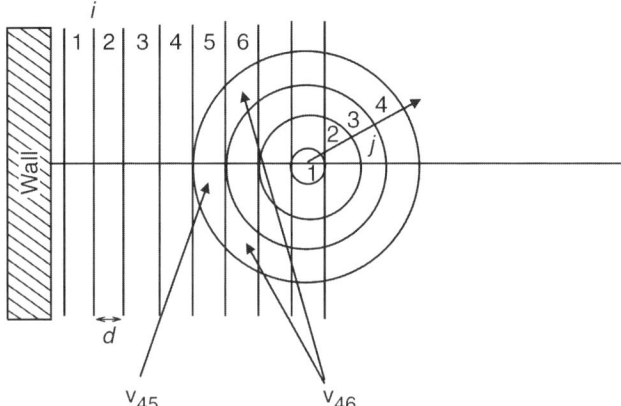

Figure 4.9. Two-dimensional representation of the overlap volumina of the planar field layers (index *i*) and the spherical kernel shells (index *j*). The overlap volumina of kernel shell 4 with planar field layer 6 (w_{46}) and of kernel shell 4 with system layer 5 (w_{45}) are indicated.

The Kohn–Sham self-consistent iteration scheme as described in Section 4.3.5 is now easily implemented into a computer program. We have used the GNU Scientific Library [35], also known under the abreviation GSL, to solve the nonlinear equations. To be more precise, we have used the 'Broyden' secant implementation of the GSL nonlinear equation solver. The performance of the GSL nonlinear equation solver is excellent. Convergence is reached after *ca* 40 iterations. In Figure 4.10 results for the equilibrium density profiles are shown for $L = 5\sigma$ and

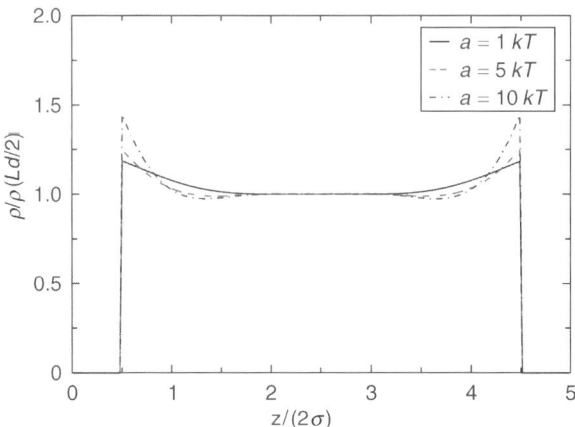

Figure 4.10. Equilibrium density profiles for a mixture between two flat hard walls using a soft-core potential as defined in equation (4.97). The following parameter values were applied: $\sigma = 40d$, ε as indicated, $L = 200d$ and $\langle\rho\rangle\sigma^3 = 1.33$. The number of kernel shells used was 100.

three different values of ε (indicated). The resolution was $d = \sigma/40$ and could be computed on an ordinary PC operated with Linux within 5 min. For higher repulsive soft-core potentials ($\varepsilon = 10kT$) we do not obtain, of course, the strong spatial ordering of the particles perpendicular to the walls as you will obtain from Monte Carlo simulations using a Lennard–Jones potential or from sophisticated density functional theories [88]. Soft-core particles do interpenetrate each other as they enclose free volume. This is exactly the reason for taking a rather high average particle density of $\langle \rho \rangle \sigma^3 = 1.33$. However, the soft-core potential defined by equation (4.97) does provide a simple means to deal with excluded volume effects for interpenetrable beads. Although we have completely neglected the correlation functional F_{cor} in this example, i.e. taking a van der Waals approach, we do capture some of the basic physics.

It is easy to extend the numerical formalism to binary mixtures of soft-core particles of different volumina, that is, different σ. As an example, we take a look at a binary mixture of soft-core particles in which particles of type 1 have a diameter twice that of particles of type 2: $\sigma_1 = 40d$ and $\sigma_2 = 20d$. The other parameters are kept at equal values, that is, $\varepsilon_1 = \varepsilon_2 = 5kT$. The volume fraction profiles of both particles in a binary mixture of 50 percent particles of type 1 and 50 percent particles of type 2 are presented in Figure 4.11. Again, we capture the basic physics in order to extend to approach to polymer mixtures. That the total volume fraction does reach a maximum beyond 1 seems strange but is not. Again, we are dealing with soft particles; that is, they can interpenetrate, and thus the volume fraction defined with respect to the particle volumina $[4\pi(\sigma/2)^3/3]$ can indeed reach locally values beyond one.

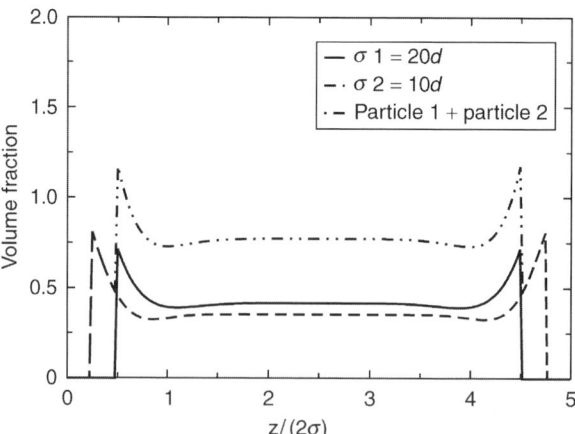

Figure 4.11. Equilibrium volume fraction profiles for a binary mixture of two soft-core particles between two flat hard walls, $\phi_1 = \phi_2 = 0.35$, using a soft-core potential as defined in equation (4.97), with $\sigma_1 = 40d$ and $\sigma_2 = 20d$. The following parameter values were applied: $\varepsilon_1 = \varepsilon_2 = 5kT$ and $\langle \rho_1 \rangle \sigma_1^3 + \langle \rho_2 \rangle \sigma_2^3 = 1.33$. The number of kernel shells used was 100. The volume fraction is defined as $\phi_i = 4\pi(\sigma/2)_i^3 \rho_i/3$.

INHOMOGENEOUS POLYMER MIXTURES 187

The examples of this section are simple systems. The density functional approach taken here is easy to implement into a computer program using the Gnu Scientific Library (GSL) as a numerical device. This sub-section should not be interpreted as a hint to replace sophisticated theoretical approaches or detailed atomistic Monte-Carlo implementations for simply fluids by the crude density functional approach introduced here.

4.3.7. Ensemble of Quasi-ideal Systems

Applying the Kohn–Sham self-consistent scheme as introduced in Section 4.3.5 to polymer mixtures would require introduction of intermolecular and intramolecular bead–bead correlations through the correlation functional $F_{cor}[\rho]$ and their expression in terms of the single bead density fields. This sounds good, but is hard to realize in practice. In this section we introduce the concept of a 'quasi-ideal' system, that is, a system including intramolecular correlations, that enables us to formulate a Kohn–Sham-like scheme that can be applied to inhomogeneous polymer mixtures. Special attention is paid to the question of how to parameterize the polymer chain model to include intramolecular correlations.

4.3.7.1. Definition. We start our derivations by defining a 'quasi-ideal' system i.e. a system in which there are no nonbonded bead–bead interactions and no intermolecular correlations; however the intramolecular correlations resulting from the interconnected beads are accounted for. In polymer density functional literature the system we call here 'quasi-ideal' is mostly referred to as 'ideal.' There are other approaches taken in the literature, for instance, Schmid [99] introduces a 'reference' system in which, in addition to the intramolecular correlations, the excluded volume of the beads are taken into account by a simple hard core repulsion.

Let $u^{qid,eff}(\mathbf{r}_{s,i})$ be the effective external potential field applied to the 'quasi-ideal' system. The probability distribution function, $\Psi^{qid}[u^{qid,eff}]$, for the 'quasi-ideal' system can be expressed as a product of single chain distribution functions, $\psi_i[u^{qid,eff}]$, as we ignore all intermolecular correlations:

$$\Psi^{qid}[u^{qid,eff}] = \prod_i \psi_i[u^{qid,eff}]^{n_i} \qquad (4.98)$$

The equilibrium bead density functions, $\rho^{qid}(\mathbf{r}_{s,i})$, are then obtained by applying the bead density operator as in equation (4.58) to the quasi-ideal probability distribution:

$$\rho_0^{qid}(\mathbf{R}_{s,i}) = \mathrm{Tr}_{cl} \Psi_0^{qid}[u] \delta(\mathbf{R}_{s,i} - \mathbf{r}_{s,i})$$

$$= \langle n_i \rangle \prod_s \left\{ \int d\mathbf{r}_{s,i} \right\} \psi_i[u^{qid,eff}] \delta(\mathbf{R}_{s,i} - \mathbf{r}_{s,i}) \qquad (4.99)$$

To apply the Kohn–Sham self-consistent scheme to polymer mixtures we must modify it by defining a set of quasi-ideal effective external potential fields, $u^{\text{qid,eff}}$, according to:

$$u^{\text{qid,eff}}[\rho_0](\mathbf{r}_{s,i}) = u(\mathbf{r}_{s,i}) - \beta^{-1} c^{\text{qid},(1)}[\rho_0](\mathbf{r}_{s,i}) \quad (4.100)$$

where the set of equilibrium density functions, ρ_0, is to be computed from equation (4.99) instead of using the 'ideal' system pendant (4.95). The 'quasi-ideal' one-particle direct correlation function is derived from its definition (4.92) by splitting the correlation functional F_{cor} into an intramolecular part $F_{\text{cor,intra}}$ and an intermolecular part $F_{\text{cor,inter}}$ and is given by:

$$c^{\text{qid},(1)}[\rho_0](\mathbf{r}_{s,i}) \equiv -\beta \left\{ \frac{\delta J[\rho]}{\delta \rho(\mathbf{r}_{s,i})} + \frac{\delta F_{\text{cor,inter}}[\rho]}{\delta \rho(\mathbf{r}_{s,i})} \right\} \quad (4.101)$$

The intramolecular correlations are accounted for by the single chain distribution functions in the density operation. The set of equations (4.100) must be solved simultaneously, either in ρ_0 or in $u^{\text{qid,eff}}$. The Kohn–Sham scheme is best implemented by taking the effective external potential fields, $u^{\text{qid,eff}}$, as the iteration variables for the nonlinear equations (4.100) and compute the corresponding density functions with equation (4.99). Numerically, this means that you will have to evaluate nested Fredholm integrals [equation (4.99)].

4.3.7.2. Gaussian Chain Model.

A well-known model for the single chain distribution function is the Gaussian chain [16] in which the statistical distribution of the end-to-end vector only depends on the local structure of the chain through the effective bond length b. The effective bond length squared is defined as the average of the various bond lengths, $b_{s,s+1,i}$ between neighboring beads squared; we will consider nonbranched polymers only,

$$b^2 = \frac{1}{N_i} \sum_{s=1}^{N_i-1} b_{s,s+1,i}^2 \quad (4.102)$$

The single chain distribution function for the Gaussian chain model is a product of normal distributions for the various bead–bead bonds in the polymer chain:

$$\psi_i[u] = \prod_{s=1}^{N_i-1} \left(\frac{3}{2\pi b_{s,s+1,i}} \right)^{3/2} \exp\left[-\frac{3}{2 b_{s,s+1,i}^2} (\mathbf{r}_{s,i} - \mathbf{r}_{s+1,i})^2 \right] \quad (4.103)$$

The Gaussian chain model implies that the bond vectors $\mathbf{r}_{s,i} - \mathbf{r}_{s+1,i}$ are independent. The mean of all bond vectors is zero and the second moment of for instance the bond between beads s and $s+1$ is derived as:

$$\langle (\mathbf{r}_{s,i} - \mathbf{r}_{s+1,i})^2 \rangle = \int \prod_{s=1}^{N_i-1} \{ d(\mathbf{r}_{s,i} - \mathbf{r}_{s+1,i}) \} (\mathbf{r}_{s,i} - \mathbf{r}_{s+1,i})^2 \psi_i[u] = b_{s,s+1,i}^2 \quad (4.104)$$

For the second moment of the end-to-end vector, $\mathbf{r}_{1,i} - \mathbf{r}_{N_i \cdot i} = \sum_{s=1}^{N_i-1}(\mathbf{r}_{s,i} - \mathbf{r}_{s+1 \cdot i})$, one derives [16] with the definition of the effective bond length, b, given in equation (4.102), that:

$$\langle(\mathbf{r}_{1,i} - \mathbf{r}_{N_i \cdot i})^2\rangle = \int \prod_{s=1}^{N_i-1}\{d(\mathbf{r}_{s,i} - \mathbf{r}_{s+1 \cdot i})\} \left\{\sum_{s=1}^{N_i-1}(\mathbf{r}_{s,i} - \mathbf{r}_{s+1 \cdot i})\right\}^2 \psi_i[u] = N_i b^2 \quad (4.105)$$

Actually, the Gaussian chain model is defined by equation (4.105) and from the central limit theorem the single chain distribution function is derived to be given by equation (4.103). The Gaussian chain model is often a good approach for describing the quasi-ideal system in case the polymer chains are long enough. The equilibrium bead density is found from the familiar Feynman decomposition of the path integrals [16]:

$$\rho_0^{\text{qid}}(\mathbf{R}_{s,i}) \propto G_{s,i}(\mathbf{R}_{s,i})\sigma[G_{s+1,i}^{\text{inv}}](\mathbf{R}_{s,i}) \quad (4.106)$$

The sets of (once integrated) Green functions $G_{s,i}(\mathbf{r})$ and $G_{s+1,i}^{\text{inv}}(\mathbf{r})$ are related by recurrence equations:

$$G_{s,i}(\mathbf{r}) = \exp[-\beta u_{s,i}^{\text{qid,eff}}(\mathbf{r})]\sigma[G_{s-1,i}](\mathbf{r}) \quad (4.107)$$

$$G_{s,i}^{\text{inv}}(\mathbf{r}) = \exp[-\beta u_{s,i}^{\text{qid,eff}}(\mathbf{r})]\sigma[G_{s+1,i}^{\text{inv}}](\mathbf{r}) \quad (4.108)$$

with $G_{1,i}(\mathbf{r}) = G_{N_i,i}^{\text{inv}}(\mathbf{r}) = 1$. The linkage operator σ is defined as a convolution with the kernel of the Gaussian chain model:

$$\sigma[f](\mathbf{r}) \equiv \left(\frac{3}{2\pi b}\right)^{3/2} \int d\mathbf{r}' \exp\left[-\frac{3}{2b^2}(\mathbf{r}-\mathbf{r}')^2\right]f(\mathbf{r}') \quad (4.109)$$

where b should be set to the appropriate bond length. There are basically two ways to implement this 'Gaussian chain quasi-ideal' system into simulation programs: by numerical integration or by Fourier transforms.

Evaluating the convolution integrals numerically on a grid means applying a trapezoidal summation scheme (as in the last section) or quadrature rules. The advantage of applying quadrature rules lies in the fact that a far coarser grid can be used in contrast to the trapezoidal integration schemes. Maurits et al. [73] have derived a simple quadrature rule to compute the Gaussian convolution integrals [equation (4.109)], including only nearest-neighbor grid nodes, a 27-point stencil operation. The quadrature rule can be implemented data parallel as shown by Maurits et al. [73], which is rather straightforward. The disadvantage of the quadrature integration method is that it is only valid for systems in which all bonds are of nearly equal length. Sok and Evers [102] applied a trapezoidal integration scheme using a very fine grid, thereby allowing for bonds of different size. This method is computationally very expensive, of course.

Another way to numerically compute the convolution integrals is by applying discrete Fourier transforms [89] since the convolution integral in real space becomes a simple product in Fourier space. To include bonds of different size, these discrete fast Fourier transforms must also be computed on a very fine grid and the procedure becomes computationally very expensive. The advantage of this Fourier transform approach is its straightforward implementation into a computer program. There exists an excellent 'public domain' software library [34] which implements a data parallel discrete fast Fourier transform. Its name is FFTW which stands for 'the fastest Fourier transform in the west.' Does this means there exists a better and faster one in the 'east'? Let's not quarel about redundancies from the past! An example of making use of this parallel DFFT implementation is the numerical self-consistent field theory simulation method of Sides and Fredrickson [106], published very recently. (As a general aside, whenever you have to deal with numerical issues you are unfamiliar with, do use the famous book *Numerical Recipes* by Press et al. [89]; it does not matter whether it is *Numerical Recipes in Fortran*, *Numerical Recipes in C*, or *Numerical Recipes in C++*.)

It is possible to prevent any evaluation of the convolution integrals; one applies the following method as outlined for instance by Doi and Edwards [16]. The suffix s of the Gaussian chain in equation (4.103) is regarded as a continuous variable, hence the summation of the bond vectors is replaced by an integral:

$$\psi_i[u] = \prod_{s=1}^{N_i-1} \left(\frac{3}{2\pi b_{s,s+1,i}}\right)^{3/2} \exp\left[-\frac{3}{2b_{s,s+1,i}^2}(\mathbf{r}_{s,i} - \mathbf{r}_{s+1,i})^2\right]$$

$$= C \exp\left[-\frac{3}{2b^2} \int_0^{N_i} ds \left(\frac{\partial \mathbf{r}_{s,i}}{\partial s}\right)\right] \qquad (4.110)$$

This distribution is known as the Wiener distribution and C is a normalization constant. The Green function $G_i(\mathbf{r}, s)$ obeys the diffusion equation [42–45, 80]:

$$\left\{\frac{1}{N_i}\frac{\partial}{\partial s} - \frac{1}{6}b^2 \nabla^2 + u^{\text{qid,eff}}(\mathbf{r})\right\} G_i(\mathbf{r}, s) = 0 \qquad (4.111)$$

with the initial condition $G_i(\mathbf{r}, 0) \equiv 1$. We have assumed that all beads in the chain experience the same quasi-ideal effective external potential field. This is not a necessary restriction; one can apply different effective external potential fields to different parts (blocks) of the chain [72]. The equilibrium bead density is then given by:

$$\langle \rho_0^{\text{qid}}(\mathbf{R}_{s,i}) \rangle = \frac{\langle n_i \rangle}{V} \int_0^1 ds\, G_i(\mathbf{R}_{s,i}, s) G_i(\mathbf{R}_{s,i}, 1-s) \qquad (4.112)$$

The advantage of this approach is clear: The diffusion equation (111) can be solved with standard numerical methods (e.g. implicit integration with the Crank–Nicolson

INHOMOGENEOUS POLYMER MIXTURES   191

scheme [13, 89]). One can also Fourier transform the diffusion equation, thereby preventing to apply finite difference operations [13]. A good example of this continuous Gaussian chain approach has been published by Schmid [99], in which excluded volume effects are included through a hard-core repulsion corrected to obtain the correct compressibility for polymer blends.

4.3.7.3. Parameterization Issues. To apply the 'quasi-ideal' system using the Gaussian chain model in practice you will need to parameterize the Gaussian chain model, which means you will need to derive rules for the segmentation of the real polymer chains into Gaussian beads and bonds.

In a recent publication, Lam and Goldbeck-Wood [64] gave a detailed discussion on deriving rules for the parameterization of the Gaussian chain model and applied these rules to Pluronic triblock copolymers ($EO_N PO_M EO_N$). They extended a procedure published earlier by van Vlimmeren et al. [111]. Gaussian chain statistics and structure factors of Gaussian chain block copolymers, introduced by Leibler [67] using a random phase approximation (RPA) approach, were used in the parameterization. To determine the bond length, b, of the Gaussian chain model, the chain dimensions of an atomistic chain and that of the corresponding Gaussian chain were compared. Lam and Goldbeck-Wood [64] applied the RIS Metropolis Monte Carlo (RMMC) method [49], as implemented in the Cerius^2 software package by Accelrys. The RMMC method was validated by comparing the atomistically simulated chain characteristics (end-to-end distance) with known PEO and PPO homopolymer values. Lam and Goldbeck-Wood parameterized the Gaussian chain model of Pluronic triblocks in order to apply it in MesoDyn [33] simulations. The EO–EO and PO–PO bonds were restricted to equal values as the MesoDyn implementation in the Cerius^2 software package by Accelrys can only deal with bonds of equal length. We will, however, give the parameterization method for the general case of different bond lengths.

We take a di-block copolymer ($PA_N PB_M$) as test case, where PA and PB denote monomer type and N and M the number of these monomers in both blocks. The end-to-end distance squared of the Gaussian chain of beads, R_l^2, and that of the corresponding atomistic chain, $R_{l,0}^2$, should be the same:

$$R_l^2 = R_{l,0}^2 \qquad (4.113)$$

Let l_{PA} and l_{PB} denote the monomer–monomer bond lengths in both blocks, PA_N and PB_M, respectively. These numbers can be obtained from atomistic simulations. It is clear that here the temperature comes into play as it influences the *cis–trans* ratio. The number of monomers, PA, making up one bead, A, is denoted by n, and we have the following constraint for n and the bead–bead bond length, b_A:

$$N l_{PA} = n b_A \qquad (4.114)$$

and of course, for the B beads,

$$Ml_{PB} = mb_B \qquad (4.115)$$

From the Gaussian chain model [equations (4.102) and (4.105)] we have for the end-to-end distance squared

$$R_l^2 = nb_A^2 + mb_B^2 = R_{l,0}^2 \qquad (4.116)$$

Substituting equations (4.114) and (4.115) for b_A and b_B in equation (4.116) for the end-to-end distance squared, we obtain a constraining relation for the number of beads (n_A and n_B):

$$m(Nl_{PA})^2 + n(Ml_{PB})^2 = nmR_{l,0}^2 \qquad (4.117)$$

Thus, we have a set of three equations (4.114), (4.115) and (4.117) in four unkowns (n, m, b_A, b_B). A solution often applied is to introduce a fixed bead–bead bond length ratio as Lam and Golbeck-Wood did, in their parameterization $b_A = b_B$. Another way is to extract the end-to-end distances squared for both blocks from the atomistic simulations, $R_{l,0,PA}^2$ and $R_{l,0,PB}^2$; this results in two expressions,

$$nb_A^2 = R_{l,0,PA}^2 \qquad (4.118)$$

$$mb_B^2 = R_{l,0,PB}^2 \qquad (4.119)$$

which together with relations (4.114) and (4.115) constitute a set of four equations in the four unknown Gaussian chain parameters (n, m, b_A, b_B).

We turn our attention again to the parameterization undertaken by Lam and Goldbeck-Wood [64] for the PEO–PPO–PEO triblocks. We will take as an example their results on Pluronic P85 ($EO_{26}PO_{40}EO_{26}$). The monomers are denoted by EO and PO, and the beads by E and P. The end-to-end distance squared for P85 is 32.4 nm^2; the monomeric bond lengths are found to be 0.2910 and 0.3394 nm, respectively, and $b_E = b_p$. Applying equation (4.117), we have $m(52*0.2910)^2 + n(40*0.3394)^2 = 32.4*nm$, $nb = 52*0.2910$ and $mb = 40*0.3394$. The Gaussian chain parameters are found by solving these three equations: $n = 13.4$, $m = 12.03$ and $b = 1.12$. It is clear that n and m need to be integers. One choice would be $n = 14$ and $m = 12$ with $b = 1.12$ nm. Hence, the mapping of the atomistic chain onto the Gaussian chain is $EO_{26}PO_{40}EO_{26} \to E_7P_{12}E_7$. However, this does not need to be the best one; $E_6P_{12}E_6$ or $E_7P_{11}E_7$ might be even better. Lam and Goldbeck-Wood show by means of the correlation function derived from equations of the RPA that actually the bead segmentation $E_6P_{12}E_6$ is the most appropriate (Figure 4.12). The peak observed is due to the 'correlation hole effect' in block copolymers: the probability of finding in the neighborhood of an A monomer, another A monomer belonging to a different chain is slightly decreased due to the

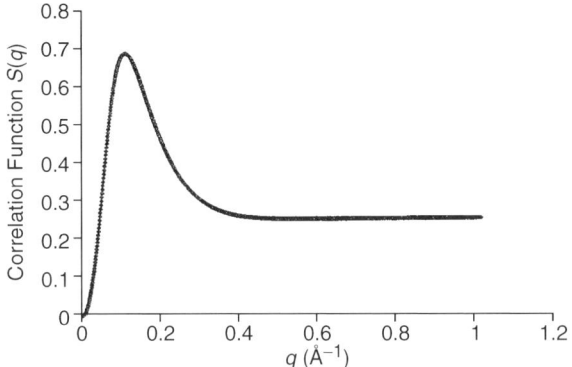

Figure 4.12. Correlation function plot for P85 with the $E_7P_{12}E_7$ structure derived from equations based the random phase approximation [67]. Reproduced from Lam and Goldbeck-Wood [64] with permission.

repulsion between the chains. At small scattering vectors, $q \to 0$, the scattering power goes to zero. The combination of both effects leads to a maximum in the intermediated range. This peak represents the end-to-end distance and is observed at 5.67 nm. In order to check the effective bead–bead bond length parameter the end-to-end distance squared obtained from the Gaussian, chain statistics [equation (116)] and that obtained from RPA is plotted in Figure 4.13. It is clearly observed that in the range $1 < b < 1.4$ nm we have a rather good agreement. That the segmentation $E_6P_{12}E_6$ is the most appropriate follows by comparing the end-to-end distances of these Gaussian chains with the atomistic simulation result and the close correspondence with the RPA results.

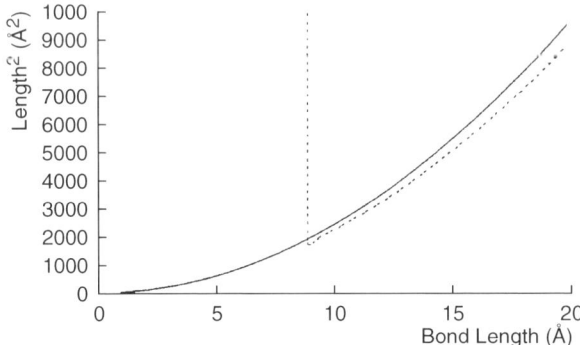

Figure 4.13. Comparison of the end-to-end distance squared as obtained from Gaussian chain statistics ($R^2 = nb^2$) with that of the peak position from the correlation function derived from equations based on the RPA. Reproduced from Lam and Goldbeck-Wood [64] with permission.

Another way to parameterize the Gaussian chain model for block copolymers is to make use of experimental values of the characteristic ratio of the corresponding homopolymers. The characteristic ratio is defined as $C_N = R_{l,e}^2/R_l^2 = R_{l,e}^2/(Nl^2)$, where $R_{l,e}^2$ is the experimentally determined end-to-end distance squared and N is the length of the polymer in number of atomistic backbone bonds with length l. The characteristic ratio is thus a measure of chain stiffness. The length of the fully stretched atomistic chain is denoted by L_N, the ratio between L_N and Nl is denoted by Λ_N. For the coarse grained beads, actually the Kuhn segments [61, 62], we thus have $nb = \Lambda Nl$ and $nb^2 = C_\infty Nl^2$, from which it follows that $b = C_\infty l/\Lambda$ and $n = \Lambda^2 N/C_\infty$. For example, poly-methylene has a characteristic ratio $C_\infty = 6.7$ and $\Lambda_\infty = 0.83$, from which it follows that $N/n = 9.7$ real C–C bonds per bead–bead bond and the bead–bead bond length itself is $b = 8.1l = 12.24$ Å. Tacticity of the polymer has, as expected, a large effect on the characteristic ratio. In case of poly-propylene we have for the atactic form $C_\infty = 6.9$ and for the isotactic form $C_\infty = 5.5$. Poly-oxythylene has a characteristic ratio of $C_\infty = 4.0$. It must be emphasized that small chain lengths have normally characteristic ratios that are different from C_∞ [31]. Taking as estimation for Λ_∞ again a value of 0.83 we obtain $N/n = 5.8$ backbone bonds and $b = 4.8l = 7.4$ Å. This leads to the result of mapping two EO monomers onto one Gaussian chain bead. Actually, this method gives an estimation of the number of monomers per Gaussian bead. It might be a rapid way to parameterize the Gaussian chain; however, the method of Lam and Goldbeck-Wood [64] is a much better approach.

4.3.8. Mean-Field Approximation

In the last section we have introduced a self-consistent Kohn–Sham scheme and concentrated on including the intrachain correlations into density functional theories by introducing a 'quasi-ideal' system. Now we turn our attention to the single-particle direct correlation functional, $c^{\text{qid},(1)}[\rho_0](\mathbf{r}_{s,i})$, which is related to the functional derivatives of the noncorrelated interaction energy functional J and the intermolecular correlation functional $F_{\text{cor,inter}}$ according to equation (4.101).

We apply the mean-field approximation to the single-particle direct correlation functional; that is, we neglect completely the intermolecular correlations ($F_{\text{cor,inter}} = 0$), so the noncorrelated interaction energy functional is replaced by a phenomenological approach as already introduced in Section 4.2.5,

$$J[\rho] = \sum_{I,J>I} \int\int d\mathbf{r}_I d\mathbf{r}_J B_{I,J}^{\text{eff}} \phi(\mathbf{r}_I) K(\mathbf{r}_I - \mathbf{r}_J) \phi(\mathbf{r}_J) \quad (4.120)$$

where the symmetric kernel $K(\mathbf{r}_I - \mathbf{r}_J)$ must be a normalized distribution and simply introduces nonlocality to the interaction functional. The indices I and J run over all bead types present in the mixture. The effective excess interaction energy parameter $B_{I,J}^{\text{eff}}$ depends on local composition and temperature. In contrast to the derivation of the chemical potential in Section 2.5 where we had to include the constraint that the sum of the volume fractions equals 1, here we have no constraints to statisfy. The functional derivative of $J[\rho]$ with respect to $\rho(\mathbf{r}_I)$ is then derived, to be given by:

$$\frac{\delta J[\rho]}{\delta \rho(\mathbf{r}_I)} = \sum_{J \neq I} \int d\mathbf{r}_J v_I B_{I,J}^{\text{eff}} K(\mathbf{r}_I - \mathbf{r}_J) \phi(\mathbf{r}_J)$$
$$+ \sum_{K,J>K} \iint d\mathbf{r}_K d\mathbf{r}_J \phi(\mathbf{r}_K) K(\mathbf{r}_I - \mathbf{r}_J) \phi(\mathbf{r}_J) \sum_{P=I;J} \frac{\partial B_{K,J}^{\text{eff}}}{\partial \phi(\mathbf{r}_P)} \frac{\partial \phi(\mathbf{r}_P)}{\partial \rho(\mathbf{r}_I)}$$
(4.121)

In case of an composition-independent interaction parameter, $B_{IJ} = (kT/v_I)\chi_{IJ}$, where v_I is the volume of beads of type I, we obtain:

$$\frac{\delta J[\rho]}{\delta \rho(\mathbf{r}_I)} = \beta^{-1} \sum_{J \neq I} \int d\mathbf{r}_J \chi_{I,J} K(\mathbf{r}_I - \mathbf{r}_J) \phi(\mathbf{r}_J) \quad (4.122)$$

Note, the bead density distributions are in units of number per m^3. This is actually the formula that is most often applied in density functional theories for inhomogeneous mixtures of polymers. As we have shown in the first section, it will mostly not be able to reproduce or predict real industrial applications. A better approach to deal with real applications is to apply the parameterization of the phenomenological approach described in detail in Section 4.2.5 and insert the resulting expression for B_{IJ}^{eff} in terms of ϕ_I and ϕ_J in equation (4.121). Unfortunately, I have not found published results on applying this phenomenological approach in density functional theories for inhomogeneous polymer mixtures. In-house applications at BASF do show the potential of such an approach. However, the implementation into density functional simulation models for inhomogeneous mixtures is far from straightforward.

The single particle direct correlation functional, $c^{\text{qid},(1)}[\rho_0]$, for the 'quasi-ideal' system applying the mean-field approach described here becomes in terms of the Flory–Huggins χ parameter:

$$c^{\text{qid},(1)}[\rho_0](\mathbf{r}_{s,i}) = -\sum_{J \neq I} \int d\mathbf{r}_J \delta_{s,i,I} \chi_{I,J} K(\mathbf{r}_I - \mathbf{r}_J) \phi(\mathbf{r}_J) \quad (4.123)$$

where we have used the chain rule

$$\frac{\partial J[\rho]}{\partial \rho(\mathbf{r}_{s,i})} = \frac{\partial J[\rho]}{\partial \rho(\mathbf{r}_I)} \frac{\partial \rho(\mathbf{r}_I)}{\partial \rho(\mathbf{r}_{s,i})} = \frac{\partial J[\rho]}{\partial \rho(\mathbf{r}_I)} \delta_{s,i,I} \quad (4.124)$$

The Dirac-delta function $\delta_{s,i,I}$ equals one if bead s of chains i is of bead type I and zero otherwise.

4.3.9. Lagrange Multiplier Fields

Most self-consistent simulation methods not only apply a quasi-ideal system approach to the Kohn–Sham scheme but in addition incompressibility is assumed. As already mentioned, Schmid [99], for instance, includes into the quasi-ideal system the bead excluded volume by hard-core interactions. Most self-consistent field analytic theories and simulation models, for example the continuous field model of Noolandi and Hong [83], the lattice model of Scheutjens and Fleer [26, 95, 96] or the orginal dynamic density functional method of Fraaije [32], are all based on complete incompressibility:

$$\sum_{i,s} \phi(\mathbf{r}_{s,i}) = 1 \tag{4.125}$$

This constraint is easily integrated into the Kohn–Sham scheme using the well known Lagrange multiplier method, thereby introducing a Lagrange parameter field, $\lambda(\mathbf{r})$. The Kohn–Sham self-consistent set of equation to be solved simultaneously in order to obtain the equilibrium density functions now becomes:

$$u^{\text{qid,eff}}[\rho_0](\mathbf{r}_{s,i}) = u(\mathbf{r}_{s,i}) + \lambda(\mathbf{r}_I) - \beta^{-1} c^{\text{qid},(1)}[\rho_0](\mathbf{r}_{s,i})$$
$$\sum_{i,s} \phi(\mathbf{r}_{s,i}) - 1 = 0 \tag{4.126}$$

This set of equations must be solved with $u^{\text{qid,eff}}[\rho_0](\mathbf{r}_{s,i})$ and $\lambda(\mathbf{r}_I)$ as the iteration variables the bead density fields are computed with equation (4.99) using an appropriate model for the single chain distributions.

4.3.10. Numerical Example: The Scheutjens–Fleer Model

The self-consistent field theory of Scheutjens and Fleer [95, 96] for inhomogeneous polymer mixtures is based on a lattice model and a popular simulation model for polymer adsorption. The entropic and enthalpic contributions to the free energy are derived from lattice statistics. However, we can directly obtain the Scheutjens–Fleer equations for multicomponent copolymer mixtures [23] from the density functional self-consistent field Kohn–Sham formalism presented in this section.

Consider a lattice between two parallel plates, as in Figure 4.14. The lattice layers parallel to the surface are numbered from one surface to the other: $z = 1, \ldots, M$. It is assumed that all lattice sites are filled with beads. The quasi-ideal self-consistent field equations [equation (4.126)] now read:

$$u^{\text{qid,eff}}[\rho_0](z_{s,i}) = u(z_{s,i}) + \lambda(z_I) - \beta^{-1} c^{\text{qid},(1)}[\rho_0](z_{s,i})$$
$$\sum_{i,s} \phi(z_{s,i}) - 1 = 0 \tag{4.127}$$

INHOMOGENEOUS POLYMER MIXTURES

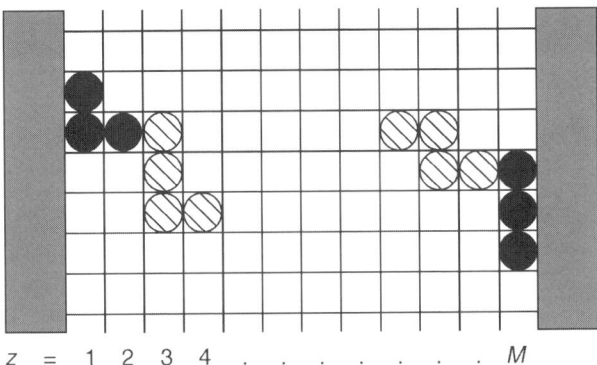

Figure 4.14. Two chains, AAABBBB, in a lattice between two surfaces.

The volume fractions $\phi(z_{s,i})$ are obtained from the Green functions $G_{s,i}(z_{s,i})$ and $G_{s+1,i}^{\text{inv}}(z_{s,i})$ as in equation (4.106):

$$\rho_0^{\text{qid}}(z_{s,i}) \propto G_{s,i}(z_{s,i})\sigma[G_{s+1,i}^{\text{inv}}](z_{s,i}) \tag{4.128}$$

The sets of (once integrated) Green functions $G_{s,i}(z)$ and $G_{s+1,i}^{\text{inv}}(z)$ are related by recurrence equations:

$$G_{s,i}(z) = \exp[-\beta u_{s,i}^{\text{qid,eff}}(z)]\sigma[G_{s-1,i}](z) \tag{4.129}$$

$$G_{s,i}^{\text{inv}}(z) = \exp[-\beta u_{s,i}^{\text{qid,eff}}(z)]\sigma[G_{s+1,i}^{\text{inv}}](z) \tag{4.130}$$

with $G_{1,i}(z) = G_{N_i,i}^{\text{inv}}(z) = 1$. The linkage operator σ is defined as a simple quadrature rule on the lattice:

$$\sigma[f](z) \equiv \lambda_1 f(z-1) + \lambda_0 f(z) + \lambda_1 f(z+1) \tag{4.131}$$

where λ_0 and λ_1 depend on the lattice type applied. For a cubic lattice, $\lambda_0 = 1/6$ and $\lambda_1 = 4/6$. The quasi-ideal one-particle direct distribution function [equation (101)] in the lattice version becomes:

$$c^{\text{qid},(1)}[\rho_0](z_{s,i}) \equiv -\beta\left\{\frac{\delta J[\rho]}{\delta\rho(z_{s,i})} + \frac{\delta F_{\text{cor,inter}}[\rho]}{\delta\rho(z_{s,i})}\right\} \tag{4.132}$$

In the Scheutjens–Fleer lattice model the intermolecular correlation functional, $F_{\text{cor,inter}}[\rho]$, is neglected and the noncorrelated interaction energy functional, $J[\rho]$, is approximated with the Flory–Huggins expression (11) locally applied on the lattice:

$$J = RT \frac{1}{2} \sum_z V \sum_{I,J} \rho(z_I) \chi_{I,J} \{\lambda_1 \phi(z_J - 1) + \lambda_0 \phi(z_J) + \lambda_1 \phi(z_J + 1)\} \qquad (4.133)$$

Often, this functional is extended with an interaction energy term for the interaction between beads adjacent to the surface and the surface using a χ_s parameter. Thus, it is possible to derive the Scheutjens–Fleer equations without making use of lattice-based derivations, rather the lattice is introduced as the basis for a very coarse numerical integration scheme to the density functional Kohn–Sham scheme. Of course, extensions to the Scheutjens–Fleer theory such as bond angle restrictions [66] do depend on lattice-based derivations. The success of the Scheutjens–Fleer theory lies in its straightforward implementation into computer programs and relatively short computation times. Although the very coarse numerical lattice integration scheme makes the simulation method inadequate for real systems containing oligomers or short surfactants, it is applicable to inhomogeneous mixtures of very long polymers and is computationally still rather inexpensive. The parameterization of the χ parameters follows exactly the scheme introduced in Section 4.2.5. The segmentation of real polymer chains into beads of equal size is straightforward if the chains are very long; again for short chains there is no way to obtain a realistic segmentation.

Although the Scheutjens–Fleer theory is mostly used to investigate physical behavior of theoretical systems, such as for instance depletion stabilization [27, 28], there are many predictions of real practical applications. See Fleer et al. [26] for an overview of applications and extensions of the Scheutjens–Fleer theory. We restrict ourselves here to two examples. The Scheutjens–Fleer theory makes it possible to study the interaction between two flat plates covered with adsorbed polymer layers. Mostly, the restricted equilibrium approach is taken; that is, the polymer molecules are not allowed to leave the gap between the plates whereas the solvent molecules are freely exchanged between surface region and bulk solution. By performing calculations at various plate distances and constant amounts of polymer, one obtains interaction energy curves between both adsorbed plates. Figure 4.15 shows typical interaction curves for different adsorbed amount in a Θ solvent. The theoretical curves are qualitatively in agreement with the experimental ones. At low adsorbed amounts there is a strong attractive minimum, which gradually disappears when the adsorption increases. The possibility to simulate the interaction between adsorbed layers of polymers makes up one of the successes of the Scheutjens–Fleer theory. For example, this self-consistent field lattice model gave evidence that attraction between adsorbed polymer layers is possible by bridging, even in good solvent [97, 98]. Measurements in Θ solvents, where osmotic forces are zero or repulsive, have confirmed this theoretical prediction [53].

At BASF, I applied together with my colleagues Gregor Ley and Erich Hädicke the Scheutjens–Fleer theory [21, 22] to investigate spherical polymer–monomer aggregates using a spherical lattice introduced by van der Schoot et al. [108]. Polymer aggregates in solution are not only of interest to academia, they play an

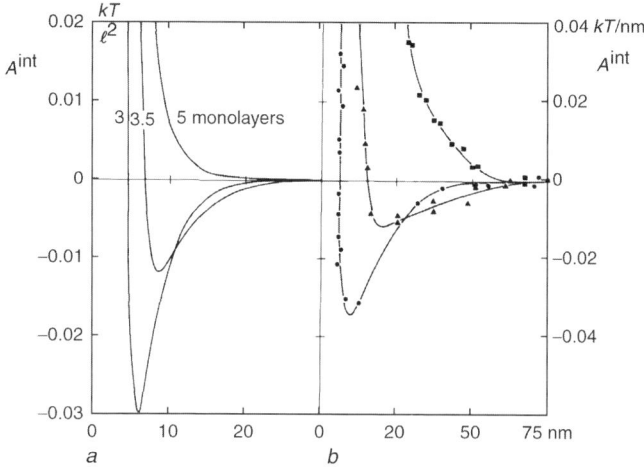

Figure 4.15. Comparison of theoretical (*a*) and experimental (*b*) interaction curves for flat plates in the presence of various amounts of polymer under Θ conditions. (*a*) Calculated with the Scheutjens–Fleer model [97, 98], the energy per surface site as a function of the plate separation (in lattice layers) for $N = 1000$, $\chi_s = 1$, $\chi = 0.5$ and $\lambda_0 = 0.5$. (*b*) Adapted from Almog and Klein [4], A^{int} in kT/nm^2; the system is monodisperse polystyrene ($M = 2000$ K) adsorbed from cyclopentane on mica. From left to right the adsorbed amount increases. Figure taken from reference [26].

important role in many industrial applications, for example, lattices. Latex particles are typically examples of mesoscopic systems with radii from 20 nm up to a few hundred nanometers. The internal structure of latex particles is of interest as it affects the mechanism and kinetics of emulsion polymerization. Casassa and Tagami [12] derived a bead density profile for a single chain inside a spherical cavity which does not touch the surface enclosing the cavity. This model was applied by Dabdub et al. [14] and Yang et al. [114] to model monomer–polymer particles. Their model is of course far from a real monomer–polymer particle as they simply model one polymer chain inside an empty spherical cavity with reflecting boundaries. In this way they obtain a depletion layer which is then interpreted as the monomer enriched shell. The Scheutjens–Fleer simulations we performed on monomer–polymer particles show a different result, the monomer enriching at the surface (see Figure 4.16), but in the core of the monomer–particle the concentration of monomers in constant. These results allowed us to interpret emulsion polymerization processes better.

4.3.11. Numerical Example: Continuous Field Model

Although the title of this section contains the phrase 'continuous field model,' we will actually apply a very fine trapezodial integration scheme for the convolution integrals in the quasi-ideal system approach decribed in Section 4.3.5.

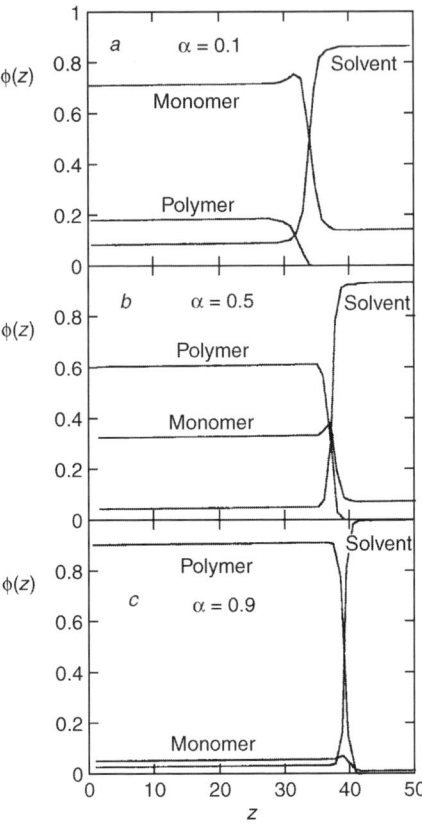

Figure 4.16. Bead volume fraction profiles of 200-mers and monomers in monomer–polymer aggregates at different polymer-to-monomer ratios [indicated, $\alpha = Nn_p/(n_m + Nn_p)$]. The solvent is a very poor solvent, $\chi_{PS} = 2.5$. The total amount of polymer beads and monomers in the system is kept at 250,000 and $M = 60$. Reproduced from Evers et al. [21] with permission.

Aggregation phenomena in systems of amphiphilic blockcopolymers in selective solvents have been the subject of a large number of studies, experimental as well as theoretical. In solvents which are selective for one of the blocks, blockcopolymers may form aggregates. Depending on, for example, the temperature, salt concentration, molecular composition, and even chain polydispersity, the aggregation behavior may vary considerably. A whole zoo of aggregate types has been found, starting from simple spherical shaped micelles, vesicles, ellipsoids, rodlike aggregates, membranes, lamella, bicontinuous phases, and so on. The critical micelle concentration (CMC), the concentration of amphiphiles where they first start to form micelles, can also be a function of all of the aforementioned parameters. Sok and Evers [102] used a self-consistent mean field model based on density functional theory to model micellization of model block copolymers, depending on their

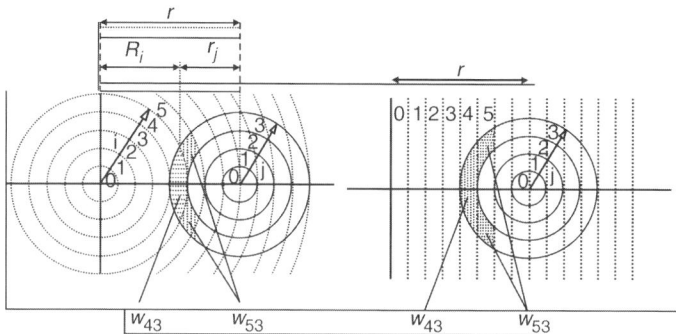

Figure 4.17. Two-dimensional representation of the overlap volume of the field layers (counter i) with the center separation r in the case of spherical symmetry (left) and planar symmetry (right). In both parts the overlap volumina of kernel layer 2 and system layer 5 ($w_{5,3}$) and of kernel layer 3 with system layer 4 ($w_{4,3}$) are indicated.

composition and parameters, applying a very fine integration scheme for the spherical geometry (Figure 4.17), as already described for the planar case in Section 4.3.6.

The small system approach we take here was originally introduced by Hall and Petica [39]; a more fundamental introduction to small system thermodynamics is given by Hill [47]. The route taken here has been extensively used in various applications of the Scheutjens–Fleer lattice theory [9, 66, 109]. We consider a system of volume V, with N_{solute}, solute molecules and N_{solvent} solvent molecules. We (hypothetically) divide this system into M subsystems each with volume V_{sub}. Each subsystem contains exactly one aggregate (one micelle) with a fixed center of mass at the center of the system. Per subsystem there are $N_{\text{solute}}^{\text{sub}}$ solute molecules (the volume fraction of solute molecules in the subsystem $\phi_{\text{solute}}^{\text{sub}}$ is equal to the overall volume fraction ϕ_{solute}). The excess Helmholtz free energy of the subsystem (with respect to the completely mixed state) is denoted by $A_{\text{ex}}^{\text{sub}}$. The total excess free energy A_{ex} is then given by:

$$A_{\text{ex}} = M A_{\text{ex}}^{\text{sub}} \tag{4.134}$$

where

$$A_{\text{ex}}^{\text{sub}} = A_{\text{exm}}^{\text{sub}} + A_{\text{tr}}^{\text{sub}} \tag{4.135}$$

$A_{\text{exm}}^{\text{sub}}$ denotes the excess free energy of the subsystem with respect to the homogeneous subsystem, and $A_{\text{tr}}^{\text{sub}}$ is the translational entropy contribution to compensate for the fixing of the aggregate. The translational entropy part is:

$$A_{\text{tr}}^{\text{sub}} = RT \ln\left(\frac{V^{\text{miscelle}}}{V^{\text{sub}}}\right) \tag{4.136}$$

where V^{micelle} is the volume of the micelle in the subsystem. It is clear that, when one uses a continuous density profile, an exact definition of micellar size becomes difficult. A number of definitions for micellar size are valid and it is possible to use different definitions for comparison with experiments. For our purpose we define a micelle (or aggregate) to end where bulk conditions set in. For example in the case of an A_xB_y amphiphile in solvent, the bulk conditions are: (i) the ratio of the A and B concentrations is x:y; and (ii) the overall concentration of amphiphile molecules is constant. In practice a 5 percent deviation from the conditions was tolerated. A typical volume fraction profile of an aggregate is shown in Figure 4.18. The radius of this aggregate, using the definition stated before, is indicated with an arrow and is roughly 20 nm. This may seem wrong, but if we look at the inset of in the figure, where the y-scale is enlarged, we see that this is exactly the radius where the bulk conditions set in. The density profile of the tails differs from the bulk value up to 20 nm. The definition of the aggregation number is straightforward: it is the total number of amphiphile molecules inside this region. If the concentration of block copolymers in the serum (i.e. outside of the micelle) were constant with changing the overall block copolymer concentration, then the serum concentration would in principle be the CMC. We have found, however, that this concentration is not constant, but rather increases with increasing overall concentration. Therefore, to evaluate the CMC we have to find the serum concentrations at several concentrations and extrapolate. The mesoscopic model we use is, however, not valid for the low concentration region and thus we should take great care when applying it to this region.

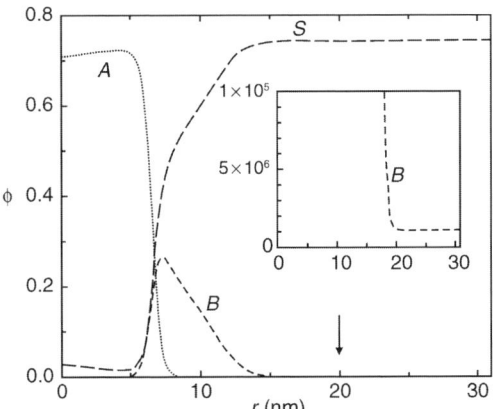

Figure 4.18. Typical volume fraction profile of an aggregate amphiphilic block-copolymers $A_{12}B_{12}$ in solvent S. The radius of the aggregate (micelle) is indicated by the arrow. The inset shows the volume fraction of the B-block on a smaller scale to show the deviation from bulk conditions up to 20 nm. The interaction parameters applied were: $\chi_{AB} = 1.0$, $\chi_{AS} = 2.0$ and $\chi_{BS} = -0.25$.

Figure 4.19 shows the excess free energy of a system of $A_{12}B_{12}$ at 5 percent total volume fraction as a function of the number of amphiphiles per subsystem (which is directly related to the sub-volume size as the volume fraction of amphiphiles is equal for all data points). The curve shows a minimum at around 46 block copolymers. Thus, the preferred aggregate size for this system, under the assumption that all aggregates of the macroscopic system are of the same size, roughly 46. For comparison, the second inset of Figure 4.19 shows the volume fraction profiles of a subsystem containing 150 amphiphiles. The volume fraction profile of the lyophilic B-block is roughly the same, but the solvent concentration in the center of the micelle has increased. Obviously, this increase in solvent concentration in the center while increasing the overall micelle size is energetically unfavorable as the free energy density has increased by 40 percent. We must note, however, that this feature is clearly favored by the monodispersity of the amphiphile chain length distribution. A polydisperse block-copolymer sample can under some circumstances show a completely different behavior.

Of course, when applying a very crude integration scheme like that of the Scheutjens–Fleer model, the same results will be obtained as with the detailed integration methods where the amphiphiles are of high molecular weight. However, if they are low molecular weight, as in the example shown here, the excess free energy as a function of the aggregation number computed with the Scheutjens–Fleer model will not be a smooth continuous curve. Instead, it shows lattice artifacts as discussed by Boehmer and Koopal [9] (see for instance Figure 4.20). To be able to extract the CMC from the Scheutjens–Fleer model they applied a third-order polynomial fit through the calculate values of A^{ex}. It is not a good idea to apply a very coarse-grained numerical integration scheme to small oligomers, which is what the Scheutjens–Fleer model is. On the other hand, it is computationally not wise to

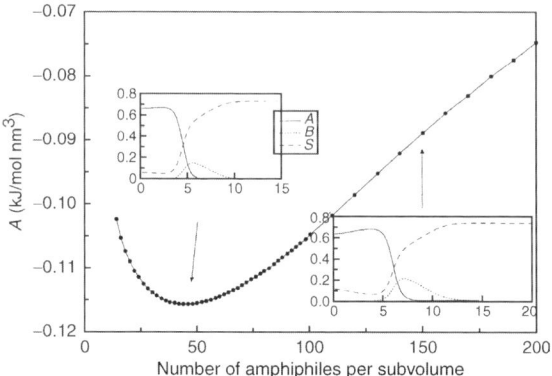

Figure 4.19. Excess free energy of a system of $A_{12}B_{12}$ at 5 percent volume fraction vs number of amphiphiles per subsystem. The two sub-figures show the volume fraction profiles of all bead types at two different numbers of amphiphiles per subsystem. Note also that the total size of the subsystem changes while the volume fraction of amphiphiles in the subsystem remains constant. The interaction parameters are the same as in Figure 4.18.

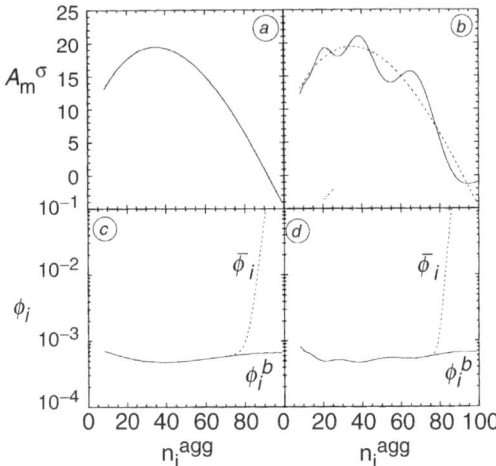

Figure 4.20. Micelle formation as computed with the Scheutjens–Fleer model by Böehmer and Koopal for a $A_{12}B_3$ copolymer in solvent. (*a*) General representation of the excess free energy of the formation of a micelle, A_{exm}^{sub}, with a fixed position in the subsystem as a function of the aggregation number. (*b*) Excess free energy of micellization, A_{exm}^{sub}, as a function of aggregation number computed with the Scheutjens–Fleer model in a 0.1 electrolyte; the dotted curve represents a fit with a third-order polynomial. (*c*) The bulk volume fraction, ϕ_p^b, and the average volume fraction in the subsystem, ϕ_p^{av}, as a function of aggregation number for $A_{12}B_3$ in a 0.1 salt solution as obtained from the smoothed curves for A_{exm}^{sub}. (*d*) Calculated curves for ϕ_p^b and ϕ_p^{av} without smoothing as a function of aggregation number. This figure was taken from Boehmer and Koopal [9] and shows how the lattice integration scheme of the Scheutjens–Fleer model introduces artifacts when applied to small copolymers.

apply a very detailed numerical integration scheme, as discussed in this sub-section, to very long polymers. The Scheutjens–Fleer model is equally accurate in this case and needs less computer resources. Hence, the choice on the self-consistent density field model to apply depends on the size of the polymers and the size of the morphologies to be expected (e.g. aggregation number in the case of micellization). In the literature, applications will be found where self-consistent density field models are applied and the polymers segmentated into beads containing, for example, a single CH_2 group. In my opinion this is nonsense. Applying such a coarse graining scheme makes the small surfactant molecule extremely flexible in the Gaussian chain model and has no relation the actual intramolecular distribution function. As shown in Section 4.3.7, the intramolecular chain distribution function in the quasi-ideal Kohn–Sham approach *must* reproduce the basic chain characteristics (characteristic ratio, end-to-end distance or even better the intramolecular pair distribution functions [111] or RPA correlation function, as Lam and Goldbeck-Wood [64] did). To conclude, be critical if too coarse numerical integration schemes are applied to too small entities like a CH_2 group. A review on modeling the self-assembly of block copolymers in selective solvents has been published by Linse [70] recently and we refer the interested reader to this paper.

4.4. DYNAMIC DENSITY FUNCTIONAL THEORY

Fraaije [32] developed his dynamic density functional theory during his time at Akzo Nobel research in the Netherlands more than a decade ago. Working in industry it is interesting to note that the dynamic density functional theory (DDFT) of Fraaije was actually developed in an industrial research department. In the course of two European research projects, CAESAR (led by British Aerospace) and MESODYN (led by BASF), the simulation method was further extended by Fraaije's group first at the University of Groningen and later at the University of Leiden. These two European projects enabled the implementation of the DDFT model into a data parallel computer program known as MesoDyn, which involved the Fraaije's group, BASF, and IBM Germany.

The basic goal of DDFT is to predict three-dimensional morphologies of complex polymer mixtures without introducing certain geometrical constraints such as we have seen in the last section (e.g. planar or spherical geometries). We first derive the basics of the dynamic density functional theory and then investigate its potential for practical applications. Special attention is paid to the 'external potential dynamics' formulation of Fraaije's DDFT.

4.4.1. The DDFT Principle

Physical transport phenomena are generally described by the Onsager principle, that is, by linear relations between fluxes J and driving forces X:

$$J = LX \tag{4.137}$$

The proportionality factor L is a materials constant. In a mixture that is not in equilibrium the gradient of the local chemical potential, $\mu^{intr}(r)$, will act as a driving force resulting in material flow, that is, $X = -\nabla \mu^{intr}(r)$. The intrinsic chemical potential, $\mu^{intr}(r)$, was derived in Section 4.3.4. Assuming local coupling, $L = \beta D \rho(\mathbf{r})$, the flux is thus given by:

$$J(r) = -\beta D \rho(r) \nabla \mu^{intr}(r) \tag{4.138}$$

with D the diffusion coefficient. Combining equation (4.138) with the continuity equation (mass conservation):

$$\frac{\partial \rho(r)}{\partial t} + \nabla \cdot J(r) = 0 \tag{4.139}$$

we obtain a diffusion equation:

$$\frac{\partial \rho(r)}{\partial t} = \beta D \nabla \cdot \rho(r) \nabla \mu^{intr}(r) \tag{4.140}$$

For an ideal gas, that is, a noninteracting fluid, the intrinsic chemical potential, which is a functional of the particle density distribution, follows from density functional theory (Section 4.3.5) and is given by:

$$\mu^{id}(r) = \beta^{-1} \ln[\Lambda^3 \rho^{id}(r)] \tag{4.141}$$

Applying the diffusion equation (4.140) to this ideal fluid results in Fick's law:

$$\frac{\partial \rho(r)}{\partial t} = D\nabla^2 \rho(r) \tag{4.142}$$

Diffusion equation (4.140) is thus a generalization of the famous Fick's law.

The basic idea of Fraaije's dynamic density functional theory [32] is to combine the generalized Fick's law [equation (4.140)] with the DFT result for the intrinsic potential [equation (4.71)], which results in a functional Langevin equation:

$$\frac{\partial \rho(\mathbf{r}_{s,i})}{\partial t} = D_{s,i} \nabla \cdot \rho(\mathbf{r}_{s,i}) \nabla \{\mu_{s,i} - u(\mathbf{r}_{s,i})\} + \eta(\mathbf{r}_{s,i}) \tag{4.143}$$

where $\eta(\mathbf{r}_{s,i})$ is a noise term that has a normal distribution with second moments dictated by the fluctuation–dissipation theorem [36, 91, 110]. We now apply a trick. Suppose the inhomogeneous system *is not* in equilibrium, but we look at it as if it is in equilibrium. This assumption is known as the local equilibrium approach. The latter can only be the case if an effective external potential field, $u^{eff}(\mathbf{r})$, is applied that satisfies the well-known Boltzmann (equilibrium) density distribution, as explained in Section 4.3.5:

$$\rho(\mathbf{r}) = \frac{\langle n \rangle}{V} \frac{\exp[-\beta u^{eff}(\mathbf{r})]}{\int \exp[-\beta u^{eff}(\mathbf{r}')]d\mathbf{r}'} \tag{4.144}$$

or directly form equation (4.77),

$$u^{eff}(\mathbf{r}_{s,i}) = \mu_{s,i} - \beta^{-1} \ln[\Lambda^3_{s,i\rho}(\mathbf{r}_{s,i})] \tag{4.145}$$

The effective external potential field, $u^{eff}(\mathbf{r})$, satisfying this Boltzmann distribution, is thus a functional of the density field, $u^{eff}[\rho]$. These considerations are valid since for an ensemble of systems in equilibrium we have a bijective, that is, a one-to-one relation between the external field and the bead density distributions. The word 'ensemble' should raise suspicions. Yes, all density functional relations derived in the last section were derived for 'ensembles' in equilibrium. How does this relate to the evolution of one system in time? The local equilibrium approach means that, at every instance of time, the density distributions in the system represent that of an ensemble of systems in equilibrium. Numerically, this means we need to take time steps that are large enough to allow the chains locally to relax. To continue our derivations, from equation (4.94) we extract for the external field, $u[\rho](\mathbf{r}_{s,i})$,

$$u[\rho](\mathbf{r}_{s,i}) = u^{\text{eff}}[\rho](\mathbf{r}_{s,i}) + \beta^{-1} c^{(1)}[\rho](\mathbf{r}_{s,i}) \qquad (4.146)$$

so that the generalized diffusion equation (4.143) becomes:

$$\frac{\partial \rho(\mathbf{r}_{s,i})}{\partial t} = D_{s,i} \nabla \cdot \rho(\mathbf{r}_{s,i}) \nabla \{ -\beta^{-1} c^{(1)}[\rho](\mathbf{r}_{s,i}) - u^{\text{eff}}[\rho](\mathbf{r}_{s,i}) \} + \eta(\mathbf{r}_{s,i}) \qquad (4.147)$$

Of course, neglecting all correlations and the interparticle particle interactions, that is, applying the 'ideal' system DFT, $c^{(1)}[\rho](\mathbf{r}_{s,i}) = 0$, you get Fick's law in return.

In order to apply the DDFT principle to inhomogeneous polymer mixtures, using the local equilibrium approximation, we need to adopt the quasi-ideal system approach (Section 4.3.7). This means that the quasi-ideal effective external potential fields, $u^{\text{rid,eff}}[\rho]$, are extracted from the bead density distributions, $\rho(\mathbf{r}_{s,i})$, by taking into account the intramolecular correlations [equation (4.99)]. Originally, Fraaije [32] used a lattice model like the Scheutjens–Fleer theory [95, 96], to extract $u^{\text{qid,eff}}[\rho]$ from $\rho(\mathbf{r}_{s,i})$ by numerical means. A few years later the theory of Fraaije was reformulated using the Gaussian chain model [33] with a numerical quadrature rule [73] to integrate the path integrals. Actually, this approach is the one implemented into the MesoDyn program included in the Cerius^2 package of by Accelrys. However, the DDFT principle as we derived it here is not at all restricted to this intramolecular correlation model. Having related $u^{\text{qid,eff}}[\rho]$ to $\rho(\mathbf{r}_{s,i})$, using some intramolecular correlation model, the diffusion equation reads:

$$\frac{\partial \rho(\mathbf{r}_{s,i})}{\partial t} = D_{s,i} \nabla \cdot \rho(\mathbf{r}_{s,i}) \nabla \{ -\beta^{-1} c^{\text{qid,(1)}}[\rho](\mathbf{r}_{s,i}) - u^{\text{qid,eff}}[\rho](\mathbf{r}_{s,i}) \} + \eta(\mathbf{r}_{s,i}) \qquad (4.148)$$

In the DDFT model of Fraaije [33] the quasi-ideal single bead direct correlation function, $c^{\text{qid,(1)}}[\rho](\mathbf{r}_{s,i})$, defined in equation (4.92) is approximated by a mean-field approach, all inter-molecular correlations are neglected ($F_{\text{cor,inter}}[\rho](\mathbf{r}_{s,i}) = 0$), and the noncorrelated interaction energy functional is replaced by a Flory–Huggins type expression (see Section 4.3.8),

$$\frac{\delta J[\rho]}{\delta \rho(\mathbf{r}_I)} = \beta^{-1} \sum_{J \neq I} \int d\mathbf{r}_J \chi_{I,J} K(\mathbf{r}_I - \mathbf{r}_J) \phi(\mathbf{r}_J) \qquad (4.149)$$

so that the quasi-ideal single bead direct correlation function is given by:

$$c^{\text{qid,(1)}}[\rho](\mathbf{r}_{s,i}) = -\beta u^{\text{int}}[\rho](\mathbf{r}_{s,i}) = -\sum_{J \neq I} \int d\mathbf{r}_J \delta_{s,i,I} \chi_{I,J} K(\mathbf{r}_I - \mathbf{r}_J) \phi(\mathbf{r}_J) \qquad (4.150)$$

For the kernel, $K(\mathbf{r}_I - \mathbf{r}_J)$, which expresses the spatial extension of the mean-field interactions, the MesoDyn implementation is a simple Gaussian distribution [33, 73]. Here we introduce for reasons of comparity the interaction potential, $u^{\text{int}}[\rho](\mathbf{r}_{s,i})$, so that the diffusion equation (4.148) can be written into the form as it is used in Fraaije's theory [33]:

$$\frac{\partial \rho(\mathbf{r}_{s,i})}{\partial t} = D_{s,i} \nabla \cdot \rho(\mathbf{r}_{s,i}) \nabla \{u^{\text{int}}[\rho](\mathbf{r}_{s,i}) - u^{\text{qid,eff}}[\rho](\mathbf{r}_{s,i})\} + \eta(\mathbf{r}_{s,i}) \quad (4.151)$$

This equation together with Gaussian chain model for the quasi-ideal system (Section 4.3.7) make up the basic principles of Fraaije's DDFT. In the local coupling model the diffusion coefficients of the beads of a certain molecule type are all equal: $D_{s,i} = D_i$.

4.4.2. Excluded Volume Effects

The equations derived in the previous section constitute the DDFT principle; however, one topic has not touched on—the excluded volume effects. Originally Fraaije [32] derived his DDFT principle based on a lattice model, thereby requiring incompressibility; that is, locally the sum of the volume fractions must equal one (see Section 4.3.9). Maurits et al. [75] added to the noncorrelated interaction energy function a Helfand's penalty function. The basic idea is that in a liquid mixture the density fluctuations are small and harmonic, so that for the purpose of calculating phase separation the bare cohesive energy interactions may effectively replaced by exchange interactions. The idea goes back on Helfand [41]. The Helfand penalty functional is given by:

$$F_{\text{pen}}[\rho] = \frac{\kappa_H}{2\beta} \int d\mathbf{r} \left\{ \sum \phi(\mathbf{r}_{s,i}) - \sum \langle \phi_{s,i} \rangle \right\}^2 \quad (4.152)$$

Thus, this penalty functional simply introduces an energy penalty in case the local sum of the bead volume fractions deviates from the average sum. Although Maurits et al. [75] call the parameter κ_H a compressibility parameter, it is formally better to call it an excluded volume regularization parameter. The parameter κ_H is a global constant; independent of composition, it can principally be related to experimental values of isothermal compressibility [113]. The functional derivative of the noncorrelated interaction energy functional now becomes:

$$\frac{\delta J[\rho]}{\delta \rho(\mathbf{r}_I)} = \beta^{-1} \left\{ \kappa_H v_I \sum_J \phi_J(\mathbf{r}_J) + \sum_{J \neq I} \int d\mathbf{r}_J \chi_{I,J} K(\mathbf{r}_I - \mathbf{r}_J) \phi(\mathbf{r}_J) \right\} \quad (4.153)$$

The big advantage of using the Helfand penalty functional instead of applying incompressibility ($\sum_{i,s} \phi_{s,i} = 1$) is that we do not need to introduce Lagrange multiplier fields. From a numerical point of view it is much better to work with regularization functional than with constraints on the iteration variables.

4.4.3. Numerical Example: Copolymer–Solvent Mixtures

Predicting the three-dimensional morphology of a real system in dependence on composition, temperature and pressure is actually nothing more than predicting the phase diagram of the system. The phase behavior of a mixture or polymer blend is of great importance to industry as it governs the market-segment into which the product can be placed. Actually, it is often desirable to bring a certain base formulation into the required morphology. The three-dimensional morphologies, for instance, are related to the rheological or viscous–elastic behavior of the formation.

For historical reasons we first deal with a triblock–water mixture. The system Pluronic L64–water is the first system for which the DDFT of Fraaije was parameterized and its phase diagram was reproduced with the MesoDyn simulation program. Pluronic is a trademark of BASF. This work was performed by van Vlimmeren et al. [111]. In general, although diblock or triblock–solvent mixtures are simple two-component mixtures, they do expose very complicated phase diagrams where the solvent is selective for one of the blocks [54]. Pluronic L64 is an $EO_{13}PO_{30}EO_{13}$ triblock copolymer that in a binary mixture with water shows in the narrow concentration range from 50 to 70 percent four mesophases as determined by Alexandridis and co-workers [1–3]: micellar, hexagonal, or cylindrical, bicontinuous and lamellar phases. van Vlimmeren et al. modeled Pluronic L64 by a Gaussian chain of three E, nine P and three E beads. All bonds were of equal length. This parameterization of the chain model was obtained by comparing atomistic simulated chain correlation functions with that of the Gaussian chain (see Section 4.3.7). The χ_{EW} and χ_{EW} parameters were extracted from the experimental vapor pressure results of Malcolm and Rowlinson [71]. The χ_{EP} parameter was estimated from group contribution methods. All simulation runs were started from a homogeneous solution. The simulation results at four different concentrations of L64 are shown in Figure 4.21 in the form of isodensity surfaces of the hydrophobic bead (P). Isodensity surfaces connect points in space at which the density equals that of a predefined value, the isolevel. The hexagonal or cylindrical phase (55 percent) is not reproduced completely. By adding a directional convective term to the DDFT diffusion equation, that is, applying shear, Zvelindovski et al. were able to simulate perfect hexagonally packed cylinders [116].

In Section 4.3.7 we described the Gaussian chain parameterization method of Lam and Goldbeck-Wood [64]; here we will discuss one of their MesoDyn simulation results. Mortensen [81] obtained experimentally the phase diagram of Pluronic P85 ($EO_{26}PO_{40}EO_{26}$). Figure 4.22 shows simulation results of Lam and Goldbeck-Wood for P85–water mixtures at a moderate temperature of 40°C and a higher temperature of 70°C. The PEO blocks of P85 become less hydrophilic with increasing temperature. At the lower temperature a rather ordered micellar morphology is obtained from the simulations whereas, at the high temperature rod-like structures are found. These results are in accordance with the experimental results of Mortensen. Lam and Goldbeck-Wood used a temperature-dependant interaction parameter.

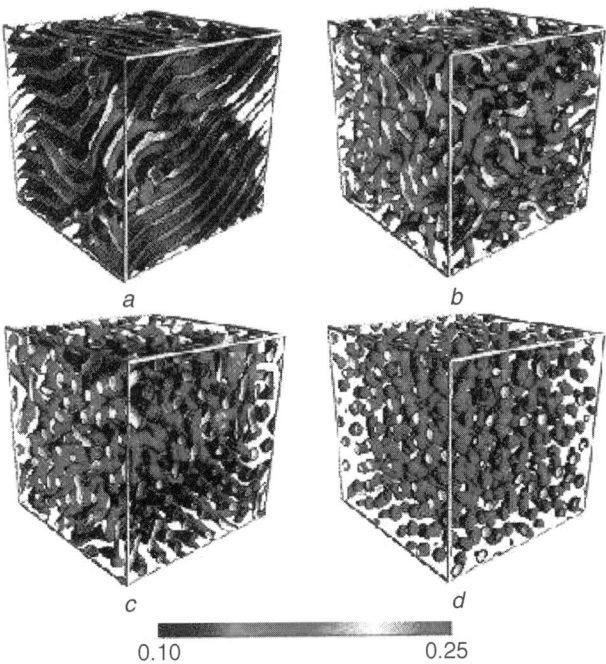

Figure 4.21. Mesoscopic morphologies of Pluronic L64–water mixtures at different concentrations of L64: (*a*) 70%; (*b*) 60%; (*c*) 55%; and (*d*) 50%. The PO iso-density surfaces are visualized with EO surface distribution in color as indicated in the color legend. Reproduced from van Vlimmeren et al. [111] with permission.

4.4.4. External Potential Dynamics

An alternative dynamical evolution strategy to the DDFT diffusion equation (4.148) is the 'external potential dynamics' model (EPD), in which the chemical potential fields are viewed as the fundamental variables rather than the density fields. This approach was proposed by Maurits and Fraaije [74].

The external potential dynamics model is not hard to understand. The derivative of the density functional $[\rho[u^{\mathrm{qid,eff}}](\mathbf{r}_{s,i})]$ with respect to the effective external potential, $u^{\mathrm{qid,eff}}(\mathbf{r}_{q,i})$, in the quasi-ideal system is the intramolecular two-body correlator $P(\mathbf{r}_{s,i}, \mathbf{r}_{q,i})$.

$$\frac{\delta \rho(\mathbf{r}_{s,i})}{\delta u^{\mathrm{qid,eff}}(\mathbf{r}_{q,i})} \equiv -\beta P(\mathbf{r}_{s,i}, \mathbf{r}_{q,i}) \tag{4.154}$$

The collective dynamics within the Rouse model [16] for polymer chains is given by [74]:

$$\frac{\partial \rho(\mathbf{r}_{s,i})}{\partial t} = \beta D_{s,i} \sum_q \nabla_{\mathbf{r}_{s,i}} \cdot \int d\mathbf{r}_{q,i} P(\mathbf{r}_{s,i}, \mathbf{r}_{q,i}) \nabla_{\mathbf{r}_q} \mu^{\mathrm{intr}}(\mathbf{r}_{q,i}) \tag{4.155}$$

DYNAMIC DENSITY FUNCTIONAL THEORY

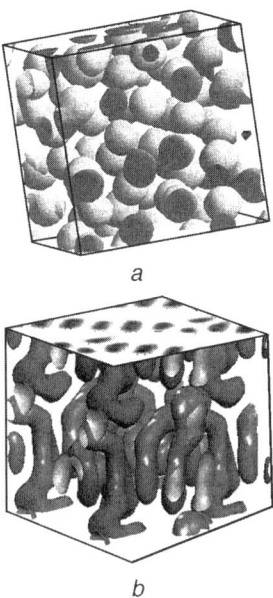

Figure 4.22. Simulated morphologies of 30 wt% $EO_{26}PO_{40}EO_{26}$ at (*a*) 40°C and (*b*) 70°C. Reproduced from Lam and Goldbeck-Wood [64] with permission.

Now let us apply the chain rule to the time derivative of $\rho(\mathbf{r}_{s,i})$, which results in:

$$\frac{\partial \rho(\mathbf{r}_{s,i})}{\partial t} = \sum_q \int d\mathbf{r}_q \frac{\delta \rho(\mathbf{r}_{s,i})}{\delta u^{\mathrm{qid,eff}}(\mathbf{r}_{q,i})} \frac{\partial u^{\mathrm{qid,eff}}(\mathbf{r}_{q,i})}{\partial t} \qquad (4.156)$$

so that, with the definition of the two-body correlator P [equation (4.154)] one obtains:

$$\frac{\partial \rho(\mathbf{r}_I)}{\partial t} = -\beta \sum_q \int d\mathbf{r}_{q,i} P(\mathbf{r}_{s,i}, \mathbf{r}_{q,i}) \frac{\partial u^{\mathrm{qid,eff}}(\mathbf{r}_{q,i})}{\partial t} \qquad (4.157)$$

Following Maurits and Fraaije we insert this expression for the time derivative of the density into the left-hand side of the collective Rouse dynamics equation (4.155):

$$\sum_q \int d\mathbf{r}_{q,i} P(\mathbf{r}_{s,i}, \mathbf{r}_{q,i}) \frac{\partial u^{\mathrm{qid,eff}}(\mathbf{r}_{q,i})}{\partial t} = -D_{s,i} \sum_q \nabla_{\mathbf{r}_{s,i}}$$

$$\times \int d\mathbf{r}_{q,i} P(\mathbf{r}_{s,i}, \mathbf{r}_{q,i}) \nabla_{\mathbf{r}_{q,i}} \mu^{\mathrm{intr}}(\mathbf{r}_{q,i}) \qquad (4.158)$$

Maurits and Fraaije point out that this equation is still exact. Since the correlator matrix, P, occurs both on the left and right-hand sides, essentially both the forces and the fluxes are transformed from u to ρ space. In the linear regime the operators ∇ and $P(\mathbf{r}_{s,i}, \mathbf{r}_{q,i})\nabla$ commute for the dot inner product since the two-body correlators are translationally invariant and equation (4.158) reduces to:

$$\sum_q \int d\mathbf{r}_{q,i} P(\mathbf{r}_{s,i}, \mathbf{r}_{q,i}) \frac{\partial u^{\text{qid,eff}}(\mathbf{r}_{q,i})}{\partial t} = -D_{s,i} \sum_q \int d\mathbf{r}_{q,i} P(\mathbf{r}_{s,i}, \mathbf{r}_{q,i}) \nabla^2_{\mathbf{r}_{q,i}} \mu^{\text{intr}}(\mathbf{r}_{q,i}) \qquad (4.159)$$

The entire relation is expressed in u space by applying the inverse operator to the integration operation on the left-hand side of equation (4.159), which exists since, for a system in equilibrium, the ρ–u relationship is bijective. The result is:

$$\frac{\partial u^{\text{qid,eff}}(\mathbf{r}_{s,i})}{\partial t} = -D_i \nabla^2 \mu^{\text{intr}}(\mathbf{r}_{s,i}) \qquad (4.160)$$

where we have introduced D_i, the diffusion coefficient of the polymer chain, as in the collective Rouse dynamics all beads have the diffusion coefficient of the chain as a whole. Equation (4.160) is also valid for some nonlinear regimes as Maurits and Fraaije proved, for instance in the case of a homogeneous phase.

The Langevin equation in the EPD approach thus becomes with equation (4.146), including the functional dependencies:

$$\frac{\partial u^{\text{qid,eff}}[\rho](\mathbf{r}_{s,i})}{\partial t} = D_i \nabla^2 \{u^{\text{int}}[\rho](\mathbf{r}_{s,i}) - u^{\text{qid,eff}}[\rho](\mathbf{r}_{s,i})\} + \eta(\mathbf{r}_{s,i}) \qquad (4.161)$$

As a consequence of the bijective relation between the effective external fields and the density distributions in the 'quasi-ideal' system, that is $\rho[u^{\text{qid,eff}}]$, we obtain the EPD diffusion equation in closed form:

$$\frac{\partial u^{\text{qid,eff}}(\mathbf{r}_{s,i})}{\partial t} = D_i \nabla^2 \{u^{\text{int}}[u^{\text{qid,eff}}](\mathbf{r}_{s,i}) - u^{\text{qid,eff}}(\mathbf{r}_{s,i})\} + \eta(\mathbf{r}_{s,i}) \qquad (4.162)$$

This diffusion equation is much easier to implement in a computer program and is an order of magnitude faster; the ∇^2 operator can be implemented very efficiently, either by quadrature rules or by discrete fast Fourier transforms. Maurits and Fraaije applied the EPD formalism to a melt of a symmetric block copolymer A_8B_8 where the A and B beads are incompatible ($\chi_{AB} = 1.0$) and compared it with the local coupling diffusion equation (4.151). Their results are shown in Figure 4.23. The EPD formalism produces larger structures and fewer defects. It is also noted that the structures are formed in fewer time steps. Both simulations were started from a homogeneous melt.

The interesting point of the external potential dynamics formalism is that, according to Reister et al. [90], it describes the dynamics of polymer mixtures better than the original DDFT diffusion equation (4.148). Reister et al. performed Monte Carlo

 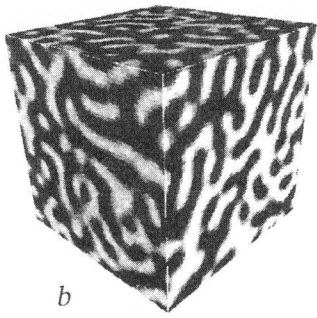

Figure 4.23. Morphology of compressible A_8B_8 block copolymer melt after 800 time steps. (*a*) Local kinetic coupling model. (*b*) External potential dynamics model. Figure taken from reference [7].

simulations during the early stage of spinodal decomposition of a symmetric polymer mixture and compared it with EPD results for the same system. The quantitative comparison of the relaxation rate of the global structure factor shows that the EPD model gives a much better agreement for the kinetic coefficient than the local coupling model. At this point we should also refer to a review article of Ganesan et al. [38] in which the potential of applying field theoretic methods to polymer melts and blends is discussed and compared with DDFT. Discussing field theoretic simulations is beyond the scope of this chapter; see a splendid article by Moreira and Netz [79]. However, the remark of Ganesan et al. [38]: 'Equilibrium structures computed with DDFT in fact represent solutions of the SCFT equations' should be taken very seriously. By SCFT is meant self-consistent field theories, as discussed in the previous section. The advantage of the DDFT approach is that no *a priori* symmetry is applied to the system of interest, such as planar or spherical symmetry. However, the question remains: do the obtained simulation trajectories have any meaning?

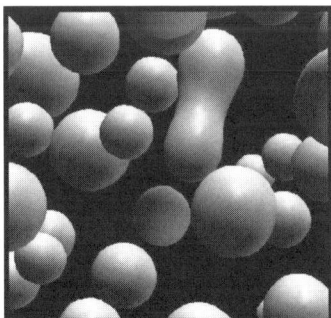

Figure 4.24. DDFT simulation result after 20,000 time steps for a simple A_2B_8–water mixture with $\chi_{Aw} = 0$, $\chi_{Bw} = 2.5$, $\chi_{AB} = 2.0$. The concentration of the block copolymer is 10%. At this instant of time we observe small micelles that are in the process of dissolving and the coagulation of small micelles.

From the work of Reister et al. [90] there seems to be one, at least for a simple block copolymer melt.

However, my personal opinion is that one should be very careful in interpreting the trajectories. I have often observed for systems in a state of dynamic equilibrium, like micelles, that DDFT (after equilibration) does produce the correct overall equilibrium dynamics of the system. After the micelles have initially been formed after a quench, for example block copolymers in water, one observes coagulation of micelles into larger aggregates, growth of micelles at the cost of smaller ones, and finally the process of splitting of the larger objects into small micelles. Figure 4.24 shows DDFT results for such a system at a certain instant of time, that is, long after equilibration. DDFT is thus able to reproduce these fundamental aspects of micellar solutions as we learned about them at school. The question is, does DDFT tell us anything about the initial stage of micelle formation from a homogeneous solution? The answer is no, of course not. The DDFT of Fraaije uses a mean-field approach to the correlation functional, so we will learn *nothing* about the dynamics in the dilute regime where two or 10 block copolymers start to aggregate into micelles. The mean-field chemical potential is completely out of order for this initial stage of micelle formation. In concentrated polymer blends the situation seems to be different, as Reister et al. [90] demonstrated. However, to conclude this section, although we cannot principally expect DDFT simulations to reproduce the real dynamics of the system under consideration, it does gives the possibility of freeing ourselves from geometrical restrictions as introduced in SCFT simulations in order to investigate equilibrium conditions. We have shown DDFT simulation results for a simple binary A_2B_8–water mixture at 10 percent block copolymer concentration (fig. 4.24). The A_2B_8 block copolymer is, however, a rather hydrophobic surfactant as it contains two hydrophilic beads (A) and eight hydrophobic beads (B). We should expect rather interesting results at higher surfactant

Figure 4.25. DDFT simulation result after 20,000 time steps for a simple A_2B_8–water mixture with $\chi_{Aw} = 0$, $\chi_{Bw} = 2.5$, $\chi_{AB} = 2.0$. The concentration of block copolymer is (left) 50% (right) 70%. At this instant of time for the 70% mixture we observe reverse micelles, that is, water droplets in the block–copolymer melt stabilized by the copolymer itself. It is interesting to note that the reverse micelles coagulate.

concentrations, for instance, at 50 or 70 percent of this block copolymer in water (Figure 4.25).

REFERENCES

[1] P. Alexandridis and T.A. Hatton. *Colloid Surface A*, **96**(1–2): 1, 1995.
[2] P. Alexandridis, U. Olsson, and B. Lindmann. *Macromolecules*, **28**: 7700, 1995.
[3] P. Alexandridis, U. Olsson, and B. Lindmann. *J. Phys. Chem.*, **100**: 280, 1996.
[4] Y. Almog and J. Klein. *J. Colloid Interface Sci.*, **106**: 548, 1985.
[5] V.A. Baulin and A. Halperin. *Macromolecules*, **35**: 6432, 2002.
[6] S. Bekiranov, R. Bruinsma, and P. Pincus. *Phys. Rev E*, **55**(1): 577, 1997.
[7] M. Bercea, M. Cazacu, and B.A. Wolf. *Macromol. Chemistry and Physics*, **204**: 1371, 2003.
[8] K. Binder. *Adv. Polym. Sci.*, **112**: 181, 1994.
[9] M.R. Boehmer and L.K. Koopal. *Langmuir*, **8**: 2649, 1992.
[10] D. Broseta, G.H. Fredrickson, E. Helfand, and L. Leibler. *Macromolecules*, **23**: 132, 1990.
[11] M. Carlsson, D. Hallen, and P. Linse. *J. Chem. Soc. Faraday Trans.*, **91**: 2081, 1995.
[12] E.M. Casassa and Y. Tagami. *Macromolecules*, **2**: 4, 1969.
[13] J. Crank. *The Mathematics of Diffusion*. Oxford Science Publications, Oxford, 1997.
[14] A. Dabdub, A. Klein, and L.H. Sperling. *J. Polym. Sci. Part B*, **30**: 787, 1992.
[15] J.E. Dennis and R.B. Schnabel. *Numerical Methods for Unconstrained Optimization and Nonlinear Equations*. Prentice-Hall, Engelwood Cliffs, NJ, 1983.
[16] M. Doi and S.F. Edwards. *The Theory of Polymer Dynamics*. Number 73 in International Series of Monographs on Physics. Oxford Science Publications, Oxford, 1986.
[17] E.E. Dormidontova. *Macromolecules*, **35**: 987, 2002.
[18] A. Eliassi, H. Modarress, and G.A. Mansoori. *J. Chem. Eng. Data*, **44**: 52, 1999.
[19] R. Evans. *Fundamentals of Inhomogeneous Fluids*.
[20] R. Evans. *Adv. Phys.*, **28**: 143, 1979.
[21] O.A. Evers, E. Hädicke, and G. Ley. *Colloids Surf.*, **90**: 135, 1994.
[22] O.A. Evers, G. Ley, and E. Hädicke. *Macromolecules*, **26**: 2885, 1993.
[23] O.A. Evers, J.M.H.M. Scheutjens, and G.J. Fleer. *Macromolecules*, **23**: 5221, 1990.
[24] F. Farassat. Introduction to generalized functions with applications in aerodynamics and aeronautics. Technical Paper 3428, NASA, Langley Research Center, Hampton, VA, 1994.
[25] M.E. Fisher. *Rev. Mod. Phys.*, **46**: 597, 1974.
[26] G.J. Fleer, M.A. Cohen-Stuart, J.M.H.M. Scheutjens, T. Cosgrove, and B. Vincent. *Polymers at Interfaces*. Chapman and Hall, London, 1993.
[27] G.J. Fleer and J.M.H.M. Scheutjens. *Croat. Chem. Acta*, **477**: 1, 1987.
[28] G.J. Fleer, J.M.H.M. Scheutjens, and M.A. Cohen Stuart. *Colloids Surf.*, **31**: 1, 1988.
[29] P.J. Flory. *J. Chem. Phys.*, **9**: 660, 1941.
[30] P.J. Flory. *J. Chem. Phys.*, **10**: 51, 1942.

[31] P.J. Flory. *Statistical Mechanics of Chain Molecules*. Oxford University Press, Oxford, 1969.

[32] J.G.E.M. Fraaije. *J. Chem. Phys.*, **99**: 9202, 1993.

[33] J.G.E.M. Fraaije, B.A.C. van Vlimmeren, N.M. Maurits, A.V Zvelindovsky, M. Postma, O.A. Evers, C. Hoffmann, P. Altevogt, and G. Goldbeck-Wood. *J. Chem. Phys.*, **106**: 4260, 1997.

[34] M. Frigo and S.G. Johnson. *Proc. ICASSP*, **3**: 1381, 1998.

[35] M. Galassi, J. Davies, J. Theiler, B. Gough, G. Jungman, M. Booth, and F. Rossi. *GNU Scientific Library*; www.gnu.org/software/gsl, 2003.

[36] C.W. Gardiner. *Handbook of Stochastic Methods*. Springer, Berlin, 1990.

[37] I.M. Gelfand and G.E. Shilov. *Generalized Functions. Volume I—Properties and Operations*. Academic Press, London, 1964.

[38] V. Ganesan, G.H. Fredrickson and F. Drolet. *Macromolecules*, **35**: 16, 2002.

[39] D.G. Hall and B.A. Petica. *Nonionic Surfactants*. Marcel Dekker, New York, 1963.

[40] J.P. Hansen and I.R. McDonald. *Theory of Simple Liquids*. Acadamic Press, London, 1986.

[41] E. Helfand. *J. Chem. Phys.*, **62**: 999, 1975.

[42] E. Helfand and A.M. Sapse. *J. Chem. Phys.*, **62**: 1327, 1975.

[43] E. Helfand and Y. Tagami. *J. Polym. Sci. B*, **9**: 741, 1971.

[44] E. Helfand and Y. Tagami. *J. Chem. Phys.*, **56**: 3592, 1971.

[45] E. Helfand and Y. Tagami. *J. Chem. Phys.*, **57**: 1812, 1972.

[46] D. Henderson. *Fundamentals of Inhomogeneous Fluids*. Marcel Dekker, New York, 1992.

[47] T.L. Hill. *Thermodynamics of Small Systems*. Benjamin, New York, 1963.

[48] T.L. Hill. *An Introduction to Statistical Thermodynamics*. Dover, New York, 1986.

[49] J.D. Honeycutt. *Theor. Polym. Sci.*, **8**(1/2): 1, 1998.

[50] M.L. Huggins. *J. Chem. Phys.*, **9**: 440, 1941.

[51] M.L. Huggins. *J. Phys. Chem.*, **46**: 151, 1942.

[52] M.L. Huggins. *J. Am. Chem. Soc.*, **64**: 1712, 1942.

[53] J.N. Israelachvili, M. Tirrel, J. Klein, and Y. Almog. *Macromolecules*, **17**: 204, 1984.

[54] B. Jönsson, B. Lindman, K. Holmberg, and B. Kronberg. *Surfactants and Polymers in Aqueous Solution*. Wiley, New York, 1998.

[55] R.P. Kanwal. *Generalized Functions—Theory and Techniques*. Academic Press, London, 1983.

[56] G.J. Karlstrom. *J. Chem. Phys.*, **89**: 4962, 1985.

[57] W. Kohn and L.J. Sham. *Phys. Rev. A*, **145**: 1133, 1965.

[58] R. Koningsveld. *Adv. Colloid Interface Sci.*, **2**: 151, 1968.

[59] R. Koningsveld, L.M. Kleintjes, and E. Nies. *Croat. Chim. Acta*, **60**: 53, 1987.

[60] R. Koningsveld, L.M. Kleintjes, and H.M. Schoffelers. *Pure Appl. Chem.*, **39**: 1, 1974.

[61] W. Kuhn. *Kolloid-Z.*, **76**: 258, 1936.

[62] W. Kuhn. *Kolloid-Z.*, **87**: 3, 1939.

[63] R.H. Lacombe and I. Sanchez. *J. Phys. Chem.*, **80**: 2568, 1976.

[64] Y-M. Lam and G. Goldbeck-Wood. *Polymer*, **44**: 3593, 2003.

[65] L.D. Landau and E.M. Lifschitz. *Statistical Physics*. Pergamon Press, Oxford, 1958.
[66] F.A.M. Leermakers and J.M.H.M Scheutjens. *J. Chem. Phys.*, **89**: 3264, 1988.
[67] L. Leibler. *Macromolecules*, **13**: 1602, 1980.
[68] H. Li, Y. Yang, R. Fujitsukad, T. Ougizawa, and T. Inoue. *Polymer*, **40**: 927, 1999.
[69] A. Lindner. *Grundkurs Theoretische Physik*. Teubner Studienbücher, 1994.
[70] P. Linse. *Amphiphilic Block Copolymers*.
[71] G.N. Malcolm and J.S. Rowlinson. *Trans. Faraday. Soc.*, **53**: 921, 1957.
[72] M.W. Matsen and M. Schick. *Curr. Opin. Colloid Interface Sci.*, **1**: 329, 1996.
[73] N.M. Maurits, P. Altevogt, O.A. Evers, and J.G.E.M. Fraaije. *Comp. Polym. Sci.*, **6**: 1, 1996.
[74] N.M. Maurits and J.G.E.M. Fraaije. *J. Chem. Phys.*, **107**: 5879, 1997.
[75] N.M. Maurits, B.A.C. van Vlimmeren, and J.G.E.M. Fraaije. *Phys. Rev. E*, **56**: 816, 1997.
[76] W.E. McMullen. *Physics of Polymer Surfaces and Interfaces*. Butterwoirth–Heinemann, Boston, MA, 1992.
[77] D.A. McQuarrie. *Statistical Mechanics*. Harper Collins, London, 1976.
[78] D.N. Mermin. *Phys. Rev. A*, 137: 1441, 1965.
[79] A.G. Moreira and R.R. Netz. *Electrostatic Effects in Soft Matter and Biophysics*.
[80] D.C. Morse and G.H. Fredrickson. *Phys. Rev. Lett.*, **73**: 3235, 1994.
[81] K. Mortensen. *Europhys. Lett.*, **19**(7): 599, 1992.
[82] P. Munk. *Introduction to Macromolecular Science*. J Wiley, New York, 1989.
[83] J. Noolandi and K.M. Hong. *Macromolecules*, **15**: 482, 1982.
[84] P. Hohenberg and W. Kohn, *Phys. Rev. B*, **136**: 864, 1964.
[85] H.-M. Petri, N. Schuld, and B.A. Wolf. *Macromolecules*, **28**: 4975, 1995.
[86] H.M. Petri, R. Horst, and B.A. Wolf. *Polymer*, **37**: 2709, 1996.
[87] H.M. Petri and B.A. Wolf. *Macromolecules*, **27**: 2714, 1994.
[88] J.G. Powels, G. Rickayzen, and M.L. Williams. *Mol. Phys.*, **64**: 33, 1988.
[89] W.H. Press, B.P. Flannery, S.A. Teukolsky, and W.T. Vetterling. *Numerical Recipies in C*. Cambridge University Press, Cambridge, 1988.
[90] E. Reister, M. Müller, and K. Binder. 2003.
[91] H. Risken. *The Focker-Planck Equation*.
[92] S. Saeki, N. Kuwahara, S. Konno, and M. Kaneko. *Macromolecules*, **6**: 246, 1973.
[93] I.C. Sanchez and R.H. Lacombe. *J. Phys. Chem.*, **80**: 2352, 1976.
[94] I.C. Sanchez and R.H. Lacombe. *Macromolecules*, **11**: 1145, 1978.
[95] J.M.H.M. Scheutjens and G.J. Fleer. *J. Chem. Phys.*, **83**: 1619, 1979.
[96] J.M.H.M. Scheutjens and G.J. Fleer. *J. Chem. Phys.*, **84**: 178, 1980.
[97] J.M.H.M. Scheutjens and G.J. Fleer. *Macromolecules*, **18**: 1882, 1985.
[98] J.M.H.M. Scheutjens and G.J. Fleer. *J. Colloid Interface Sci.*, **111**: 504, 1986.
[99] F. Schmid. *J. Chem. Phys.*, **104**: 9191, 1996.
[100] N. Schuld and B.A. Wolf. Polymer—solvent interaction parameters. In *Polymer Handbook*. Wiley, New York, 1999.

[101] L. Schwartz. Theorie des distributions. In *Actualities Scientiques et Industrielles*, Vols I and II. Hermann and Cie, Paris, 1957.

[102] R.M. Sok and O.A. Evers. to be published.

[103] I. Stakgold. *Boundary Value Problems of Mathematical Physics*, Vol. II. Macmillan, New York, 1968.

[104] H.E. Stanley. An *Introduction to Phase Transitions and Critical Phenomena*. Oxford University Press, Oxford, 1971.

[105] S. Stryuk and B.A. Wolf. *Macromol. Chem. Phys.*, **204**: 1948, 2003.

[106] S.W. Sides and G.H. Fredrickson. *Polymer*, **44**: 5859, 2003.

[107] Z. Tong, Y. Einaga, T. Kitagawa, and H. Fujita. *Macromolecules*, **22**: 450, 1989.

[108] P.P.A.M. van der Schoot and F.A.M. Leermakers. *Macromolecules*, **21**: 1876, 1988.

[109] B. van Lent and J.H.M.H. Scheutjens. *Macromolecules*, **22**: 1931, 1989.

[110] B.A.C. van Vlimmeren and J.G.E.M Fraaije. *Comp. Phys. Commun.*, **99**: 21, 1996.

[111] B.A.C. van Vlimmeren, N.M. Maurits, A.V. Zvelindovsky, G.J.A. Sevink, and J.G.E.M. Fraaije. *Macromolecules*, **32**: 646, 1999.

[112] B.A. Wolf. *Macromol. Chem. Phys.*, **204**: 1381, 2003.

[113] D.T. Wu, G.H. Fredrickson, J-P. Carton, A. Ajdari, and L. Leibler. *J. Polym. Sci. Part B: Polym. Phys.*, **33**: 2373, 1995.

[114] S.I. Yang, A. Klein, L.H. Sperling, and E.F. Casassa. *Macromolecules*, **23**: 4582, 1990.

[115] Y. Zhang, W. Li, B. Tang, S. Ge, X. Hu, M.H. Rafailovich, J.C. Sokolov, D. Gersappe, D.G. Peiffer, Z. Li, A.J. Dias, K.O. McElrath, M.Y Lin, S.K. Satija, S.G. Urquhart, and H. Ade. *Polymer*, **42**: 9133, 2001.

[116] A.V. Zvelindovsky, B.A.C. van Vlimmeren, G.J.A. Sevink, N.M. Maurits, and J.G.E.M. Fraaije. *J. Chem Phys.*, **109**: 8751, 1998.

5

PREDICTION OF MECHANICAL PROPERTIES OF SEMICRYSTALLINE POLYMERS

A. RAPHAEL AND I. ALIG
Deutsches Kunststoff-Institut, Schlossgartenstr. 6, 64289 Darmstadt, Germany

M. KRÖHN
Siemens VDO Automotive AG, VDO-Str. 1, 64832 Babenhausen, Germany

CONTENTS

5.1. Introduction	219
5.2. Force Field Simulations	221
5.3. Preparation of Crystalline and Amorphous Model Cells	224
5.3.1. Crystal Cell	224
5.3.2. Amorphous Cell	225
5.4. Simulation of Mechanical Properties	226
5.4.1. Elastic Constants	226
5.4.2. Simulation Approaches	228
5.5. Mechanical Properties of Isotactic Polypropylene	231
5.5.1. Mechanical Properties of the Crystalline Phase	231
5.5.2. Mechanical Properties of the Amorphous Phase	233
5.5.3. Combination of Molecular Simulation and Micromechanical Models	235

Molecular Simulation Methods for Predicting Polymer Properties, Edited by Vassilios Galiatsatos.
ISBN 0-471-46481-3 Copyright © 2005 John Wiley & Sons, Inc.

5.1. INTRODUCTION

The simulation of structure and properties of polymers on a molecular level has made tremendous progress in recent decades. However, the simulations using realistic molecular models, for example, on a quantum mechanical level, are still limited to relatively small systems. These methods can only predict properties which are defined on the length scale of the molecular structure used for the simulations. On the other hand, it is well known that many of the macroscopic properties of polymers are determined by its superstructure. For instance a large number of polymers of technical importance form semicrystalline structures. The crystalline lamellae of those polymers are arranged in superstructures, for example, the well-known spherulites, which are embedded in an amorphous matrix of the same polymer. The knowledge of the mechanical and thermal properties of semicrystalline polymers is of major importance for the simulation of technical processes, like injection or compression molding, including shrinkage and warpage effects. (e.g. by finite element programs).

Even for semicrystalline polymers of high technical relevance, like polyethylene (PE) or polypropylene (PP), the structure–property relation is not yet fully understood, because the morphology of semicrystalline polymeric materials is extremely dependant on the preparation (temperature, pressure, shear rate, cooling rate, etc.). Furthermore microscopic anisotropies resulting from processing (e.g. deformed spherulites, shish kebab structures) have not yet been included in macroscopic finite element calculations because of the lack of experimental data for mechanical or thermal properties depending on prehistory. Molecular simulations in combination with micromechanical models can assist in providing detailed data for a large number of different structures. Since the direct experimental determination of the mechanical properties of the crystalline phase is difficult, this opens up an interesting field for atomistic simulation.

To model the composite structure of crystalline domains in an amorphous matrix, one would have to cover an area of approximately several cubic micrometers. Although very recently the possibility of simulating structure formation in semicrystalline polymers using coarse-grained models has been proved [1], quantitative simulations of macroscopic properties of the semicrystalline superstructure on an atomistic level are still beyond the scope of current computer capacities. An alternative way to predict the mechanical behavior of semicrystalline polymers is to simulate the amorphous and crystalline phases separately at the length scale of several chains or chain fragments and employ micromechanical mixing rules to calculate the properties of the 'composite.'

In the past, several models have been developed to calculate the mechanical properties of lamella structures, uniaxial fibers in a matrix [2, 3] and spherical inclusions in a matrix [4]. While the fiber–matrix model is capable of describing the final spherulitic material [5], the spherical inclusion model can be applied to describe the growth of spherulites in the polymer melt [6]. For the prediction of the mechanical properties of semicrystalline polymers, the knowledge of the elastic constants of amorphous and crystalline phase is necessary. By combining micromechanical models for the superstructure with mechanical properties calculated for the crystalline and amorphous phase on a molecular level, it is possible to predict the macroscopic mechanical

FORCE FIELD SIMULATIONS

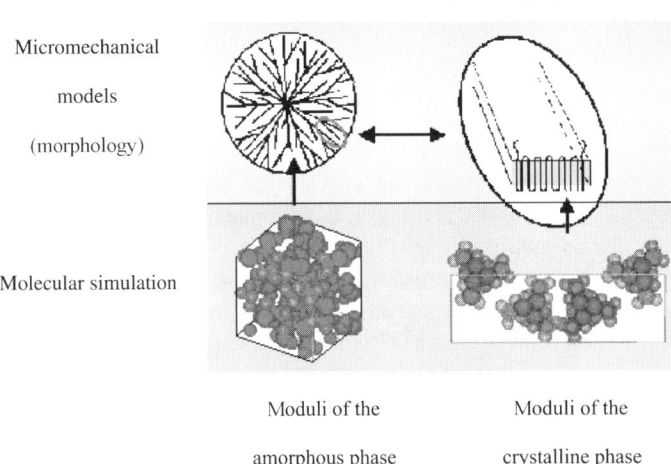

Figure 5.1. Schematic illustration of the combination of molecular simulation and micromechanical models. After the amorphous and crystalline unit cells have been constructed and elastic constants have been calculated separately for the crystalline and amorphous cells, the prediction of the mechanical properties of the semicrystalline material is performed by a successive calculation of the elastic constants of the lamella, the spherulite, and finally the composite.

properties in a straightforward manner. The procedure is schematically illustrated in Figure 5.1.

In this way, combination of molecular simulations and micromechanical models can extend the databases for process simulations, including prehistory effects. This may be less cost-intensive than experiments on a large number of materials prepared under different processing conditions. Although until now boundary effects and interphases have not been included, we feel that the combination of molecular simulation, micromechanical models, and process simulations can help to close the gap between modeling on a molecular level and the need for material properties data for engineering.

After a short introduction on forcefield simulations, the preparation of amorphous and crystalline cells is described. Next the basics of simulation of elastic properties and the different methods are presented. The application of these methods is demonstrated for the crystalline and amorphous phases of polypropylene. Finally the elastic constants derived by molecular simulation are included in micromechanical models to predict the mechanical properties of semicrystalline polymers.

5.2. FORCE FIELD SIMULATIONS

At the most basic simulation level, all nuclear and electronic degrees of freedom must be treated quantum mechanically. Nevertheless, when simulating large systems such as condensed polymers, the goal is often to extract statistical properties that depend on an average over a set of atomic configurations. Under these circumstances the details of electronic structure are lost in the averaging processes. Therefore, bulk properties can often be extracted if a good approximation of the energy potential in which the atoms move is available.

The purpose of a *force field* is to describe this energy potential with reasonable accuracy. Chapter 2 gives a detailed explanation to the subject. Briefly, and for the purposes of this chapter, a force field describes molecules by a combination of intramolecular interactions depending on bond distances, d, bond angles, Θ, dihedral torsion angles, Φ, and intermolecular interactions (van der Waals, electrostatic) depending on the distances, r, between the atoms. The atomic coordinates of a given structure combined with a forcefield create a so-called 'energy expression.' This energy expression describes the potential energy surface of a particular structure as a function of its atomic coordinates.

$$U = U_{\text{bond}}(d) + U_{\text{angle}}(\theta) + U_{\text{torsion}}(\phi) + U_{\text{coulomb}}(r) + U_{\text{vdW}}(r) \quad (5.1)$$

An important method in molecular simulation is the minimization with respect to potential energy of the system being examined. The minimization finds configurations that are stable points on the potential surface. The energy expression must be defined and evaluated for a given conformation. The conformation is then adjusted to minimize the value of the energy expression.

Once an energy expression and, if necessary, a minimized structure have been defined for the system of interest, a molecular dynamics (MD) simulation can be performed in order to obtain information on its time evolution. The principle in MD is the numerical integration of the classical Newton's equation of motion:

$$F_i(t) = m_i a_i(t) \quad (5.2)$$

The force on atom i can be computed directly from the derivative of the potential energy, V, with respect to the coordinates, r_i:

$$-\frac{\partial V}{\partial r_i} = m_i \frac{\partial^2 r_i}{\partial t^2} \quad (5.3)$$

During simulation, the equations of motion for the particles have to be integrated forward in time. The differential equation (5.3) is solved numerically by the finite-difference method. Starting from the initial coordinates and velocities at time t, the positions and velocities at time $t + \Delta t$ are calculated.

Although the initial coordinates are determined, the initial velocities are randomly generated at the beginning of a dynamics run, according to the desired temperature. The macroscopic temperature, T, is related to the microscopic description through the kinetic energy, E_{kin}, which is calculated from the velocities of the atoms, $\mathbf{v}_i$.

$$\tfrac{3}{2} N k_B T = E_{\text{kin}} = \sum_i^N \tfrac{1}{2} m_i |\mathbf{v}_i|^2 \tag{5.4}$$

The distribution of velocities in a system is given by the Maxwell–Boltzmann equation, which expresses the probability that a molecule i has a velocity of $\mathbf{v}_i$ when the system is at temperature T. This distribution does not remain constant as the simulation continues. During molecular dynamics, kinetic and potential energy are exchanged and as a consequence the temperature changes. To maintain the correct temperature, the computed velocities have to be rescaled periodically.

In order to simulate a virtually infinite system, periodic boundary conditions have to be defined. The cell is replicated in three dimensions, thus the simulated cell should have translational periodicity. The molecules near the surface now interact with molecules in adjacent cells.

If periodic boundary conditions are applied to a molecular system, volume and pressure can be defined. Pressure is calculated using the virial theorem. Pressure, p, temperature, T, volume, V, and internal virial can be related in the following way:

$$p = \frac{1}{V} \left(2 \sum_i^N \tfrac{1}{2} m_i \mathbf{v}_i \mathbf{v}_i^T + \sum_i^N \mathbf{r}_i \mathbf{f}_i^T \right) \tag{5.5}$$

where $\mathbf{f}_i$ is the force acting on atom i, with position $\mathbf{r}_i$ and velocity $\mathbf{v}_i$.

From the definition of the pressure tensor p, the instantaneous hydrostatic pressure p is calculated as one-third of the trace of the pressure tensor. Sometimes the stress tensor $\sigma = -p$ is used instead of the pressure tensor. The pressure can be changed by changing the coordinates of the particles and the size of the cell under periodic boundary conditions.

Depending on which state variables (pressure, p, temperature, T, volume, V, etc.) are kept fixed, different statistical ensembles can be generated. A variety of structural, energetic, and dynamic properties can then be calculated from the averages or the fluctuations of these quantities for the actual ensemble. The constant-volume, constant-temperature ensemble (NVT) is suitable for simulating low-density model cells during the initial stages of system setup. It is also the ensemble of choice when local properties such as interface energies or diffusion coefficients are concerned. Cell vectors are kept constant while temperature is controlled by coupling velocities to an external heat bath. The constant-pressure, constant-temperature ensemble (NPT) simulations allow changes in cell size and shape. Not only the structure of the model cell contents but also its volume and density can equilibrate. This yields more realistic structures than the NVT ensemble. In addition, one can easily compute thermodynamical properties such as heat capacity or mechanical moduli from density fluctuations.

Our simulations were done with the InsightII^R molecular modeling system and the Discover^R program, using the *pcff* and *cvff* forcefields.

5.3. PREPARATION OF CRYSTALLINE AND AMORPHOUS MODEL CELLS

5.3.1. Crystal Cell

Modeling of the elastic constants of the crystalline cells is based on exact knowledge of the crystalline structure, because errors in intermolecular distances and cell parameters of the crystal cell may result in large errors in the simulated moduli. Until now the elastic constants have been modeled in this way only for a limited number of polymeric crystals [7].

In order to determine the structure of the crystalline unit cell, one can use wide angle X-ray scattering. Because of the isotropic orientation of the unit cells in spherulite structures, which are typical for the supermolecular arrangement of the crystalline phases in many semicrystalline polymers, the data gained from scattering are comparable to powder diffraction and deliver only a small number of characteristic peaks.

Construction of a crystal unit cell is a two-step process: First, an elementary cell has to be found, which fits the peak positions in the X-ray diagram. In the next step the molecular basis and intermolecular distances have to be arranged and adjusted to match intensities. While a first guess on cell shape and the approximation of atom positions is usually obtained rapidly, the process of adjusting the calculated intensities to the experiment by varying atom positions is more time-consuming.

According to Bragg's theory and with knowledge of the atomic form factors, a geometrical structure factor can be calculated from the positions of the atoms. The structure factor:

$$F_{hkl} = \sum_{j=1}^{n} f_j(\Delta \mathbf{k}) \exp(i\Delta \mathbf{k} \mathbf{d}_j) \tag{5.6}$$

depends mainly on the relative positions, $\mathbf{d}_j$, of the atoms, while the atomic form factor, $f_j(\Delta \mathbf{k})$, is the Fourier transform of the electron density distribution function, $\rho(\mathbf{r})$.

$$f_j(\Delta \mathbf{k}) = -\frac{1}{e} \int d\mathbf{r} \exp(i\Delta \mathbf{k} \mathbf{r}) \rho(\mathbf{r}) \tag{5.7}$$

where $\Delta \mathbf{k}$ is the difference in wave vector of the penetrating radiation.

As an example Figure 5.2 shows the final result of such a refinement for polyether-ether-ketone (PEEK). The calculated X-ray pattern of the final structure is in very good agreement with the experimental one.

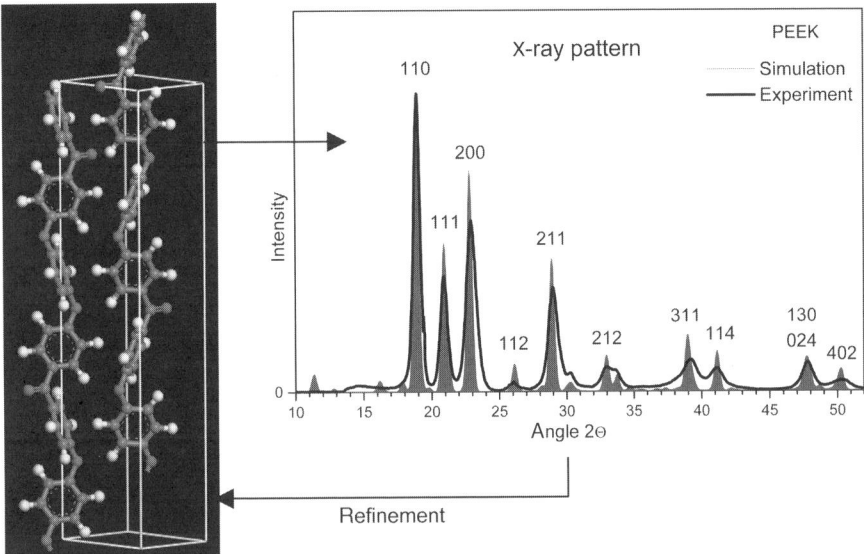

Figure 5.2. Crystal cell for PEEK (left) and experimental and simulated X-ray pattern (right) after refinement of the cell.

5.3.2. Amorphous Cell

Constructing the initial configuration of the simulation cell of an amorphous polymeric system requires some special considerations in contrast to crystalline or low molar mass systems. In the case of polymers, it is important to start with a configuration in which the global characteristics of the chain for the material under consideration are represented.

The conformational statistics of unperturbed chains are well described by the rotational isomeric state (RIS) theory [8], which can be used as a basis for generating chains in solution or bulk systems. All degrees of freedom except torsions in the backbone are neglected. In a first step, the first bond is placed in the cell in some random orientation. Then the chain is generated by stepwise adding step by step new bonds to the backbone. Unfortunately, the RIS description of a molecule does not explicitly prohibit overlapping of atoms separated by more than a few backbone bonds. Theodorou and Suter [9] overcome these limitations for generating an 'amorphous cell' while retaining the advantages of RIS, by taking long-range interactions into account.

Building cells at or near the known experimental density may result in intermolecular overlaps of a chain and its images due to the periodic boundary conditions. In such a case, the initial construction of the cell is performed at a reduced density, followed by compression to the experimental density, using an NPT molecular dynamics with stepwise increasing pressure.

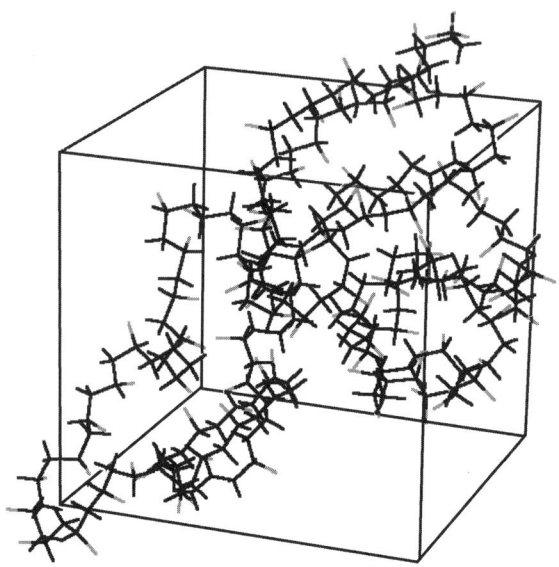

Figure 5.3. Amorphous cell of isotactic polypropylene containing one chain consisting of 100 monomers.

In the final stage, it is essential to equilibrate the system by the use of molecular dynamics to achieve uniform space filling. Figure 5.3 shows an amorphous cell containing one isotactic polypropylene chain consisting of 100 monomers, built in the manner described above.

5.4. SIMULATION OF MECHANICAL PROPERTIES

5.4.1. Elastic Constants

The application of stress to a body results in a change in the relative positions of the particles within the body. For small deformations, the relationship between the stress, σ, and strain, ε, can be expressed by a generalized Hooke's law:

$$\sigma = C\varepsilon \tag{5.8}$$

where σ is the stress tensor, ε is the strain tensor and C is the tensor of the elastic stiffness coefficients, from which all elastic constants can be derived. As both the stress and strain tensor are symmetric, it is possible to simplify this expression to a vector notation, the so-called Voigt notation:

$$\sigma_i = C_{ij}\varepsilon_j \tag{5.9}$$

SIMULATION OF MECHANICAL PROPERTIES 227

Here σ_i and ε_i are the six components of the stress and strain vectors. Elements 1–3 describe the tensile stress–strain and elements 4–6 describe the shear stress–strain. The 6×6 stiffness matrix C_{ij} is also symmetric, so a maximum of 21 coefficients are required for a full description to describe the stress–strain behavior.

An anisotropic crystalline solid (in our case the crystalline phase of a semicrystalline polymer) requires the whole set of independent elastic coefficients for calculation of the mechanical properties. Often the inverse stiffness matrix, which is called the compliance matrix, is used:

$$\mathbf{S} = \mathbf{C}^{-1} \tag{5.10}$$

The components of the compliance matrix $\mathbf{S}$ are related to the components of the elastic moduli for different directions.

$$E_{ii} = \frac{1}{S_{ii}}, \quad i \in [1, 2, 3]$$

$$G_{jk} = \frac{1}{S_{ii}}, \quad i \in [4, 5, 6], \; i - 3 \neq j, k \tag{5.11}$$

$$v_{ij} = -\frac{S_{ij}}{S_{jj}}, \quad i, j \in [1, 2, 3]$$

A given crystal symmetry can reduce the number of independent constants. Equations (5.12) and (5.13) show the stiffness matrix for a monoclinic and an orthorhombic symmetry:

$$\mathbf{C}_{\text{monocl.}} = \begin{pmatrix} C_{11} & C_{12} & C_{13} & 0 & C_{15} & 0 \\ C_{12} & C_{22} & C_{23} & 0 & C_{25} & 0 \\ C_{13} & C_{23} & C_{33} & 0 & C_{35} & 0 \\ 0 & 0 & 0 & C_{44} & 0 & C_{46} \\ C_{15} & C_{25} & C_{35} & 0 & C_{55} & 0 \\ 0 & 0 & 0 & C_{46} & 0 & C_{66} \end{pmatrix} \tag{5.12}$$

$$\mathbf{C}_{\text{orthorh.}} = \begin{pmatrix} C_{11} & C_{12} & C_{13} & 0 & 0 & 0 \\ C_{12} & C_{22} & C_{23} & 0 & 0 & 0 \\ C_{13} & C_{23} & C_{33} & 0 & 0 & 0 \\ 0 & 0 & 0 & C_{44} & 0 & 0 \\ 0 & 0 & 0 & 0 & C_{55} & 0 \\ 0 & 0 & 0 & 0 & 0 & C_{66} \end{pmatrix} \tag{5.13}$$

For an isotropic material, the stress–strain behavior can be fully described by only two independent coefficients, the so-called 'lame constants':

$$C_{iso} = \begin{pmatrix} \lambda + 2\mu & \lambda & \lambda & 0 & 0 & 0 \\ \lambda & \lambda + 2\mu & \lambda & 0 & 0 & 0 \\ \lambda & \lambda & \lambda + 2\mu & 0 & 0 & 0 \\ 0 & 0 & 0 & \mu & 0 & 0 \\ 0 & 0 & 0 & 0 & \mu & 0 \\ 0 & 0 & 0 & 0 & 0 & \mu \end{pmatrix} \quad (5.14)$$

From the lame constants all elastic moduli (Young's modulus, E, bulk modulus, B, shear modulus, G, and Poisson's ratio, v) can then be derived:

$$E = \frac{3\lambda + 2\mu}{\lambda/\mu + 1} \quad (5.15)$$

$$v = \frac{\lambda}{2(\lambda + \mu)} \quad (5.16)$$

$$B = \lambda + \frac{2\mu}{3} \quad (5.17)$$

$$G = \mu \quad (5.18)$$

5.4.2. Simulation Approaches

We focus here only on two simulation methods, the static deformation and the fluctuation method, to derive the elastic constants of the crystalline and amorphous phase of semicrystalline polymers. The first uses molecular mechanics and the second molecular dynamics. Both methods are schematically shown in Figure 5.4 and described in detail in the following.

5.4.2.1. Static Deformation. The first approach, described in the work of Theodorou and Suter [10, 11], uses a completely static method, without molecular dynamics. It considers the temperature only in an indirect way by using the correct density appropriate to the simulation temperature. Effects originating from changes in configurational entropy on deformation are ignored.

Starting with an energy-minimized structure, the cell is deformed by setting one of the components of the strain vector to some small value, ε, while keeping all other components fixed at zero. For the deformed cell the potential energy [equation (5.1)] and the induced stress [equation (5.5)] are calculated. This procedure is performed independently for all six components of strain; that is, the cell is subjected to six deformations, three uniaxial tensions and three pure shear deformations.

The elastic constants can then be calculated from the ratio of the applied strain to the resulting internal stress or from the second derivatives of the potential energy

SIMULATION OF MECHANICAL PROPERTIES

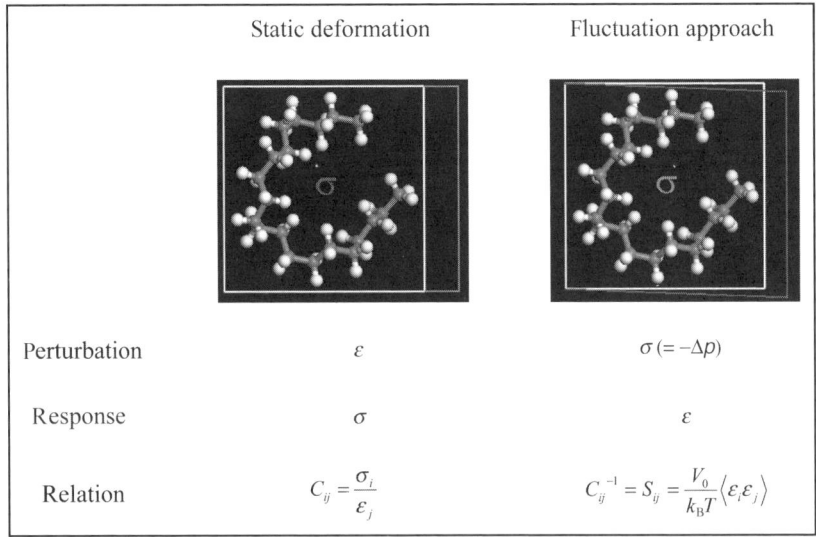

Figure 5.4. Comparison of the static deformation method and the fluctuation approach.

with respect to the strain. The first derivative of the potential energy, U, with respect to strain, ε, is the internal stress, σ, and the second derivative represents the stiffness matrix [10]:

$$C_{ij} = \frac{1}{V} \frac{\partial^2 U}{\partial \varepsilon_i \partial \varepsilon_j} = \frac{\partial \sigma_i}{\partial \varepsilon_j} \tag{5.19}$$

where V is the volume of the undeformed system.

It should be noted that the method to obtain the elastic stiffness coefficients by estimating the second derivatives of the deformation energy with respect to strain is limited. Since the six different deformations were performed independently, the potential energy depends only on a single variable and only the diagonal components can be obtained:

$$C_{ii} = \frac{1}{V} \frac{\partial^2 U(\varepsilon_i)}{\partial \varepsilon_i^2} \tag{5.20}$$

Therefore this method is only practicable for isotopic materials, where the elastic constants can be obtained from two independent coefficients. To determine the elastic properties of an anisotropic material, like crystals, all coefficients of the stiffness matrix are needed. Therefore, we have to use the stress–strain relation for the simulation of crystalline phase:

$$C_{ij} = \frac{\Delta \sigma_i}{\Delta \varepsilon_j} \tag{5.21}$$

For an isotropic amorphous solid (in our case the glassy amorphous phase) equation (5.20) suggests that two different deformations, for example, uniform hydrostatic compression and pure shear, are sufficient to fully characterize the elastic behavior. Hydrostatic compression decreases the cell lengths without changing the angles of the cell. The cell remains cubic in shape. ε is defined as the negative value of the dilatation, that is, $(-\Delta V/V)$. For simple shear the deformation is taken as the shear strain γ. The bulk (compression) (B) and the shear (G) modulus are

$$B = \frac{1}{V}\left(\frac{\partial^2 U}{\partial \varepsilon^2}\right) \quad \text{and} \quad G = \frac{1}{V}\left(\frac{\partial^2 U}{\partial \gamma^2}\right) \tag{5.22}$$

From the bulk modulus and shear modulus Young's modulus and Poisson's ratio can then be derived by equations (5.15)–(5.18). We chose this method, first described by Theodorou and Suter [10, 11] for our simulations of the amorphous systems.

5.4.2.2. Fluctuation Approach.

The fluctuation approach was developed by Parinello and Rahman [12] and applied to polymers for the first time by Gusev et al. [13]. This method makes use of the fact that the elastic constants can be obtained from the fluctuations of the cell dimensions during a molecular dynamics run. At thermal equilibrium the parameters of the cell, that is, the spatial continuation vectors, fluctuate about their mean values, due to the thermal motion of the atoms. These fluctuations are smaller, the stiffer the material is. Therefore they can be used to determine the elastic moduli. The main advantage of the fluctuation approach is that temperature and entropic effects are implicitly included.

The free (Helmholtz) energy, F, can be written as a function of the components of strain ε [14]:

$$F = F_0 + \frac{1}{2}V\sum_{i,j=1}^{6} C_{ij}\varepsilon_i\varepsilon_j \tag{5.23}$$

where V is the mean volume of the system.

The probability for a particular deformation of the cell is given by:

$$P(\varepsilon_i) \propto \exp\left\{-\frac{F-F_0}{k_B T}\right\} = \exp\left\{-\frac{V}{2k_B T}\sum_{i,j=1}^{6} C_{ij}\varepsilon_i\varepsilon_j\right\} \tag{5.24}$$

where T is the mean temperature of the system.

Averaging the product of the strain components ε_i and ε_j over all possible deformations gives:

$$\langle \varepsilon_i \varepsilon_j \rangle = \frac{\int \varepsilon_i \varepsilon_j P(\varepsilon_i, \varepsilon_j) \, d\varepsilon_i \, d\varepsilon_j}{\int P(\varepsilon_i, \varepsilon_j) \, d\varepsilon_i \, d\varepsilon_j} = \frac{k_B T}{V} C_{ij}^{-1} \tag{5.25}$$

or

$$S_{ij} = C_{ij}^{-1} = \frac{V}{k_B T}\langle \varepsilon_i \varepsilon_j \rangle \qquad (5.26)$$

where $\langle \varepsilon_i \varepsilon_j \rangle$ represents an ensemble-averaging of the different deformations. For an adequate long-molecular-dynamic run, during which the system can explore the whole 'shape' space, this ensemble averaging can be replaced by a time averaging.

5.5. MECHANICAL PROPERTIES OF ISOTACTIC POLYPROPYLENE

As an example we will show the combination of molecular simulation of the elastic constants and micromechanical models for isotactic polypropylene (i-PP). Polypropylene is one of the most important technical polymers, with a market that is still increasing. Starting with the pioneering work of Ziegler and Natta, its structure was intensively investigated in the 1950s and 1960s [15, 16]. With the development of metallocene catalysis in the 1980s, allowing polymerization of PP of defined tacticity, it again became attractive to an increasing number of researchers [17]. So far PP is one of the best investigated polymers.

5.5.1. Mechanical Properties of the Crystalline Phase

Isotactic polypropylene is known to crystallize in 3/1 helix form [17]. The helices can vary in handedness and sidegroup orientation. There are three crystalline modifications known to exist. The monoclinic α-phase, the hexagonal β- and the triclinic γ-phase. The occurrence of these modifications strongly depends on the sample preparation. We focus here on the monoclinic α-phase, which is the most common and stable one.

The cell dimensions used for the simulations were taken from the literature [17–19]. The simulation cell was then filled with preoptimized fragments of i-PP, which consist of nine monomer units. Each of these fragments has three helix turns, as shown in Figure 5.5. Starting from an initial guess according to the

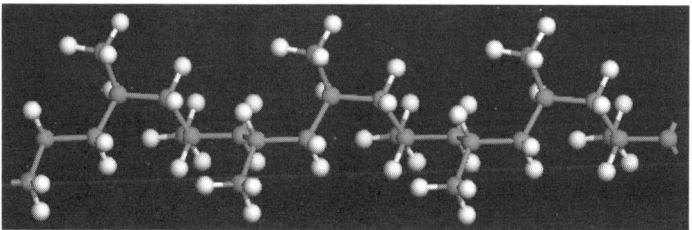

Figure 5.5. Helix of isotactic polypropylene.

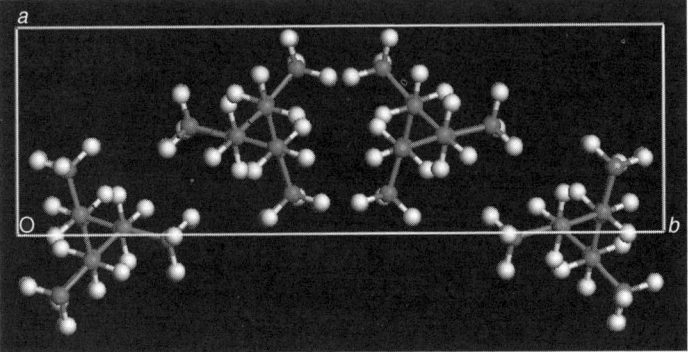

Figure 5.6. Crystal cell of the α-form of isotactic polypropylene after refinement ($a-b$ plane).

3/1 helix geometry, the fragments were energy-minimized and four fragments were then positioned symmetrically in the cell, as shown in Figure 5.6. The refinement of the cell on the basis of experimental X-ray data was carried out as described above. Both simulation approaches, the static deformation and the fluctuation method were applied on the crystal cell thus generated.

For the static method, three tensile and three pure shear deformations of magnitude 0.0005 were applied to the energy-minimized undeformed cell. The internal stress tensor was then calculated using equation (5.5) and used to obtain the six columns of the elastic stiffness coefficients matrix by equation (5.21). For the fluctuation method an NPT molecular dynamic run of 300 ps at 1 bar and $T = 250$ K was carried out. This was an adequate time to achieve a good statistic. The elastic stiffness coefficients were then calculated according to equation (5.26). The stiffness matrices obtained by the static deformation and the fluctuation approach of the α-form of isotactic polypropylene are shown in equations (5.27) and (5.28), respectively. The elements not listed are below 0.1 GPa. The monoclinic symmetry predicted by equation (5.12) is well represented.

$$C = \begin{pmatrix} 6.4 & 3.1 & 6.2 & & <1 & \\ 3.1 & 7.1 & 2.7 & & <1 & \\ 6.2 & 2.6 & 58 & & 6.4 & \\ & & & 2.5 & & \\ <1 & <1 & 6.4 & & 5.2 & \\ & & & & & 2.0 \end{pmatrix} \text{GPa} \qquad (5.27)$$

Static deformation

$$C = \begin{pmatrix} 7.5 & 2.4 & 5.7 & & & 0.3 \\ 2.4 & 6.5 & 1.6 & & & 0.7 \\ 5.7 & 1.6 & 66 & & & 10.0 \\ & & & 2.2 & & \\ & & & & & \\ 0.3 & 0.7 & 10.0 & & & 6.3 \\ & & & & & 1.9 \end{pmatrix} \text{GPa} \quad \text{Fluctuation method} \quad (5.28)$$

In order to test the reliability of the two simulation approaches described above, we compared the simulated values of the crystalline phase with experimental data. One can determine the Young's modulus of the crystalline regions of a semicrystalline polymer by X-ray scattering of a sample under constant load, because the positions of the X-ray peaks depend only on the crystal lattice dimensions. The strain ε in the crystal regions is given by $\varepsilon = \Delta d/d_0$, where d_0 denotes the initial lattice spacing, and Δd is the change in lattice spacing induced by a constant stress. The experimental Young's moduli can then be calculated by $E = \sigma/\varepsilon$. The simulated Young's moduli are calculated according to equations (5.10) and (5.11). The simulated and experimental values of the Young's moduli along the three unit cell vectors are shown in Table 5.1. In particular, the modulus E_{33} in the chain direction is in good agreement with experimental values for both methods.

TABLE 5.1. Simulated Young's Moduli of the α-Form of Isotactic Polypropylene in Comparison with Experimental Data from the Literature

	Static Deformation	Fluctuation Approach	Experimental Values [20, 21]
E_{33} (GPa)	40 ± 1.6	46 ± 1	41
E_{22} (GPa)	5.5 + 0.2	5.6 ± 0.3	3
E_{11} (GPa)	3.9 ± 0.2	6.0 ± 0.3	3

5.5.2. Mechanical Properties of the Amorphous Phase

To predict amorphous behavior, an ensemble of five unit cells was constructed. The cells were generated with low densities ($\rho = 0.4$ g cm^{-3}) to avoid unphysical overlaps. A molecular dynamics run was exerted on the systems, leaving only temperature, T, and external pressure, p, constant (NPT-MD). By raising the pressure stepwise, the structures were annealed until the final density of $\rho = 0.89$ g cm^{-3} was reached. This is the density of glassy atactic polypropylene, well below its glass transition temperature of $T_g = 253$ K, which was assumed to be suitable for the amorphous isotactic phase as well.

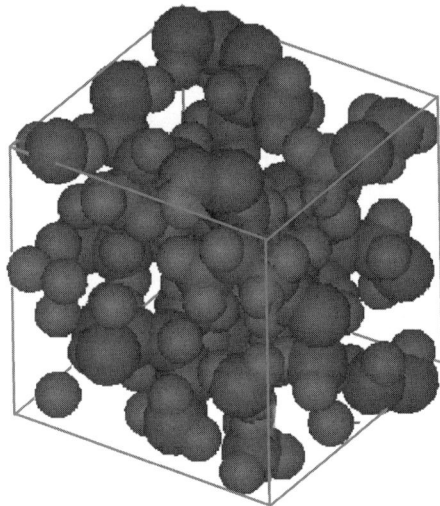

Figure 5.7. Amorphous cell of polypropylene containing a single chain of 20 monomers. The length of the cubic cell is $l = 1.18$ nm.

The microsystems were equilibrated in a 100 ps molecular dynamics run and minimized by a conjugate gradient procedure to a local energy minimum. Again the static deformation and the fluctuation approach were applied on the systems thus generated.

For the static method the strain was increased stepwise. The increments were chosen to be very small $[\varepsilon = O(10^{-3})]$, guaranteeing the continuity of the deformation sequence. After each deformation step an energy minimization was performed. The new structure was then taken as a starting point for the next deformation step. With this procedure, deformations of up to 1 percent were simulated. The elastic constants, bulk, and shear moduli were then calculated using equation (5.22). From these two constants all other moduli were then derived. For the fluctuation method an NPT molecular dynamic run of 300 ps at 1 bar and $T = 253$ K was carried out. After an equilibration period of 200 ps the data collection was carried out during the last 100 ps. The elastic stiffness coefficients were then calculated according to equation (5.26).

Table 5.2 shows the mechanical moduli for the simulations using the static deformation and the fluctuation approach. The values obtained by the static deformation are twice as high as the results from the fluctuation approach. For the other amorphous polymers (polysulfon, polyethersulfon, epoxy resin) we investigated, the moduli obtained by the static approach were also considerably higher than expected. This is in agreement with the results of other authors [22], which report that the static approach only gives an upper limit for the elastic moduli. In contrast to the simulation of the crystalline phase, temperature and entropic effects, ignored by the static method, are not negligible for amorphous polymers. It should be mentioned

TABLE 5.2. Mechanical Moduli of Amorphous Isotactic Polypropylene Derived by the Static Deformation and the Fluctuation Approach

	Static Deformation	Fluctuation Approach
Shear modulus, G (GPa)	2.8	1.2
Young's modulus, E (GPa)	7.6	3.2
Bulk modulus, K (GPa)	8.6	3.1
Poisson ratio	0.35	0.33

that Theodorou and Suter [10] obtained good results for atactic polypropylene using the static deformation method (Table 5.3).

TABLE 5.3. Comparison of the Mechanical Moduli of Amorphous Atactic Polypropylene Derived with the Static Deformation by Theodorou and Suter [10] with Experimental Data from the Literature

	Simulation [10]	Experiment [10, 23]
Shear modulus, G (GPa)	1.1	0.97
Young's modulus, E (GPa)	3.0	2.65
Bulk modulus, K (GPa)	3.5	3.4
Poisson ratio	0.35	0.37

5.5.3. Combination of Molecular Simulation and Micromechanical Models

The elastic constants calculated above refer to homogenous crystalline or amorphous materials. To determine composite behavior of semicrystalline polymers, one has to employ models for the specific morphology. Under homogenous conditions (e.g. no external deformation, temperature gradients, or epitactic effects) the crystalline phase in many polymers grows in the form of lamellae arranged in a supermolecular structure of spherical shape. These spherulites themselves consist of highly branched crystalline lamellas with amorphous regions in between. For simplicity we assume that the mechanical properties of these amorphous inclusions do not differ considerably from those of the free amorphous phase. It should be noted here that this is a rather strong assumption since many authors [24] assume the existence of a 'rigid amorphous' phase due to the restrictions of the chains close to the lamellae. A model capable of reproducing the spherulite structure was adopted from laminate theory by Halpin and Kardos [2, 3]. They suggested that crystalline polymers behave, with respect to their mechanical properties, as a two-phase composite consisting of an amorphous matrix reinforced by short fibers. Therefore,

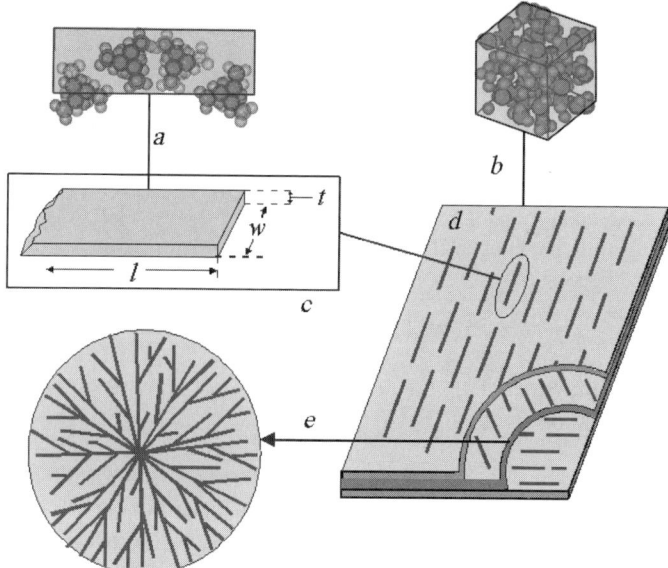

Figure 5.8. Laminate model of spherulites: The spherulite is considered to consist of crystalline lamellae (*c*) in an amorphous matrix (*b*). The spherulites are modeled by a laminate system consisting of layers of unidirectional orientated lamellae (blue fibers) in an amorphous matrix (*d*). The layers are rotated with respect to each other (*e*). *l* is the length, *w* is the width, and *t* is the thickness of the lamella.

spherulites could be treated as a composite of randomly distributed and orientated short fibers (lamellae) embedded in an amorphous phase. By applying lamination theory, the isotropic mechanical properties of the spherulites can be mathematically modeled by an orientation averaging of a laminate system consisting of n layers of unidirectional orientated lamellae in an amorphous matrix.

This procedure was used to predict the elastic moduli of semicrystalline materials from experimental data for the moduli of the crystalline and amorphous phase [5]. Here the experimental data are replaced by the results of the molecular simulation.

Below we show the combination of molecular simulation and micromechanical models for isotactic polypropylene. After the amorphous (Figure 5.8*b*) and crystalline (Figure 5.8*a*) unit cells have been constructed and elastic constants have been calculated separately for the crystalline and the amorphous cell, the prediction of the mechanical properties of the semicrystalline material is performed by a successive calculation of the elastic constants of the lamella (Figure 5.8*c*), the layers of unidirectional orientated lamellae in an amorphous matrix (Figure 5.8*d*) and finally the spherulite by orientation averaging (Figure 5.8*e*).

The crystalline lamella is formed of parallel chains which are eventually folded back. The moduli parallel to the lamella axis $E_{\parallel,L}$ can be calculated averaging the two unit cell directions perpendicular to the helices:

$$E_{\parallel,L} = \frac{E_{11} + E_{22}}{2} \tag{5.29}$$

Because of lamella twisting, the average value of the moduli perpendicular to the lamella axis $E_{\perp,L}$ can be represented by a parallel model suggested by Kardos and Raisoni [2]:

$$E_{\perp,L} = \frac{1}{2}(E_{\parallel,L} + E_{33}) \tag{5.30}$$

We use here a more accurate method, by averaging over all possible orientations of the crystalline unit cell [25]:

$$E_{\perp,L} = \frac{1}{2\pi} \int_0^{2\pi} \left\{ \left(\frac{\cos[\alpha]}{E_{\parallel,L}}\right)^2 + \left(\frac{\sin[\alpha]}{E_{33}}\right)^2 \right\}^{-1/2} d\alpha \tag{5.31}$$

The direction-dependent elastic moduli of an amorphous matrix with unidirectional oriented lamellas (only one of the layers in Figure 5.8d) can now be calculated using the theory of Halpin and Kardos:

$$E_{11,u} = \frac{1 + \xi\eta\phi_c}{1 - \eta\phi_c} E_m, \quad \text{with } \eta = \frac{E_{\parallel,L}/E_m - 1}{E_{\parallel,L}/E_m + \xi}, \quad \xi = 2\frac{l}{t} \tag{5.32}$$

$$E_{22,u} = \frac{1 + \xi\eta\phi_c}{1 - \eta\phi_c} E_m, \quad \text{with } \eta = \frac{E_{\perp,L}/E_m - 1}{E_{\perp,L}/E_m + \xi}, \quad \xi = 2\frac{w}{t} \tag{5.33}$$

$$G_{12,u} = \frac{1 + \xi\eta\phi_c}{1 - \eta\phi_c} G_m, \quad \text{with } \eta = \frac{G_L/G_m - 1}{G_L/G_m + \xi}, \quad \xi = 1 \tag{5.34}$$

where $E_{11,u}$ and $E_{22,u}$ are the Young's moduli of the layer, parallel and perpendicular to the lamellas; $G_{12,u}$ is the shear modulus of the layer; ϕ_c is the degree of crystallinity. The indices L and m define the lamella and amorphous matrix, respectively, and the index u the unidirectional layer, with lamellas of width w, length l and thickness t. The shear modulus G_L of the lamella was taken to be G_{12}, which is perpendicular to lamella axis.

From the moduli derived by equations (5.32)–(5.34), the elastic coefficient matrix of the two-dimensional layer with unidirectional orientation can be calculated:

$$C_{11}^u = \frac{E_{11,u}}{1 - v_{12,u}v_{21,u}}$$
$$C_{22}^u = \frac{E_{22,u}}{1 - v_{12,u}v_{21,u}}$$
$$C_{12}^u = C_{21}^u = \frac{v_{12,u}E_{11,u}}{1 - v_{12,u}v_{21,u}} = \frac{v_{21,u}E_{11,u}}{1 - v_{12,u}v_{21,u}} \qquad (5.35)$$
$$C_{66}^u = G_{12,u}$$

The Poisson ratio was derived assuming a simple mixing rule, $v_{ij,u} = v_{ij}\phi_c + v_m\phi_m$.

To describe the whole spherulite and not only one unidirectional layer, Halpin and Kardos proposed an orientation averaging. Following this idea, the isotropic mechanical properties of the spherulites can be mathematically modeled by a laminate system consisting of n layers of unidirectionally orientated lamellae in an amorphous matrix, as shown in Figure 5.8e. The layers are oriented at angles of π/n with respect to each other. The averaged moduli only depend on the invariants (with respect to rotation) of the elastic coefficient matrix. The invariants are expressed in terms of the elastic coefficients.

$$U_1 = \frac{1}{8}(3C_{11}^u + 3C_{22}^u + 2C_{12}^u + 4C_{66}^u)$$
$$U_5 = \frac{1}{8}(C_{11}^u + C_{22}^u - 2C_{12}^u + 4C_{66}^u) \qquad (5.36)$$

The isotropic elastic moduli of the spherulite can then be expressed as:

$$\bar{G} = U_5$$
$$\bar{E} = 4\frac{U_5}{U_1}(U_1 - U_5) \qquad (5.37)$$
$$\bar{v} = \frac{U_1 - 2U_5}{U_1}$$

TABLE 5.4. Mechanical Moduli of the α-Form of Isotactic Polypropylene Derived from Simulation

	Static Deformation	Fluctuation Approach
E_{11}, GPa	3.9	6.0
E_{22}, GPa	5.5	5.6
E_{33}, GPa	40	46
G_{12}, GPa	2.0	1.9
v_{12}	0.49	0.33
v_{21}	0.39	0.36

TABLE 5.5. Mechanical Moduli of Amorphous Polypropylene Derived from Simulation

	Static Deformation	Fluctuation Approach
E, GPa	7.6	3.2
G, GPa	2.8	1.2
ν	0.35	0.33

For the calculation we need the simulated elastic moduli of the crystalline phase of i-PP (Table 5.4). We also need the simulated elastic moduli of the amorphous phase of i-PP (Table 5.5).

The ratios of length-to-thickness and width-to-thickness of the lamellae were approximated as $l/t = 100$ and $w/t = 20$, which is reasonable for most polymers [2]. Both values enter the geometry factor ξ.

For equations (5.32)–(5.34) the degree of crystallinity ϕ_c has to be known. Therefore an i-PP sample (Novolen® 1000) was investigated by wide-angle X-ray scattering (WAXS) using a Siemens D 500 diffractometer (Figure 5.9). After subtraction of the underlying amorphous halo, the amount of the crystalline phase was found to be 54 percent, suggesting that integral intensities of the amorphous halo and the Bragg peaks are proportional to the mass of the two phases.

Regarding the four dominant peaks, the second is due to the β-phase only (010), while the others are related to the α-phase (110), (040) and (130). The amount of β-phase is about 25 percent. Since the experimental moduli of the β-phase differ only slightly from those of the α-phase [26], we consider in our simulation only the α-phase.

In order to compare the simulation with experimental data, the elastic moduli of the i-PP sample were determined by dynamical mechanical analysis (DMTA). The

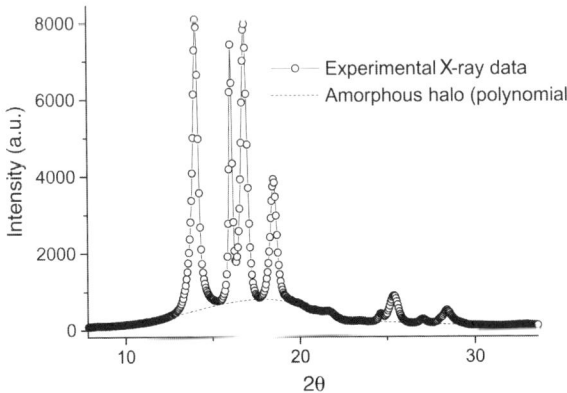

Figure 5.9. X-ray scattering pattern of isotactic polypropylene after Lorentz, polarization and Compton corrections.

TABLE 5.6. Comparison of the Elastic Moduli
Calculated from Micromechanical Models with
Experimental Data

	Static Deformation	Fluctuation Approach	Experiment
E, GPa	7.5	5.9	5.0
G, GPa	2.7	2.2	1.7
ν	0.37	0.35	—

real (G') and imaginary (G'') parts of the complex shear modulus, $G^* = G' + iG''$, were obtained using an ARES rheometer in torsional mode. From this experiment a storage modulus, G', of about 1.7 GPa was determined for temperatures of $-20°C$, which is below T_g. The Young's modulus was measured with Seikos Exta 4700 in the same temperature regime. The real part of the modulus was $E' = 5.0 \pm 0.5$ GPa.

Table 5.6 compares the results of micromechanical calculations on basis of both simulation methods with experimental data from the measurements described above. Whereas the predictions from the fluctuation approach are in quite good agreement with the experiment, the moduli values obtained by the static deformation are slightly higher. This is due to the overestimation of the moduli of the amorphous phase by the static method, which ignores temperature and entropic effects. For the crystalline phase both simulation methods gave good results. For amorphous systems molecular dynamics simulations tend to be more accurate than molecular mechanics.

The agreement of the moduli obtained by combination of molecular simulation and micromechanical models with experimental data, when using the fluctuation approach, opens up the possibility of applying this procedure to different polymers.

Furthermore, the random orientation of the lamellas can be replaced by other structural models and the results can be taken as parameters for microscopic finite element analysis, gaining more accurate descriptions of moldflow injection processes and thermic behavior, for example shrinkage, of semicrystalline polymers.

ACKNOWLEDGMENT

Financial support from the Federal Ministry of Economics and Labour [Bundesministerium für Wirtschaft und Arbeit (BMWA)] through the Federation of Industrial Cooperative Research Associations "Otto von Guericke" [Arbeitsgemeinschaft industrieller Forschungsvereinigungen e.V. (AiF)] is gratefully acknowledged.

REFERENCES

[1] Meyer, H., Müller-Plathe, F., *J. Chem. Phys.*, **115**, 7807 (2001).
[2] Kardos, J.L., Raisoni, J. *Polym. Engng. Sci.*, **15**, 183 (1975).

REFERENCES

[3] Halpin, J.C., Kardos, J.L., *J. Appl. Phys.*, **43**, 2235 (1972).
[4] Kerner, E.H., *Proc. Phys. Soc.*, **69B**, 808 (1956).
[5] Alig, I., Tadjbakhsch, S., *J. Polym. Sci. B*, **31**, 2949 (1998).
[6] Alig, I., Tadjbakhsch, S., Floudas, G., Tsitsilianis, C., *Macromolecules*, **31**, 6919 (1998).
[7] Le Bail, A., *Trends in Structure Determinations by Powder Diffractometry, ECM-18 Proceedings* (1998).
[8] Flory, P.J., *Statistical Mechanics of Chain Molecules*. Hanser, New York (1989).
[9] Theodorou, D.N., Suter, U.W., *Macromolecules*, **18**, 1467 (1985).
[10] Theodorou, D., Suter, U.W., *Macromolecules*, **19**, 139 (1986).
[11] Theodorou, D., Suter, U.W., *Macromolecules*, **19**, 379 (1986).
[12] Parrinello, M., Rahman, A., *J. Chem. Phys.*, **76**, 2662 (1982).
[13] Gusev, A.A., Zehnder, M.M., Suter, U.W., *Macromol. Symp.*, **90**, 85 (1995).
[14] Landau, L.D., Lifschitz, E.M., *Course of Theoretical Physics*, Vol. VII.
[15] Natta, G., Corradini, P., Cesari, M., *Rend. Accad. Naz. Lincei*, **21**, 365 (1956).
[16] Turner Jones, A., Aizlewood, J.M., Beckett, D.R., *Makromol. Chem.*, **75**, 134 (1964).
[17] Cheng, S.Z.D., Janimak, J.J., Rodriguez, J., *Polypropylene: Structure, Blends and Composites* (1995).
[18] Natta, G., Corradini, P., *Nuovo Cim.*, **15**(Suppl.), 40 (1960).
[19] Tashiro, K., Kobayashi, M., Tadokoro, H., *Polym. J.*, **24**, 899 (1992).
[20] Sakurada, I., Kaji, K., *J. Polym. Sci. C*, **31**, 57 (1970).
[21] Sakurada, I., Taisuke, I., Katsuhiko, N., *J. Polym. Sci. C*, **15**, 75 (1966).
[22] Raaska, T., Niemelä, S., Sundholm, F., *Macromolecules*, **27**, 5751 (1994).
[23] Koppelmann, J., Leder, H., Royer, F., *Colloid Polym. Sci.*, **257**, 637 (1979).
[24] Cheng, S.Z.D., Cao, M.-Y., Wunderlich, B., *Macromolecules*, **19**, 1868 (1986).
[25] Kröhn, M., Computersimulation mechanischer Eigenschaften von Polymeren, Ph.D. Thesis, p. 96 (2002).
[26] Karger-Kocsis, J., Varga, J., *J. Appl. Polym. Sci.*, **62**, 291 (1996).
[27] Landau, L.D., Lifschitz, E.M., *Course of Theoretical Physics*, Vol. V.
[28] Parrinello, M., Rahman, A., *J. Appl. Phys.*, **52**, 7182 (1981).

6

CROSSLINKING SIMULATIONS IN POLYMER DESIGN

ROBERT T. JOHNSTON
Ethylene Elastomers R&D, DuPont Dow Elastomers LLC, Freeport, Texas, USA

CONTENTS

6.1. Introduction	244
6.2. Review of Theory	247
6.2.1. Network Formation (Gelation)	247
6.2.2. Network Structure	250
6.2.3. Rubber Elasticity	253
6.3. Why a Simulation?	253
6.4. A Taxonomy of Crosslinking and Scission Simulations	256
6.4.1. Empirical Models	257
6.4.2. Mathematical Theoretical Models	259
6.4.3. Scission of Linear and Branched Polymers	259
6.4.4. Crosslinking Without Spatial Effects or Molecular Motion	260
6.4.5. Crosslinking and Scission Without Spatial Effects or Molecular Motion	261
6.4.6. Crosslinking With Spatial Effects Without Molecular Motion	262
6.4.7. Crosslinking With Spatial Effects and Molecular Motion	264
6.4.8. Simulation of Network Formation and Deformation (Rubber Elasticity)	267
6.5. Application of Simulations to Rational Design	268
6.5.1. Refinement of Theories	268
6.5.2. Refinement of Analytical Methods	269

Molecular Simulation Methods for Predicting Polymer Properties, Edited by Vassilios Galiatsatos.
ISBN 0-471-46481-3 Copyright © 2005 John Wiley & Sons, Inc.

6.5.3. Effects of Polymer Structure and Chemistry	269
6.5.4. General Approach to Developing Validated Simulations	270
6.6. Conclusion	272

6.1. INTRODUCTION

Many first time users of computer modeling and simulation software have the expectation that it will serve as a 'black box' into which they can enter desired product properties and obtain as output the polymer structure required to obtain those properties, in short, a computer-enabled 'first principles' basis for rational design. Is this expectation realistic? The focus of this chapter is on using crosslinking simulations for design of polymer products, including what simulations can and cannot do.

Long after the Stone, Bronze and Iron ages have passed, we are now in an age in which nontraditional advanced materials, including plastics and rubber as well as new versions of the traditional ceramics, glasses, and metals, are being used in a wide variety of increasingly sophisticated and demanding applications. Whether or not we live in the Polymer Age is debatable (one could easily argue for the Nano Age or the Bio Age), but few would disagree that the moniker 'Information Age' also applies. The superpositioning of virtual and material worlds has abetted the ongoing revolution in materials science. In polymers as well as other materials, computer technology has created new opportunities for rational design.

Polymers have not always been designed rationally. Historically, serendipity has played a major role in polymer technology development, starting with Goodyear's discovery of natural rubber vulcanization [1] in 1839 and continuing right through to the discovery of crystalline polypropylene by Hogan and Banks [2]. After Staudinger's publication of the macromolecular nature of polymers (certain 'colloids') in 1920, Carrothers reasoned from the macromolecular hypothesis that dihydroxides and diacids could be reacted together in 1:1 stoichiometry to obtain polyesters. This success eventually led to the discovery of nylon [3]. Nonetheless, rational design took a backseat to serendipity in many of the most important industrial polymer developments, and polymer science principles have often proved more useful for explaining and optimizing breakthrough innovations than in finding them in the first place [4].

The potential for rational design of polymers has increased significantly since the days of Carrothers. Macromolecules and polymer networks are better understood today and, equally importantly, the number of synthetic approaches for creating them has vastly expanded. Within the field of polyolefins, for instance, the past 15 years have seen a revolution in design capability brought about by metallocene catalysts. These homogeneous catalysts have been used to produce well-defined polymer structures that are highly controlled by catalyst selection, reactor design and operating conditions. Molecular weight (MW) and molecular

INTRODUCTION

weight distribution (MWD), comonomer distribution and tacticity are all controlled, and can now be 'dialed in' by plant operators with an unprecedented degree of predictability, thanks to well-defined catalyst behavior and computerized kinetic models.

These capabilities certainly do not extend to all classes of polymers, and practical synthetic methods still do not exist for many desirable complex structures. For example, within the family of polyolefins, considerable research effort continues to be focused on creating block copolymers, including stereo block copolymers, and the ability to add polar functionality during polymerization of linear polyolefins remains a challenge. Nevertheless, within several polymer product families, the ability to controllably manufacture a target polymer structure exists today. The larger gap for rational design appears to be on the applications side: What should the target structure be?

One of the reasons polymers have been so successful is that they are versatile. A few major industrial polymer classes have been able to function in a wide variety of applications. This has allowed these polymers to be manufactured in large volumes with resultant economies of scale, which has led to low overall costs. The versatility of polymers comes not only in the variety of molecular weights, compositions, and so on, that can be produced in a given product family, but also the many ways they can be formulated and processed. Polymers are formulated with a variety of fillers, plasticizers, pigments, stabilizers, and functional additives such as coupling agents. Surface properties are modified with additives that bloom to the surface, such as slip or antiblocking additives, additives that react at the surface to change its properties, such as UV graftable light stabilizers, or with corona treatments or coatings. Polymer structures are themselves altered via controlled scission, crosslinking, and grafting with various functional agents. Polymers may be blended or reacted with other polymers, producing complex morphologies. Polymers are also processed in a wide variety of ways, including a variety of compounding and mixing processes, and a variety of shaping or fabrication processes, including solution casting, film and profile extrusion, injection molding, blow molding, and thermoforming. Variations in macromolecular orientation, morphology, crystallization, degradation, and so on, result from the different processes and process operating conditions, which may vary in temperature, pressure, shear rate, and residence time. The formulation, polymer blending, or modification can take place before, during, or after these compounding and fabrication processes.

This versatility is also what makes rational design such a challenge. The number of variables and the number of physical laws and relationships that must be understood to predict product properties in a formulated and fabricated polymer article is enormous. To predict product properties, one must first be able to predict the interaction of the polymer structure and compounding or fabrication processes. Rheological properties, compatibility with compounding ingredients, and processing stability are examples of the properties to be predicted. If the polymer structure is changing during processing, as, for example, during dynamic vulcanization or scission, the interaction of process variables must be with this changing structure. Even if this is successfully predicted, the final article must meet a defined set of performance requirements in order to be successful. Examples of product properties that may be important to predict include hardness, stiffness, tensile strength, tear resistance, permeability, chemical resistance, temperature resistance, and weatherability. One must be able to predict the relationship between the processed structure and these properties.

Unfortunately, the physical laws governing polymer structure–property relationships are in many cases too poorly understood to predict the properties of typical formulations after processing. Many models have been created to explain these phenomena, but most are limited to simple cases or are not 'first principles' models that derive properties directly from the molecular structure. This does not negate the value of empirical models, but their utility for general prediction is limited. The same can be said of process models which have been developed for some simplified processes or empirically for more complex processes but which must be adapted to each specific case, generally requiring experimental effort.

As a result of this complexity, a comprehensive 'first principles' model that can be applied generally to a wide variety of polymers to predict structures that will produce desired sets of properties is not available, nor are there any immediate prospects. Dynamic atomistic or quantum mechanical simulations on even small collections of atoms take massive computer power for even short simulations (a few nanoseconds). A realistic and general simulation of a large number of high molecular weight polymer chains in a crosslinking and scission reaction in the presence of other functional additives and fillers with prediction of the properties resulting therefrom has not yet been achieved.

These points are important to remember when considering what a computer model or simulation can realistically be expected to do (and not do). The widespread availability of computers has led some to the unrealistic expectation that 'black box' software will eliminate the need for human intelligence. This is unlikely. In offices, computers have been found to function best as *tools* for human intelligence, creating higher value jobs for human 'knowledge workers' who are skilled in using computers to acquire, manage, and analyze information; the same may be expected in laboratories. Thus, our enthusiasm for computer models and simulations should be tempered by the realization that they are not likely to replace serendipity as a source of breakthrough innovations, they do not eliminate the need for skilled polymer scientists, and we most likely will use them as analytical or search tools rather than directly as 'invention machines' or as polymer science 'black boxes.'

Having hopefully by now disabused the reader of any notion that there is a software/hardware system that will allow a chemist to directly predict crosslinked product properties from a given initial polymer structure or, in reverse, to find an optimal polymer structure that will produce a set of properties, we may now ask: What, then, can a chemist realistically hope to achieve today by using crosslinking simulations?

We have found that computer simulations are most useful not for making direct predictions, but for developing understanding. Whereas we initially thought considerable laboratory effort could be spared by using computer simulations, we found that laboratory experiments had to be performed to effectively use the simulations to develop new knowledge. The simulations functioned as *analytical tools* to *understand* what took place in the laboratory. In most cases, the simulations and accompanying laboratory experiments have been done on simple systems. Complex systems (e.g. industrially relevant polymer formulations) could not be directly modeled in many cases, but the insights gained from simple simulations were useful for understanding the behavior of the more complex systems.

In this chapter, our goal is to present a taxonomy of crosslinking models and simulations with an emphasis on computer applications, compare their characteristics, and explain how they may be used. The intention is not to provide a comprehensive review of the rather extensive literature on crosslinking models and simulations, but to highlight major categories and examples that have proved useful as analytical tools for rational design of polymers for crosslinked applications.

6.2. REVIEW OF THEORY

It is beyond the scope of this chapter to review theories of polymer networks and rubber elasticity in detail. Classical theories are described in polymer science textbooks [3, 5], and recent developments have been reviewed by Erman and Mark [6]. In this section, we only briefly highlight key concepts that are necessary for an understanding of crosslinking simulations, especially in relation to network structures.

6.2.1. Network Formation (Gelation)

To understand network–property relationships, one must understand the nature of polymer networks. Polymer networks comprise linear or branched polymer chains that have been crosslinked* together until the extent of crosslinking exceeds the gel point, the critical crosslink density where an 'infinite' network is formed. Statistically, this is the point above which there is a finite probability that a randomly selected primary polymer chain will belong to the network. Paradoxically, at the gel point, there is an infinitely small amount of 'infinite' molecular weight polymer. The paradox arises from statistical arguments. Borrowing from percolation theory, the gel point can be considered the point where there arises a continuous pathway from one side of the container to the other, for example, a *spatially* infinite network, albeit comprising an insignificant fraction of the total polymer mass at the gel point. At crosslink densities above the gel point, the fraction of polymer in the gel network increases rapidly.

The gel point is a critical point analogous to the critical temperature in liquid–vapor equilibria, below which the condensed phase density increases and the vapor phase density decreases. Above the gel point in a randomly crosslinked system, the molecular weight of the gel increases while the molecular weight of the sol decreases. This behavior follows from the statistical probabilities of randomly selected crosslinkable units belonging to high molecular weight chains vs

---

*We use the term 'crosslink' here, but the concept also applies to other mechanisms of linking such as endlinking, grafting, or copolymerization with branching agent monomers.

low molecular weight chains, and the resultant preferential incorporation of mass from high molecular weight molecules into the gel.

Defining the gel point in a simulation, which necessarily involves a finite number of polymer chains, can be somewhat tricky. One commonly used approach is to define the 'reduced weight average molecular weight' as the weight-average molecular weight calculated from all polymer molecules except the largest one. This parameter peaks at the gel point and then falls rapidly at higher crosslink densities in randomly crosslinked systems. By graphing the reduced weight average molecular weight in a simulation versus crosslink density, the gel point can be estimated.

The partitioning of mass between sol and gel depends on crosslink density, crosslink distribution (random or otherwise), crosslink functionality, initial polymer molecular weight, and molecular weight distribution. Various statistically based analytic expressions have been developed to describe the crosslinking behavior of various systems, generally assuming random crosslinking. These are collectively referred to as sol–gel partition theories. Examples include Flory–Stockmayer [7], Charlesby–Pinner [8], Saito–Inokuti [9, 10] and Miller–Macosko [11–13].

The number of crosslinks per molecule is usually the parameter of greatest interest when considering the effects of crosslinking on polymer solubility. Three quantities that have been found to be particularly useful are the crosslink density (q), which is the fraction of total monomer or repeat units involved in a crosslink (two units per crosslink); the crosslinking index (γ), or the number of crosslinked units per initial number average (M_n) molecule; and the crosslinking coefficient (δ)—the number of crosslinked units per initial weight average molecule (M_w). It follows that $\delta/\gamma = M_w/M_n$.

The Flory–Stockmayer relation derives from probabilistic arguments for randomly crosslinked polymers of arbitrary molecular weight distribution [7]. With P_w being the weight average degree of polymerization (equal to M_w/m where m is the weight of the repeating unit), the condition for the gel point is when the probability that any one monomer unit is crosslinked is equal to the reciprocal of the weight average degree of polymerization:

$$(P_c)_{crit} \approx 1/P_w \qquad (6.1)$$

or

$$\delta = 1 \qquad (6.2)$$

From the definition of γ and δ at the beginning of this section, it follows that:

$$\delta = (M_w/M_n)\gamma \qquad (6.3)$$

To describe the gel fraction at any crosslink density, not just at the gel point, various relations have been defined. A popular one is that derived by Charlesby and Pinner for randomly crosslinked and/or scissioned polymers initially having the most probable distribution of molecular weights [8]. The equation was derived in units related

REVIEW OF THEORY

to radiation crosslinking:

$$s + s^{0.5} = p/q + 1/q_0 r P_n \tag{6.4}$$

where s is the sol fraction, p is the number of scissions per monomer unit, q is the proportion of monomer units which are crosslinked, r is the radiation dose, $q_0 = q/r$ (e.g. the proportion of crosslinked monomer units per unit dose), and P_n is the number average degree of polymerization. By simply plotting $s + s^{0.5}$ vs $1/r$, one can obtain the ratio of scissions to crosslinked units from the intercept.

A modified form of the Charlesby–Pinner (C-P) equation suitable for chemically crosslinked systems was published by Lal and McGrath [14], who applied it to a study of peroxide crosslinked polyvinyl butyl ethers:

$$s + s^{0.5} = p/q + 1/2mEP_n[i] \tag{6.5}$$

where s, p, q, and P_n retain their previous definitions, $[i]$ is the concentration of decomposed peroxide (mol/g), E is the crosslinking efficiency (mol crosslinks/ mol decomposed peroxide), and m is the weight of the repeat unit. They plotted $s + s^{0.5}$ vs $1/[i]$ and observed linear plots for all their samples, except a sample of polypropylene oxide, which became linear after the 5 phr peroxide level. Crosslink efficiencies and crosslink/scission ratios were obtained directly from the graphs. Note that E in equation (6.5) is the *crosslinking* efficiency, not the initiator efficiency, since $E[i] = q$ and does not include the p term, thereby not counting initiator expended on scission reactions.

Polymers not having an initial most probable distribution of molecular weights are sometimes assumed to reach it if they undergo simultaneous scission, but this assumption can be risky, as discussed below. Preferably, an alternative model is used that describes the sol–gel partitioning for any arbitrary distribution of molecular weights. An example is the model of Saito and Inokuti [9, 10]:

$$\begin{aligned} 1 - g = \frac{1}{(2g+\lambda)^3} \Bigg\{ & \frac{4\lambda g}{y} + 2g\lambda^2 + \lambda^3 - \frac{4P_n \lambda g}{y} G_0\left[\frac{y(2g+\lambda)}{P_n}\right] \\ & - 4(2g^3 + g^2\lambda) \frac{\partial G(z)}{\partial z}\bigg|_{z=\frac{y(2g+\lambda)}{P_n}} \Bigg\} \end{aligned} \tag{6.6}$$

$$G(z) = \int_0^\infty m(P) e^{-zP} dP \quad (z \geq 0)$$

where $m(P)$ is normalized such that

$$\int_0^\infty P m(P) dP = 1$$

and y = number of crosslinks per initial number average chain (e.g. half of Flory's crosslinking index γ); x = number of scissions per initial number average chain; and $\lambda = x/y$, or the ratio of scissions to crosslinks.

The author has adapted the Saito–Inokuti model to numerical distributions of molecular weight data from gel permeation chromatography so that it can be directly applied to experimentally determined molecular weight distributions as well as arbitrary distribution functions. Figure 6.1 shows log normal distributions based on 10,000 M_n with polydispersity up to 10. Figure 6.2 shows the calculated Saito–Inokuti models graphed in the form of the Charlesby–Pinner equation, assuming random crosslinking and no scission. The gel point ($s + s^{0.5} = 2$) occurs at lower crosslink density (y) the broader the MWD or higher M_w, as predicted by Flory–Stockmayer. Figure 6.3 shows the effect of simultaneous crosslinking and scission. As λ increases, the y-intercept on the Charlesby–Pinner plot increases, but otherwise the curves are congruent. The graph also illustrates the error introduced in this case by assuming a most probable distribution (for which $M_w/M_n = 2$) rather than the log–normal distribution using the true M_w.

6.2.2. Network Structure

Networks can be described as follows.

Network—a system of interlinked polymer chains (in this context, we restrict our definition to covalently bonded links, for example, chemical crosslinks) that is spatially 'infinite,' traversing the dimensions of the sample. Practically, one can define networks in other ways, especially for mathematical convenience in simulations, but the main idea is that they are interlinked chains where one chain is linked to another, which is linked to another, and so on, *ad infinitum*, in a branched, nonlinear fashion.

Gel—the network portion of a network-containing polymer sample. This material is insoluble in solvents provided degradation does not occur.

Sol—the nonnetwork portion of a network-containing polymer sample. This material typically can be extracted or dissolved by suitable solvents. Typically, oils, fillers, and so on, are excluded from the definition, with only the nonnetwork *polymeric* material counted as sol.

Elastically active chain—the polymer chain segment between two elastically active junctions in a network.

Elastically active junctions—the connection points for elastically active chain segments. These are junctions that have more than two elastically active chains emanating from them. That is, the junction has at least three paths leading away from it that are independently attached to the network [15, 16].

Dangling chain end—a pendant chain end that is attached to the gel but is elastically inactive due to the fact that one end is not attached to an elastically active junction.

Loop—although there can be elastically active loops, what is generally referred to as a loop is an elastically inactive loop; that is, a loop is a path from a junction point that leads back to the same junction point without the path containing other junctions that make it elastically active.

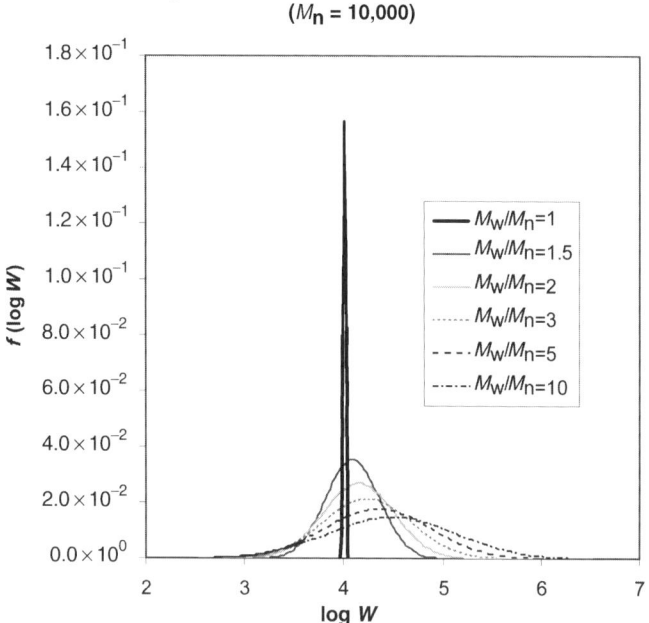

Figure 6.1. Log–normal distribution functions with various M_w/M_n ratios.

Perfect network—a network having no dangling chain ends, ineffective junctions (e.g. not elastically active), or loops.

In practice, analysis of a real network can proceed by first reducing it to a perfect network. This may be done by recursively removing all dangling chain ends, all

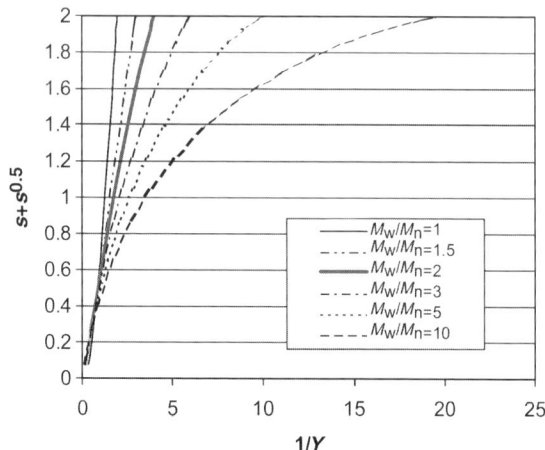

Figure 6.2. Effect of M_w/M_n on sol–gel partitioning of log–normal distributions ($M_n = 10,000$) according to the Saito–Inokuti model.

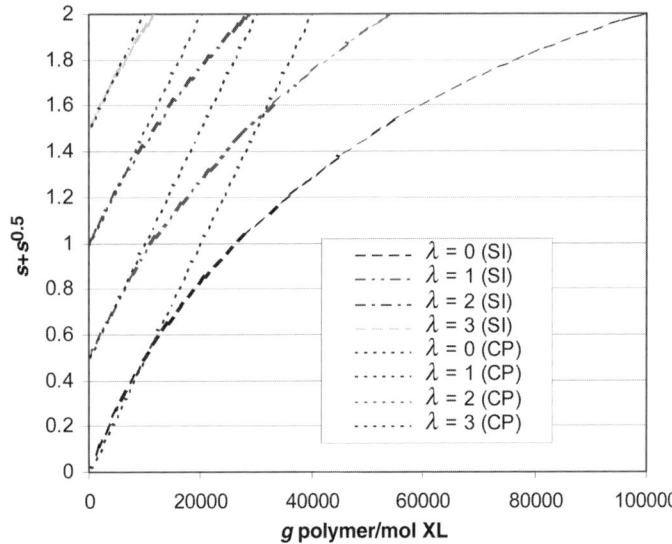

Figure 6.3. Effect of various scission:crosslink ratios on sol–gel partitioning of a log–normal distribution having $M_n = 10{,}000$ and $M_w = 50{,}000$ according to the Saito–Inokuti equation (SI), and the most probable distribution having $M_n = 10{,}000$ and $M_w = 20{,}000$ according to the Charlesby–Pinner equation (CP).

junctions with functionality of less than 3 (e.g. creating one chain from two that are linked by a two-functional junction), and all elastically inactive loops [17]. Once the real network is reduced to a perfect network, the topology of the network can be described by the following parameters: ϕ = average junction functionality (e.g. the average number of network chains emanating from a junction); ν = number of network chains; μ = number of junctions; ξ = cycle rank, or the number of network chains that must be cut to reduce the network to a spanning tree with no closed cycles (e.g. the number of redundant links that could be cut without dividing the network molecule into two or more parts; illustrated in Figure 6.4); M_c = number average molecular weight between junctions (note that the literature contains alternative definitions).

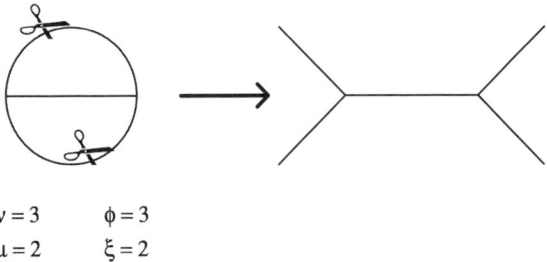

Figure 6.4. Illustration of network parameters using a simple 'network.'

WHY A SIMULATION?

Note that *for a perfect network* [18],

$$\mu = 2\nu/\phi \tag{6.7}$$

$$\xi = \nu - \mu + 1 \approx \nu - \mu = (1 - 2/\phi)\nu \tag{6.8}$$

$$\xi/V_0 = \frac{(1 - 2/\phi)\rho}{M_c/N_A} \tag{6.9}$$

where V_0 is the volume of the network in the state of formation, ρ is the corresponding density, and N_A is Avagadro's number. Equations (6.7) and (6.8) can be applied to imperfect networks if they are first converted to perfect networks as described above. However, equation (6.9) cannot be used in that case, and is applicable only to a 'true' perfect network. This is because the network density includes the imperfections that are excluded from the perfect network.

6.2.3. Rubber Elasticity

There are numerous models of rubber elasticity, and many debates about the role of entanglements, relative contributions of chains vs junctions, and so on. Among the most commonly used models (excluding phenomenological models like Mooney–Rivlin) are the affine model [equation (6.10)] and the phantom model [equation (6.11)] [5], along with various combinations such as the constrained junction model [19–21].

$$G = \nu kT/V_0 \tag{6.10}$$

$$G = \xi kT/V_0 \tag{6.11}$$

where G is the equilibrium shear modulus, k is the Boltzmann constant, and T is absolute temperature. Pearson and Graessley [22] described a model that includes trapped entanglements:

$$G = G_c + T_e G_e^{max} \tag{6.12}$$

where G_c is the contribution from chemical crosslinks (e.g. from rubber elasticity theory) and G_e^{max} is equated with the plateau modulus, G_N^0. The trapping factor T_e is equal to the square of the weight fraction of elastically active chains, W_{EANC}. Application of this model requires significant knowledge about the network structure, either from a statistical model or from a simulation.

6.3. WHY A SIMULATION?

It is difficult to develop the science of something one cannot measure. While certain of their properties can be measured, crosslinked polymers suffer from their inability to dissolve and therefore to be analyzed by chromatographic and fractionation techniques. The science of noncrosslinked polymers such as polyethylene, polypropylene, polystyrene, and so on, has been advanced by the development and application of gel permeation chromatography (GPC) and fractionation techniques such as automated temperature rising elution fractionation (ATREF). These methods have undergirded the understanding of structure–property relationships by allowing scientists to relate molecular weight and monomer composition and distribution variables to observed processing and property characteristics.

Crosslinked networks, however, cannot be directly analyzed in this fashion. The best that can be done is to measure secondary properties such as tensile properties, swelling, or sol–gel partitioning, and then, using *theoretical* relationships between these properties and network structure, infer the network structure. While one might argue that even methods such as GPC rely on theoretical relationships between molecular weight and secondary properties such as hydrodynamic volume and retention time, the relationship is more direct and amenable to calibration with absolute methods such as osmometry. In contrast, the theoretical relationships used to relate elastomer properties to network structure are more limited and rely on assumptions that cannot always be tested independently.

Thus, for example, it is common for rubber technologists to refer to M_c as a fundamental parameter of crosslinked networks which relates to swelling or tensile properties through relationships such as the Flory–Rehner equation (swelling) or the affine or phantom modulus (tensile). Commonly, these equations are used in reverse. That is, one doesn't know M_c and thus, instead of using it to predict tensile or swelling properties, one measures the equilibrium stress–strain properties or swelling properties of a crosslinked elastomer and then applies the appropriate equation in reverse to obtain M_c. Beyond the problem of limitations in the accuracy of the models themselves, a problem with this approach is that the equations being applied assume a certain network functionality that is generally not known in advance, and they assume a perfect network (no dangling chain ends or loops). Corrections are sometimes applied to adjust for the lack of network perfection, but the correction itself is just an assumption, not based on an absolute measure of the number of defects. Even when the more fundamental term, v/V, the density of network chains, is used rather than M_c, this approach is still limited by the lack of ability to directly measure the fundamental property and to know the network functionality. Technologists can use the relationships to make useful predictions, but the absolute value of the terms involved is suspect. This in turn hampers efforts to develop reliable structure–property relationships.

A form of circular reasoning can develop in which one assumes a perfect network (or applies an assumed correction), assumes a network functionality (e.g. tetrafunctional junctions), and assumes a certain model of rubber elasticity (e.g. affine deformation of Gaussian chains), and then derives M_c from theory and experimental data. The M_c thus obtained for a series of samples describes the properties of the rubber and even enables reasonable predictions, and thus the assumptions are believed to be reasonable. However, all that happened here was that M_c was essentially used as an adjustable parameter to fit data without ever knowing what M_c truly was or if the assumptions were correct.

This situation is akin to one encountered in the literature of sol–gel partition theories. Many authors have used the Charlesby–Pinner equation to estimate the number of crosslinks and the crosslink/scission ratio from sol–gel partition data (such as sol fraction and gel fraction from hot xylene extraction experiments). In doing so, they implicitly assume random crosslinking and/or scission and an initial most probable distribution of molecular weights. Some authors have justified application of the Charlesby–Pinner equation to other initial molecular weight distributions based on the tendency for random scission to eventually produce a most probable distribution. By linear regression through experimental data and extrapolation to nonzero y-intercepts on a graph of $s + s^{1/2}$ vs $1/\text{dose}$ or reciprocal initiator concentration, these authors have rationalized that scission did occur at significant rates and thus their use of the Charlesby–Pinner equation was justified. This is circular reasoning, however. As can be seen in Figure 6.2, the broader the initial distribution, the higher the y-intercept that would be obtained from linear extrapolation of low to intermediate dose (or initiator concentration) data. Thus, for example, one study [23] of low density polyethylene (LDPE) concluded that $p/q = 0.16-0.36$, despite the fact that LDPE typically has polydispersity significantly greater than 2. In fact, it is unlikely that LDPE could have had significant scission during peroxide crosslinking.

Our purpose here is not to criticize other authors, but to point out that the difficulty in directly analyzing the molecular-level structure of crosslinked networks has led to problems and confusion in the development of structure–property relationships. These difficulties arise directly from the need to interpose theories and assumptions between experimental data and a knowledge of network structure. These difficulties extend across the spectrum of theories and properties commonly used, including sol–gel partitioning, swelling, and mechanical deformation. An even more critical assessment of the state of network characterization methods may be found in a review by Funke [24].

What is the solution to this situation? Perhaps some day direct methods of analyzing network structure will become available. In the meantime, however, computer simulations are a useful way to bridge the gap. If the chemistry describing a network can be adequately described (and the definition need not be all that precise—the primary point is to be able to estimate the rates of scission and crosslinking at each polymeric site), then networks can be built in virtual reality rather than in the laboratory. A virtual network has the advantage that one can readily analyze it within the computer in almost any way imaginable (provided one has the patience and skill to write the algorithm for a particular analysis). Thus, one can 'directly' measure M_c, for example, in a virtual network, whereas one cannot do so in a real network. In fact, one needn't be confined to averages like M_c; one can determine the *distribution* of molecular weights between crosslinks.

Without a simulation, one can *measure* a physical property and then using a theory *estimate* a network structure *derived* from an initial polymer composition.

Using a simulation, one can create a virtual network structure that simulates the real one, and this simulation can be validated in a number of redundant ways, using comparisons to not one but multiple physical properties via theory, and also by predicting measurable properties such as the monomer compositional partitioning between sol and gel. Once a virtual network can be built that is congruent with a real network, then one can proceed to develop structure–property relationships based on the network structure itself rather than the uncured polymer properties. Overall, this approach is less dependent on assumptions and is less susceptible to circular reasoning than the approaches described above.

Ideally, a simulation should provide enough output properties that the validity of the simulation can be tested in several complementary ways. For example, a simulation could provide sol–gel partitioning information that can be compared with experimental extraction data, network structural parameters that can be used to calculate rubber elasticity by various models for comparison with experiments, and in the case of nonrandom copolymers or blends, partitioning of monomers between sol and gel fractions for comparison with spectroscopic (NMR or FTIR) characterization data. Such a simulation would not depend on the assumptions of a single theory or phenomenon.

6.4. A TAXONOMY OF CROSSLINKING AND SCISSION SIMULATIONS

Many simulations use Monte Carlo techniques. 'Monte Carlo,' derived from the European casino-rich city-state with that name, refers to a simulation that uses random sampling techniques to obtain approximate solutions to complex mathematical or physical problems. Randomization plays an important role in determining such things as the reaction type and reacting monomers for each step of the simulation, and depending on the simulation, the initial composition or distribution of molecular chain segments, including spatial distributions and segment motions if applicable. The complexity of simulations increases as they move from simple to more complex molecular structures, and as they move from statistical to more realistic spatially distributed models, and ultimately, to molecular dynamics models (models that take into account the motion of the monomer units and chains by simultaneously solving equations of motion for all particles in the system). This hierarchy is described below.

Durand [25] reviewed several Monte Carlo simulation studies and grouped them into two classes, A and B. Class A includes those Monte Carlo methods that generate tree-like polymer structures and allow intramolecular reactions inside the largest cluster identified as the gel. In class A simulations all the molecular species are assumed to have infinite mobility just as in the classical theories (such as Flory–Stockmayer). Class B Monte Carlo methods place monomers in fixed positions on a finite lattice and then generate cyclic polymer structures from them. Monomers are assumed to have zero mobility. In this review, we additionally define class C methods as those that not only place monomers on a lattice or other array, but also utilize some form of molecular motion (including molecular dynamics*) to move them around. Within this review, we refer to class B and C simulations as 'spatially distributed' since they position polymer chains in space and keep track of positional information in addition to the reaction and connectivity tracking of class A simulations.

If we consider the range of reactions of interest and the range of modeling and simulation approaches described in the literature, we can construct hierarchies of both reaction complexity and simulation complexity. With increased simulation complexity may come increased accuracy (although not necessarily, since sometimes applying a more fundamental type of simulation adds more unknown parameters and approximations than higher level simulations coupled with application of underlying classical models do), but at the expense of increased computer resources. These hierarchies are summarized in Tables 6.1 and 6.2.

6.4.1. Empirical Models

Although empirical models are not the subject of this chapter, it should be noted that they have a major place in rubber chemistry and engineering. The Mooney–Rivlin model of rubber elasticity, for instance, is widely used for finite element analysis and design of rubber parts. Empirical equations are not as general as models based on molecular structure and proven theoretical principles, but for the particular data set they are applied to, they may describe the data very well. One of the reasons empirical models will likely play a significant role for the foreseeable future is the complexity of industrial polymer formulations noted in the introduction. First principles models have not been developed with sufficient complexity to handle the large number of factors affecting network formation and performance. Thus,

TABLE 6.1. Complexity of Reaction Processes (In Order of Increasing Modeling Complexity, with the Top of the List Being Least Complex)

Random scission of linear chains
Nonrandom scission of linear chains
Random scission of branched chains
Nonrandom scission of branched chains
Random crosslinking
Random scission and crosslinking
Nonrandom crosslinking
Nonrandom scission and crosslinking

---

*In this chapter, 'molecular motion' refers to movement of molecules or atoms within a simulation. This term is used to distinguish between general methods of molecular motion, including Monte Carlo based movements, and 'molecular dynamics,' which has a more specific meaning within molecular simulation literature, referring to systems in which laws of motion are solved for each particle in a system.

TABLE 6.2. Complexity of Modeling Methods (In Order of Increasing Modeling Complexity, with the Top of the List Being Least Complex)

Method	Characteristics
1. Empirical models (regression analysis of designed experiments, neural networks, etc.)	General applicability but poor predictability outside design space. Models are mathematically simple
2. Mathematical theoretical models (such as Flory–Stockmayer, Saito, etc.)	Simple for random scission, increasingly difficult as complexity increases according to Table 6.1. These models assume no spatial effects
3. Simulation of linear chain scission	Simple algorithms and minimal tracking required. Spatial effects unlikely to be important
4. Simulation of branched chain scission	Tracking of branched structures requires more complex memory structures and algorithms. Spatial effects unlikely to be important
5. Simulation of crosslinking (or crosslinking/scission) without spatial or molecular motion effects	Tracking and analysis of network structures adds complexity. A simplified scheme may introduce errors, especially in densely crosslinked systems (e.g. thermoset resins) or highly dilute systems. Errors may be greater in systems without migratory free radicals or small molecule crosslinkers. Useful for relating experimental and modeling results to understand complex systems and to develop predictive models
6. Simulation of crosslinking with spatial effects and no molecular motion	Adds more complexity and computer resources. Various schemes such as lattice chains or off-lattice chains with capture spheres introduce spatial effects, but these are limited approximations of real spatial effects
7. Simulation of crosslinking with spatial effects and molecular motion	Increased complexity and greatly increased computing resources due to molecular motion, such as that incorporated via athermal bond fluctuation algorithms
8. Future directions	• First principles kinetics models based on force fields and realistic chain mobility models (molecular dynamics simulations) even for highly entangled imperfect networks. • Direct simulation of network properties via rubber elasticity of non-Gaussian network chains, including failure properties. • Incorporation of real-world variables such as mixing aspects, multiphase systems, presence of shear stresses, diffusion of competing species (e.g. oxygen), reactive fillers, etc.

the simulations described later can be useful for providing insight into the performance of simple formulations, but application to more complex formulations may require empirical extensions.

6.4.2. Mathematical Theoretical Models

In this class are the various analytic expressions described above for sol–gel partitioning and rubber elasticity in relation to crosslinking or scission phenomena, for example, Saito–Inokuti, Miller–Macosko and the like. These expressions can be extremely useful in simple cases, but their usefulness decreases as systems become more complex and underlying assumptions fail to be met, as discussed above. One practical use of these models for simulations, however, is as a method of validating simulation performance. By running simulations on model polymers and reactions that meet the assumptions of a particular theory, such as the Charlesby–Pinner equation, and then comparing simulation output with theoretical predictions, simulation performance can be tested. This is a useful exercise not only for the software developers, but also for users, since an understanding of the effect of various parameters on simulation performance can be gained by this approach.

6.4.3. Scission of Linear and Branched Polymers

There has been considerable interest in modeling random scission processes because of the commercial importance of peroxide-induced scission of polypropylene for rheology modification.

Workers at Texas A&M University and Fina Oil & Chemical Company recently published the results of a personal computer based model of polypropylene scission [26]. This model was not a Monte Carlo simulation, but used probability concepts to convert an input-discretized MWD to an output-discretized MWD. The model allowed input of experimental MWDs of rheology modification experiments, and then iteratively solved for the peroxide efficiency giving the best match between model and experimental results. On an Intel 486 CPU, DX2 66 MHz system, it took 3 min to obtain results. They studied a series of peroxide modification processes with varied extruder injection systems/sequences, temperatures, and peroxide concentrations [27]. They found peroxide efficiencies of approximately 60 percent when 2,5-dimethyl-2,5-di(t-butyl peroxy) hexane (Lupersol 101, Atochem) was injected into an extruder at 200°C, but only 25 percent when injected at 230°C. This work illustrates the value of modeling for understanding real commercial scale processes via the technique of comparing experimental and modeling results to obtain useful parameters such as efficiency to improve one's understanding of real processes and methods for process optimization. It also illustrates the fact that real processes have many variables that make it difficult to do realistic full kinetics simulations. Perhaps the greatest utility of modeling is in simplified modeling schemes where the kinetic parameters are not calculated from first principles, but are supplied from experiments and then used in simplified models to improve understanding of more complex real systems.

Huang et al. [28] developed a more sophisticated simulation for polypropylene scission that directly incorporated kinetics effects via a Monte Carlo simulation and a stochastic model in which time intervals are used. At each time interval a chain and then bond are selected, and then the probability of that bond being broken in that time interval is determined and the outcomes managed. This approach has the advantage of obtaining a time-dependence of reactions directly from the simulation itself (and kinetic parameters), but at the cost of considerable wasted computing resources since chains and bonds may be selected and then no reaction done. This approach also requires a large number of initial chains ($\sim$100,000) to dampen fluctuations. Computing times were on the order of 17 h on an IBM RISC 6000.

Giudici and Hamielec [29] studied the random scission of branched chains, a process that is difficult to model mathematically (in contrast to the scission of linear chains). They used the following Monte Carlo scheme: (1) define the initial population of chains with a given molecular weight and structure distribution; (2) pick a random number between 0 and 1 and multiply by the total number of bonds; (3) by a simple counting procedure define the chain and the bond that undergo scission; (4) determine the scission products (chain size and morphology); and (5) modify the population (one chain disappears and two new ones are created). Because of the nature of branched polymers, the simulation was required to store not just the chain lengths (as in a simulation of linear polymer scission), but also the branching information, including junction points, junction functionality, and so on. Only regular star and comb structures were considered in the simulation; their polymer definition scheme did not allow for dendritic structures. One thousand initial molecules were found to be sufficient to represent the process. They noted that fluctuations due to the Monte Carlo simulation were most pronounced in high MW tails of the distribution, and that these effects were best handled by averaging several experiments and by grouping the averaged results into several intervals of MW. One of the action items they noted in the conclusion of their paper was the need for a way to describe a generic branched structure and the scission products in a general and efficient way.

It should be noted that these scission simulations did not consider spatial effects as some crosslinking simulations do. Because scission only breaks bonds and does not form network structures, whereas crosslinking forms higher ordered network structures in addition to forming crosslinks, it is probable that spatial effects are relatively unimportant in scission processes in the absence of crosslinking.

6.4.4. Crosslinking Without Spatial Effects or Molecular Motion

Tobita and coworkers published a series of reports describing Monte Carlo simulations and their use in calculating MWDs after various processes. The ability to calculate MWDs rather than just MW averages and gel fractions was a significant advantage of simulations over classical theories. In a 1993 paper [30], Tobita showed that personal-computer-based Monte Carlo simulations could be used to calculate the long chain branch formation and MWDs of polymers produced in nonlinear free-radical polymerization processes. He also demonstrated good agreement with the Flory–Stockmayer equations. Tobita found that a population of

1000 polymer molecules was sufficient for good quality simulations. In 1994 Tobita and coworkers published a report [31] describing the simulation of the MWD produced by random crosslinking of polymers. They defined the gel point in their work by defining the gel molecule as a polymer that consists of more than 100 generations. Their model was a combination of Monte Carlo and statistical techniques. They claimed that this combination overcomes the limitations of a finite population via statistical sampling techniques. In their simulation methodology, they selected the reactive unit first, and then found the polymer chain it was on via complicated statistical methods [32]. The complexity of this method suggests that the reverse might prove easier to manage.

Cheng and Chiu [33] studied network formation from small molecules (amine cured epoxies) using a general Monte Carlo simulation approach. They used kinetics to first select a reaction, and then they selected the reacting molecules, did the book-keeping, and calculated product molecular weights. The simulation included no spatial effects, and in this sense was a simulation equivalent to classical network/sol–gel theories. They used the method of equating the conversion at the maximum value of the reduced weight average molecular weight with the gel point. They found that approximately 100,000 initial epoxy prepolymers were required to obtain less than 1 percent discrepancy from the theoretical results from the generating function method of determining the gel point. These results, in comparison with those of Tobita and others, suggest that, if the starting molecules/prepolymers are small, the number of units required for an accurate simulation is higher than for a macromolecular prepolymer.

Hendrickson et al. [34] studied the effects of substitution (i.e. changes in reaction rate as polyfunctional reactants become increasingly substituted) in condensation reactions of silicon alkoxide systems. They did not embed the reactants in space, but left them in a general population pool. A Monte Carlo algorithm was used to select the reaction, and then the reactive species (of various forms of substitution) were chosen and the reactions done. A maximum in the reduced weight average molecular weight was related to the gel point. This paper provides a good description of the method of probability-based reaction selection and linked list connectivity data structures, explaining the flexibility and advantages of linked lists in these kinds of simulations (as compared to standard data arrays).

A commercially available software package in this category is DryAdd, from Oxford Materials Ltd. From the available product literature [35], this simulation appears to be intended for epoxies, polyurethanes, polyesters, siloxanes, and other thermoset systems that are built from small molecules. It does not appear to be suitable for polydisperse macromolecular systems, which require more complex data structures and tracking systems.

6.4.5. Crosslinking and Scission Without Spatial Effects or Molecular Motion

It is common in the case of simultaneous scission and crosslinking to first apply scission to the initial polymer molecular weight distribution, then use the resultant distribution and apply the crosslinking calculations or simulation. However, this approach may not be suitable for nonrandom crosslinking and scission, especially where scission produces a reactive endgroup for crosslinking. In addition, if one wishes to follow a set of properties as a function of extent of reaction, the simulation must be run multiple times rather than just once.

Simulations that simultaneously track crosslinking and scission may involve a significant leap in computational time and effort. The reason for this is that, once an extensive network has formed with a high degree of interconnectivity, a simulation must recursively trace the entire network each time a scission reaction occurs to be sure that the scission reaction did not sever the network molecule in two. Galina and Lechowicz [36] recently described an algorithm for modeling degradation of polymer networks. They reported that tracking molecular identities, and the exhaustive search required after scission, were time-consuming. Even with only 5000 units in the network, scission to below the gel point required several days of CPU time.

The author has recently described algorithms that overcome this limitation [37]. The resultant simulation software, SimPolyMod^{TM}, simulates nonrandom (and random) scission and/or crosslinking of arbitrary initial molecular weight and monomer compositional distributions in macromolecules. The software calculates both network structural parameters and sol–gel partitioning of mass and monomers. A demonstration of its integrated approach to modeling rubber elasticity was recently published, showing the utility of simulations for understanding the effect of comonomer concentration on the performance of peroxide crosslinked ethylene octene copolymers [38]. By comparing simulation results with conventional industrial methods for characterizing rubber cure, for example, curemeters such as oscillating disk rheometers (ODR) or moving die rheometers (MDR), common errors in interpretation of curemeter data were shown [39]. Rather than curemeter torque being proportional to crosslink density, as is often assumed, torque was shown to depend on unrelaxed network defects, crosslink density and trapped entanglements, with each contribution varying with cure time or extent of cure, as well as varying with starting polymer structure. This example illustrates how simulations can complement traditional laboratory analytical techniques. Virtual and physical laboratory techniques used in concert lead to a more complete understanding than either method alone.

6.4.6. Crosslinking With Spatial Effects Without Molecular Motion

As noted by Eichinger and Akgiray [17], the nonspatial statistical models or simulations result in network structures that cannot be mapped in three-dimensional space. Despite this unrealistic aspect, the theories and simulations have been surprisingly successful. Presumably, this is because one typically uses the statistical output of the simulation or model rather than the three-dimensional structure itself. Thus, parameters such as fraction of dangling chains, number of elastically active chains, and so on, are apparently close to the values that would be obtained if the network structure was physically realistic.

When simulating crosslinking of small monomers instead of high polymers, and when reacting to high crosslink density, nonspatial models may be inadequate since they do not account for steric or diffusion limitations on extent of reaction. When polymerizing highly functional monomers, it may be that only the first two or three reactive sites are capable of reacting before chain constraints limit further reaction. Trapping of sol loops within the gel is another potential source of error that a spatial model can address. Yet another example where nonspatial simulations fall short is crosslinking of dilute solutions of high polymers, where a higher degree of intermolecular crosslinking takes place, resulting in different network structures (and elastic properties) than when the polymers are crosslinked in the dry state. These are examples of where a more sophisticated model may be required, one involving some type of spatially realistic model. Such models may potentially describe the topological relationships between network chains and the effect of topology on reactivity or properties.

To overcome the above limitations, various simulation schemes have been studied that involve arranging polymer units on a lattice or deploying polymer chains in a three-dimensional box and then crosslinking. An example of such a simulation is a coarse-grained Monte Carlo simulation developed by Eichinger and coworkers [40, 41] and commercially available from Accelrys as the Networks Module component of Insight II polymer modeling software [42]. This simulation defines reactive sites and reaction probability parameters, together with crosslinkers, then uses incrementally larger 'capture spheres' to search around a reactive site for another site with which to react. When one capture sphere overlaps another, the two reactive sites are allowed to react. Capture spheres are intended to simulate molecular diffusion. This approach is not a true kinetics approach, but does provide time and space dependence to the simulation. One advantage of this is that the spatial aspects of the simulation allow one to observe increases in relative amounts of intramolecular (with loop formation) vs intermolecular reactions when the polymer is in a dilute solution [17]. There is no volume exclusion in the system (two or more chains can share the same space), so the layout of chains in the box is unrealistic in that respect. The software does not support block copolymers or simultaneous crosslinking and scission reactions. This software requires a Silicon Graphics workstation, indicative of the increasing computational power required as simulation complexity increases.

Lee and Eichinger [41] argued that the classical gelation theory of Flory [5] underestimates the extent of reaction at the gel point because it does not take into account loops formed before the gel point. They also observed that a simulation of polyurethane network formation produced a higher sol fraction than was observed experimentally. They observed that the simulation produced many double-loop molecules, and suggested that in real systems this results in entrapment of sol in the gel phase, resulting in higher experimental gel fractions than predicted by the simulation. (This should be less important in simulations of crosslinking of high molecular weight polymers, where it is less probable that a prepolymer would have loop formation without also becoming directly attached to the network.)

In a study of radiation-cured polyethylene, Galiatsatos and Eichinger [43] found from simulation results that radiation produced a ratio of 4.5 crosslinks per chain scission, or nine crosslinked units per scission, in bulk amorphous polyethylene, compared with a ratio of three crosslinks per chain scission established via the Charlesby–Pinner equation. Another significant finding was that the level of loop formation was extremely small. This work was extended by Castner and Galiatsatos, showing the expected reduction in percentage of loops and dangling ends with increase in prepolymer molecular weight [44].

These results show how a spatial simulation model can overcome deficiencies in classical models and give insight into the performance of industrially important polymer networks. Details on the algorithms and theoretical background for the Insight II networks module have been published elsewhere [17, 40, 45].

6.4.7. Crosslinking With Spatial Effects and Molecular Motion

Durand [25] reviewed a number of approaches to network formation modeling. In reviewing several experimental results with dilute systems, he noted that mechanical stirring reduces the diluent effect (by increasing intermolecular contact and thereby reducing the extent of intramolecular reactions). He also noted that very high dilutions could result in precipitation of microgel clusters from the solution. Thus, modeling dilute systems is a challenge indeed, and these special situations may contribute to discrepancies even with a spatial model if it does not take into account molecular motion. Increasing realism in simulations requires force field or quantum mechanical approaches to intra- and interchain interactions, accurate space filling models, and time-dependent molecular positioning. Thus, the next level of simulation complexity introduces molecular motion.

Binder [46] reviewed the difficulties of simulating polymer chains at an atomistic level. The length scales for polymer systems require large populations of atoms for realistic simulations, on the order of 10^6 atoms or more for some phenomena, yet interactions at the pairwise atomic level strongly influence behavior. Times involved in polymer relaxation and dynamics range from very short to geological time scales, thus requiring small simulation time steps over long times, and consequently massive computational resources. The complexity of interactions is much greater than in simple liquids, and even for simple polymers like polyethylene there is disagreement about the parameters to use. Thus, quantum mechanical simulations of large polymer systems are out of the question at this time, and even coarse-grained models require considerable computational resources. Thus, the simulations in this class are generally of small groups of molecules in idealized networks, using massive amounts of computer resources for verifying specific aspects of rubber elasticity theories.

Kremer and Grest [47] reviewed entanglement effects in simulations of polymer melts and networks and highlighted the technical challenges molecular dynamics simulations of these systems pose. To study entanglement effects, chain lengths should be much longer than the entanglement chain length (or entanglement molecular weight, M_e), but relaxation times for polymer melts have been found experimentally to increase exponentially (to the 3.4 power) with chain length. These also translate into long relaxation times in computer simulations. In addition, the long chain lengths mean that many atoms or monomer units are involved, and thus either very few chains can be studied or computational times are further lengthened. Thus, there is a contradiction between the chain length requirements for studying topological effects on network properties and the computer resource requirements of long chain length molecular dynamics simulations. Because of this, for the near future, molecular dynamics simulations of long chain networks must use coarse-grained models, sometimes assisted by Monte Carlo methods. Kremer and Grest reviewed several network simulation studies and concluded that entanglements play a significant role in networks and, while the precise mechanism is not yet clear, the magnitude of the effect is related to the entanglement length (or M_e), a conclusion consistent with the model of Pearson and Graessley (M_e being related to G_N^0).

Sommer [48] described the Monte Carlo simulation of randomly crosslinked polymer networks for polymers placed on a lattice with the ability to move about via an athermal bond fluctuation algorithm. Radicals were randomly distributed along the polymer backbones, then reacted as they contacted one another through the polymer movement. In a related paper, Trautenberg et al. [49] described how this simulation was used to compare diffusion-controlled and reaction-controlled crosslinking processes. They were able to simulate in a pseudokinetics (Monte Carlo time) approach the change in reaction rate with increased functionality of crosslink sites (due to decreased mobility and increased polymer density). These studies included spatial effects (though restricted to a lattice) and molecular motion (via the athermal bond fluctuation algorithm). In this way, they demonstrated several effects of developing network structure and functionality on reaction rates. These effects, while not quantified in real terms, are nonetheless likely to play a role in the rate/kinetics of real systems. Thus, accurate kinetics of crosslinking and other reactions in real systems must take into account changes in reaction rates due to changing polymer crosslink density. It should be noted that the simulations of diffusion-controlled reactions took 90 CPU-h of Cray YMP-8 computer time for three samples.

Grest and Kremer [50] studied coarse-grained bead and spring continuum (e.g. not lattice-based) models, where each bead represented several monomer units and in which the beads interacted with a repulsive Lennard–Jones potential. An equilibrated melt was first achieved via a molecular dynamics simulation, and then randomly selected beads were instantly crosslinked with nearest neighbors. Alternatively, endlinking was performed. Effects on elastically active loop formation were discussed. Key network structural parameters were found to be similar to those output by the Eichinger simulation. This simulation was performed on a Cray supercomputer.

In a later study by the same authors [51], equilibrated systems of 30,000–50,000 monomer units distributed in 2500–500 chains, respectively, were endlinked in a simulated time-dependent reaction where kinetics and structural development were followed as crosslinking proceeded on the dynamic chains. Several tricks to increase simulation speed were used to drive the reaction to less than 2 percent dangling chains. The resultant networks were then studied for static properties. One important conclusion was that fluctuations during crosslinking did not alter the random chain conformation. Different rubber elasticity theories were compared with simulation results based on various assumptions, thereby elucidating the limitations of classical models. Entanglement contributions were found to be especially significant when the network chain length exceeded the entanglement molecular weight, and approximately 2.2 entanglement molecular weight chain segments were needed to form the equivalent of one crosslink. The study required over 1500 h of Cray YMP CPU time. Grest et al. [52] subsequently studied the stress–strain behavior of dry and swollen networks using a molecular dynamics simulation of freely jointed bead-spring chains. They found that simulation results agreed best with models that constrain monomer units in a tube formed by neighboring chains. These examples illustrate the value of molecular dynamics space-filling models for enhancing fundamental understanding and choosing between competing rubber elasticity models. They also illustrate the computing time required, which is why these simulations are not yet practical for widespread industrial use. Had a randomly crosslinked system been used with many more dangling chain ends, relaxation times, and thus computation times, would have been considerably longer.

A comprehensive summary of recent molecular dynamics simulations of network formation and behavior was published by Doherty et al. [53] These authors carried out a fully atomistic simulation of poly(methacrylate) network formation on a Cray C90 computer. Force fields were used, not quantum mechanical methods, but the chain structures were truly atomistic, with molecular structures created using molecule-building software. The simulation was applied to a typical dental resin comprised of two monomers, bisphenol A digycidylmethacrylate and triethyleneglycol dimethacrylate. The simulation approach involved placing 42 monomers on a lattice, then equilibrating them using molecular dynamics through a sequence of simulations to arrive at a constant volume cell. The monomers were subsequently polymerized using a capture sphere type algorithm but with molecular dynamics, and energy minimization after each reaction, as well as a final minimization after the end of network formation. The authors foresee several advantages to fully atomistic models, including the ability to calculate cohesive and mechanical properties, resistance to permeability and diffusion of small molecules, and adhesive properties. However, this atomistic simulation involved only 42 small monomers yet required 11 h of CPU time for the entire simulation.

Techniques that use molecular dynamics space-filling simulations to conduct reactions under dynamic conditions require enormous computing power, as the above examples show. To minimize computer time, simulations are often conducted at low polymer densities, where relaxation and equilibration happen faster. In contrast to these 'reactive molecular dynamics' simulations, Faulon [54] recently described a network builder method that combines random walk and pseudo-reptation on a diamond lattice, followed by molecular dynamics equilibration, in which the density, end-to-end distance and radius of gyration of chains is realistic. X-ray scattering, He diffusion, and elastic properties calculated from the simulated structures agreed with published results. Trapped entanglements or topological loops were shown to be major contributors to elasticity when the network chain length exceeds the entanglement length, in agreement with Graessley and Pearson.

This simulation method appears promising for eventual use in practical industrial simulations, since (a) CPU time was less than 10 min on an SGI workstation in contrast to days of CPU time on a Cray YMP for reactive molecular dynamics simulations, and (b) the simulation time scaled linearly with number of atoms as compared with molecular dynamics methods where simulation time is proportional to the number of atoms cubed. Polyisobutylene structures of up to 1,400,000 atoms were prepared via this method, in contrast to much smaller numbers for most molecular dynamics simulations. One limitation of this technique is that a separate network must be constructed for each crosslink density; for example, the development cannot be followed sequentially. Thus, multiple simulations would be required to study network properties as a function of crosslink density. It remains to be shown how well this method works for unsaturated hydrocarbon polymers where a diamond (or other fixed dimension) lattice would not fit nontetrahedral or varied carbon bond angles.

Given the computational requirements for bead- and spring-type simulations, quantum mechanical atomistic simulations should not be expected before major advances in computing technology, probably utilizing massive parallelism and/or quantum computing.

6.4.8. Simulation of Network Formation and Deformation (Rubber Elasticity)

The simulations discussed above focused primarily on the simulation of network structure under equilibrium conditions, although several simulations did examine topological properties of the network structures in relation to stress–strain behavior. Several simulations have particularly emphasized the stress–strain behavior of networks, including estimation of their ultimate properties as a function of network structure. Examples are briefly reviewed in this section.

Termonia and Smith [55, 56] described a simulation of tensile deformation of an entanglement network that produced qualitative visual output of the morphologies of semicrystalline polyethylene at various strains. This model involved laying out the chains of an entanglement network on a grid, then applying various slip, bond breakage, and so on, parameters to the junctions and then deforming this grid in an interative fashion with bond breakage, slippage, and relaxation processes at each step.

Later, Termonia extended this model to simulate network formation and deformation of elastomers [57]. He added crosslinked nodes in place of some of the entanglement nodes and then studied the deformation of the lattice. The crosslinking process itself did not constitute a significant aspect of the simulation since it simply connected monodisperse prepolymer chains to a predefined grid of junctions having fixed functionality. The resultant crosslinked network was analyzed and its defects removed, then a simulated tensile deformation was applied and the iterative bond breakage/slippage/relaxation processes were applied as previously. Termonia concluded from this study that entanglements were a significant contributor to rubber elasticity. Termonia's simulation took 4 h of CPU time on an IBM 3090 to generate and test the properties of a 60 × 60 node network (e.g. 3600 junctions/ entanglements).

A fascinating online publication of a simulation of network stress–strain behavior shows via video presentation the effect of topological constraints (linked loops) on stress–strain behavior [58]. Studying model trifunctional networks, Everaers found good agreement with entanglement models such as Pearson and Graessley's and higher moduli than predicted by classical models. The simulation videos show how topologically connected shortest path chains bear the brunt of the load.

Stevens [59, 60] recently described simulation of crosslinked adhesives and the effect of network structure and crosslink functionality on failure properties. Crosslinked networks were created between two rigid planes that were then separated at a defined deformation rate (simple tension or shear). He found that ordered networks could be designed with higher extensibility than random networks, and that the number of bonds at the interface vs the interior determined adhesive vs cohesive failure. The failure strain was determined by the average minimum path through the network chains connecting the two surfaces.

6.5. APPLICATION OF SIMULATIONS TO RATIONAL DESIGN

From the preceding review, it is clear that simulations have been used in several ways that are relevant to rational design of polymers for applications involving crosslinked networks. Without belaboring the point with additional examples of 'practical' applications, a brief summary of some of the ways simulations have and will continue to impact rational design is provided in this section.

6.5.1. Refinement of Theories

One of the most important ways simulations provide practical benefits for rational design has been validation and interpretation of nonsimulation (mathematical) models of crosslinking, network formation, and rubber elasticity. Practically, we are still quite far from having a 'bench chemist/hardware-friendly' molecular level simulation that is realistic in the sense of properly oriented molecular chains with quantum mechanically based atomistic modeling of motions and interactions. The computational requirements are too severe. However, pioneering versions of such sophisticated simulations have been used to help discriminate between competing rubber elasticity theories, for example.

The 'trickle down' effect of these findings allows the implementation of better theoretical (mathematical) models into less sophisticated simulations that *can* be run on today's desktop computers. Thus, for example, there can now be little doubt that topological effects, for example, entanglements, are significant in many crosslinked systems and must be accounted for. The Pearson–Graessley relation has been verified in several studies. At high extensions or dynamic (nonequilibrium) stress–strain conditions, models that are more complex can be selected using guidance from simulation results. The mathematical models and parameters thus selected may be used in conjunction with less computationally demanding simulations to obtain practically useful results.

6.5.2. Refinement of Analytical Methods

Another rather general way simulations have proven useful is for increasing our understanding of important analytical methods. Proper understanding of analytical and characterization methods and results is critical for successful rational design. Commonly used methods of characterizing the properties of network polymers rely on key assumptions and theories. Simulations have been used to demonstrate how such characterization methods can best be used, how their results should be interpreted, and the validity of assumptions in particular cases.

Thus, for example, simulations have shown that the assumption of a most probable distribution of molecular weights in the Charlesby–Pinner model is not met in many randomly crosslinked polyolefin systems. Simulations have been used to interpret and refine testing procedures for sol–gel analysis (extraction methods), oscillating disk rheometry (or other dynamic rheological methods), equilibrium stress–strain methods, and relaxation behavior.

6.5.3. Effects of Polymer Structure and Chemistry

As noted in the introduction, rational design of polymers for crosslinked applications involves development of polymer structure–network structure–property relationships. By using the relationships between network structure and properties described directly by simulations and by mathematical models validated by simulations (see above), we can probe the effects of a variety of polymer structure parameters on network properties using relatively simple simulations.

By using simulations, one need not make as many assumptions when comparing different products. For example, the author has noted instances where industrial chemists compared two different polyolefins having nominally similar Mooney viscosity or melt flow indices (a simple measurement of rheological behavior commonly used in industry as a proxy for molecular weight) and then concluded, based on ODR torque differences after cure, that crosslinking efficiency was higher in one material than the other. In at least one case, this led to an extensive search for chemical impurities that may have affected crosslinking efficiency. Upon application of a simulation approach, however, it was found that the higher torque was due to differences in MWD rather than crosslinking efficiency.

This example is repeated many times over in industrial laboratories focused on polymerizing, compounding, and fabricating crosslinked polymer systems. The knife cuts both ways: A failure to understand physical differences leads to a search for nonexistent chemical differences, or a failure to understand chemical differences leads to a search for nonexistent physical differences. In both cases, a simulation coupled with appropriate analytical methods may explain the phenomenon, or at least eliminate some possibilities.

When a series of designed experiments is done, varying polymer structural parameters and then comparing results with simulation and adjusting simulation parameters until the simulation is calibrated, one creates a 'simulation system' which can be used to run simulations that explore the effects of various combinations of structural parameters. Thus, for example, once one quantifies the relative reactivity of ethylene and octene monomer units towards dialkylperoxides, the relative rates of competing reactions such as scission and crosslinking, and the overall initiator yield from peroxides, then one can model the crosslinking behavior of ethylene octene copolymers. Using this information, one can then explore the effects of MW, MWD, blends, various compositions, various comonomer sequence distributions, and peroxide concentration on network structure and properties. The simulation output such as network structural parameters enable comparison of different experimental polymers. As noted in Section 6.1, these kinds of analyses are often more useful for building understanding of the relative importance and role of each structural variable than for direct prediction of performance in actual applications and formulations.

The same approach sheds light on the effects of various chemistries. For example, comparison of di- and tri-functional crosslinking agents can be made, or the effects of endlinking vs random crosslinking can be compared. By comparing experimental results with simulations based on certain assumptions regarding the reactions involved, one can test the soundness of those assumptions.

6.5.4. General Approach to Developing Validated Simulations

Development of a forward- and backward-integrated simulation, i.e. one that can be used both to predict properties and to explain properties, requires both simulation and experimental work. The following general approach has been found to be helpful when developing SimPolyMod™ simulation models (Table 6.2, method 5):

(1) Characterize the initial polymer analytically via several complementary methods, including characterization of MW, MWD, monomer composition, monomer sequence distribution (including presence of blocks), branching, and chemical analysis of antioxidants, additives and trace impurities.

(2) Identify key network properties to be tested, preferably including complementary techniques and principles. Especially helpful are sol–gel partitioning of mass (extraction testing), rubber elasticity (equilibrium stress–strain), stress relaxation or dynamic properties, network composition (FTIR, NMR) and compositional partitioning of components between sol and gel.

(3) Prepare materials with several levels of crosslinking agent, cure them, then measure the identified properties.

(4) Input the initial polymer composition, and making starting assumptions for unknown parameters such as monomer selectivity/reactivity and crosslinking or scission efficiency, run a simulation (or average a set of simulation runs).

(5) Use the outputs of the simulation to compare with the predicted properties for which experimental data exist (e.g. sol–gel content). Iteratively adjust the simulation input parameters (step 4) until the outputs agree with the experimental results. The more complementary methods are used, the more accurate the resulting model should be.

(6) Once a robust set of parameters has been identified, the simulation can be run for other related systems with a high degree of confidence in the results. Now validate the simulation model with experimental data covering a range of initial compositions.

(7) When experimental results and simulation results are in good agreement over a range of initial compositions and properties, the 'simulation system' of simulation model plus parameters plus input datasets can be considered validated. At this point, the output properties that cannot be measured experimentally are considered to have a high probability of being correct.

(8) With a robust system, the output that cannot be measured experimentally but which is obtained from the simulation can be used for development of structure–property relationships. For example, the dangling chain fraction can be obtained; this can then be used to relate to dynamic properties, and thus polymer structure relationships to dynamic properties may be developed.

As previously noted, this type of approach works best for simple systems. With complex formulations, it may prove most helpful to run simulations on simpler formulations first, to understand the effects of the major variables in that system, before attempting to extend it to the more complex formulation.

As noted in the review above, simulations involving high molecular weight initial polymers can generally be done on a relatively few chains, perhaps 500–1000, while obtaining statistically useful results. Simulations involving low molecular weight prepolymers, such as those used for thermoset polymers, generally require many more initial molecules to obtain statistically reliable results. Gel point determinations can be made via any of the methods described, but with simulations involving a sufficient number of chains, can be estimated from the peak reduced weight average molecular weight. With fewer initial chains, this may overestimate the gel point, but sol–gel data past this point can be extrapolated back to zero gel fraction to obtain a gel point estimate.

For now, simulations are more or less analytical tools. In the future, as 'simulation systems' are developed for particular formulation families, one may envision the development of iterative procedures in which parameters are adjusted within optimization software supersystems so that the optimal polymer structure or crosslinking chemistries are applied to achieve a given set of desirable properties. The supersystem could employ traditional statistical methods, neural networks or genetic algorithms. It would optimize polymer design parameters based on a set of desirability functions. At that point, we will be closer to the elusive 'black box' approach to polymer design. However, for the foreseeable future a universal 'black box' is not attainable due to the computational limitations on achieving true 'first principles' molecular models.

6.6. CONCLUSION

There is a hierarchy of complexity and sophistication in the crosslinking simulation literature, ranging from simulations that make similar assumptions regarding spatial effects as the classical theories of network formation and rubber elasticity do, to simulations that incorporate spatial (topological) and molecular motion effects. Choosing the appropriate level of complexity for a simulation project involves weighing the trade-offs between accuracy and computing resources, while also factoring in experimental parameter requirements and knowledge, ease-of-use and other indirect factors. Commercially available simulation software applications do not offer the most realistic simulation models for this reason, but the compromises made have not precluded agreement with experimental results when simulations are used in combination with appropriate theoretical models.

In this review, we have not emphasized application examples, but rather have attempted to summarize for the reader a general theoretical and experimental approach that has been found useful in conjunction with simulations to build an understanding of key parameters affecting polymer structure–network structure–property relationships. However, crosslinking simulations are widely applicable, with publications describing crosslinking simulations for polymers ranging from thermoset resins to rubber, chemistries ranging from free radical random crosslinking to functionalized endlinking, and end-use applications from dental resins to adhesives to tires.

Improved fundamental understanding of network structure–property relationships is a major goal of simulation studies. Competing rubber elasticity theories have been examined, and a significant contribution of simulations has been to highlight the importance of topological constraints or trapped entanglements. In general, application of simulations to gaining understanding of underlying fundamental aspects of crosslinking behavior appears to be a more fruitful application than use as a direct predictive tool for complex industrial formulations.

It is appealing to think of simulations as a designer's black box into which one can input desired properties and receive as output a polymer or formulation design. Unfortunately (fortunately from a chemists' employment perspective), this is not the case. Computers have proven adept at carrying out human instructions for routine tasks, but relatively poor at creativity and innovation. (One exception is in the area of search methods [61] of discovery, where computers do excel at searching a large solution space quickly, either by brute force searching an entire region, time-limited searches using genetic algorithms, or by refined searches using statistical optimization techniques.) Thus, computer simulations and models are not a panacea, nor a replacement for human intelligence. Serendipity has and will continue to play a role in the development of crosslinking chemistry and related phenomena. Computers likely will be most helpful: (a) in 'search' methods for discovery; (b) as analytical tools to characterize polymers; (c) as tools to develop fundamental understanding; and (d) in traditional computer applications for instrumentation control, data acquisition, management, and analysis. In these ways, computers will be tools for human intelligence, not replacements for it.

As tools, computers and simulation software must be used intelligently: properly aimed and focused. 'Garbage in, garbage out' certainly applies here. Promising directions for commercial simulations as computing power becomes more affordable and user-friendly include full volume exclusion models with molecular dynamics, directly simulated kinetics, and simulated network deformation. Increasing computer power will enable creation of true molecular level models that are applicable to a wider variety of systems and to formulations that are more complex. However, for the foreseeable future, simulations will be a powerful tool in the hands of skilled practitioners, not a 'black box.' Breakthroughs will happen at the interface of man and machine, not within the machine itself.

REFERENCES

[1] Subramaniam, A., Natural rubber, in *The Vanderbilt Rubber Handbook*, 13th edn, Ohm, R.F., ed. (R.T. Vanderbilt, Norwalk, CT, 1990), p. 23.

[2] Conoco-Phillips Inc., Teaching tools: serendipity; www.teachingtools.com/Slinky/serendipity.html, downloaded 28 July 2003.

[3] Sperling, L.H., *Introduction to Polymer Science*, 3rd edn (Wiley-Interscience, New York, 2001), pp. 17–18.

[4] Coleman, M.M., The incredible world of polymers: tales of innovation, luck and perseverance, in *Society of Plastics Engineers, Annual Technical Conference*, San Francisco, CA, May 5–9, 2002, from lecture notes taken by R.T. Johnston.

[5] Flory, P.J., *Principles of Polymer Chemistry* (Cornell University Press, Ithaca, NY, 1953), pp. 432–494.

[6] Erman, B. and Mark, J.E., *Structures and Properties of Rubberlike Networks* (Oxford University Press, New York, 1997).

[7] Stockmayer, W.H., *J. Chem. Phys.*, **12** (1944), 125.

[8] Charlesby, A. and Pinner, S.H., *Proc. R. Soc. Lond., Ser. A*, **249** (1959), 367.

[9] Saito, O., *J. Phys. Soc. Jpn*, **13** (1958), 1451.

[10] Inokuti, M., *J. Chem. Phys.*, **38** (1963), 2999.

[11] Macosko, C.W. and Miller, D.R., *Macromolecules*, **9** (1976), 199.

[12] Miller, D.R. and Macosko, C.W., *Macromolecules*, **9** (1976), 206.

[13] Miller, D.R. and Macosko, C.W., *J. Polym. Sci.: Part B: Polym. Phys.*, **26** (1987), 1.

[14] Lal, J. and McGrath, J.E., *J. Polym. Sci.: Part C*, **16** (1967), 33.

[15] Scanlan, J., *J. Polym. Sci.*, **43** (1960), 501.

[16] Case, L.C., *J. Polym. Sci.*, **45** (1960), 397.

[17] Eichinger, B.E. and Akgiray, O., Computer simulation of polymer network formation, in *Computer Simulation of Polymers*, Colbourn, E.A., ed., (Longman, Harlow, 1994), pp. 263–302.

[18] Mark, J.E., and Erman, B., *Rubberlike Elasticity: a Molecular Primer* (Wiley, New York, 1988).

[19] Flory, P.J., *J. Chem. Phys.*, **66** (1977), 5720.

[20] Flory, P.J. and Erman, B., *Macromolecules*, **15** (1982), 800.

[21] Flory, P.J., Networks, in *Encyclopedia of Polymer Science and Engineering*, Vol. 10 (Wiley, New York, 1987), pp. 95–112.

[22] Pearson, D.S., and Graessley, W.W., *Macromolecules*, **13** (1980), 1001.

[23] Narkis, M., and Miltz, J., *Polym. Engng. Sci.*, **9** (1969), 153.

[24] Funke, W., Polymer networks and their characterization, in Elias, H.-G. and Pethrick, R.A., eds, *Polymer Yearbook—1*, (Hardwood Academic, New York, 1984), pp. 101–111.

[25] Durand, D., Network formation—from basic theories towards more realistic models, in Pethtrick, R.A., Zaikov, G.E., eds, *Polymer Yearbook—3*, (Harwood Academic, New York, 1986), pp. 229–253.

[26] Bonilla-Rios, J., Darby, R., and Sosa, J.M., *Society of Plastics Engineers, ANTEC '95*, p. 1625.

[27] Bonilla-Rios, J., Darby, R., and Sosa, J.M., *Society of Plastics Engineers, ANTEC '95*, p. 1630.

[28] Huang, C., Tzoganakis, C., and Duever, T.A., *Polym. Reaction Engng.*, **3** (1995), 43.

[29] Giudici, R. and Hamielec, A.E., *Polym. Reaction Engng.*, **4** (1996), 73.

[30] Tobita, H., *Polym. Reaction Engng.*, **1** (1992–1993), 379.

[31] Tobita, H., Yamamoto, H.Y., and Ito, K., *Macromol. Theory Simul.*, **3** (1994), 1033.

[32] Tobita, H., *Polymer*, **36** (1995), 2585.

[33] Cheng, K.-Ch. and Chiu, W.-Y., *Macromolecules*, **27** (1994), 3406.

[34] Hendrickson, R.C., Gupta, A.M., and Macosko, C.W., *Comput. Polym. Sci.*, **4** (1994), 53.

[35] Oxford Materials Ltd, DryAdd professional software; www.oxmat.co.uk/Business/products/prodcat/dryadd.htm, downloaded 11 July 2003.

[36] Galina, H. and Lechowicz, J., *Comput. Chem.*, **22** (1998), 39.

[37] Johnston, R.T., paper no. 47, presented at a *Meeting of the Rubber Division, American Chemical Society*, Dallas, TX, 4–6 April 2000.

[38] Johnston, R.T., *Rubber Chem. Technol.*, **76** (2003), 174.

[39] Johnston, R.T., Modeling peroxide crosslinking in polyolefins, in *Society of Plastics Engineers Annual Technical Conference*, San Francisco, CA, 5–9 May, 2002.

[40] Galiatsatos, V. and Eichinger, B.E., *Rubber Chem. Technol.*, **61** (1988), 205.

[41] Lee, K.J. and Eichinger, B.E., *Polymer*, **31** (1990), 406.

[42] Accelrys Inc., Insight II for coarse grained simulation of polymers; www.accelrys.com/insight/I2_coarsegrained.html#Networks, downloaded 4 August 2003.

[43] Galiatsatos, V. and Eichinger, B.E., *J. Polym. Sci.: Part B: Polym. Phys.*, **26** (1988), 595.

[44] Castner, E.S. and Galiatsatos, V., *ANTEC '95*, p. 2547.

[45] Akgiray, O., *Makromol. Chem., Macromol. Symp.*, **76** (1993), 211.

[46] Binder, K., in *Monte Carlo and Molecular Dynamics Simulations in Polymer Science*, Binder, K., ed., Chap. 1, (Oxford University Press, New York, 1995), pp. 3–46.

[47] Kremer, K. and Grest, G.S., in *Monte Carlo and Molecular Dynamics Simulations in Polymer Science*, Binder, K., ed., Chap. 4, (Oxford University Press, New York, 1995), pp. 194–271.

[48] Sommer, J., *Macromol. Symp.*, **81** (1994), 139.

[49] Trautenberg, H.L., Sommer, J., and Göritz, D., *Macromol. Symp.*, **81** (1994), 153.

REFERENCES

[50] Grest, G.S. and Kremer, K., *Macromolecules*, **23** (1990), 4994.
[51] Duering, E.R., Kremer, K., and Grest, G.S. *J. Chem. Phys.*, **101** (1994), 8169.
[52] Grest, G.S., Pütz, M., Everaers R., and Kremer, K., *J. Non-Cryst. Solids*, **274** (2000), 139.
[53] Doherty, D.C., Homes, B.N., Leung, P., and Ross, R.B., *Comput. Theor. Polym. Sci.*, **8** (1998), 169.
[54] Faulon, J.-L., *J. Comp. Chem.*, **22** (2001), 580.
[55] Termonia, Y. and Smith, P., *Macromolecules*, **20** (1987), 835.
[56] Termonia, Y. and Smith, P., *Macromolecules*, **21** (1988), 2184.
[57] Termonia, Y., *Macromolecules*, **22** (1989), 3633.
[58] Everaers, R., *New J. Phys.* (1999), Article 12. Downloaded from www.iop.org, 6 August 2003.
[59] Stevens, M.J., *Macromolecules*, **34** (2001), 1411.
[60] Stevens, M.J., *Macromolecules*, **34** (2001), 2710.
[61] Thagard, P., *Studies in History and Philosphy of Science*, 1997 (prepublication version, from http://cogprints.ecs.soton.ac.uk/archive/0000067/00/ulcers.htm, downloaded 28 July 2003).

INDEX

Ab initio polymer quantum theory:
  band structure and wave-functions, 4–20
    Bloch's theorem, 4–5
    Hartree-Fock structure calculations, 8–12
      integration procedures, 9–10
      lattice summations, 8–9
      quasi-linear dependencies, 10–12
    helical symmetry, 12–13
    LCAO Hatree-Fock methodology, 5–7
    Møller-Plesset and wave-function-based electron correlation, 14–20
    unrestricted Hartree-Fock approach, 13
  force field parametrization, 53–54
  future research issues, 36–38
  polymer properties and applications, 20–36
    band structure, density of states, and photoelectron spectroscopy, 20–23
    electronic polarizabilities and hyperpolarization, 28–32
    excitation energies: band gaps and excitons, 23–28
    structural geometries, vibrationala and mechanical properties, 32–36
  research background, 2–4
Aggregation phenomena, inhomogeneous polymers, mesoscopic simulation, continuous field model, 200–204
Alaylioglu, Evans, and Hyslop (AEH) quadratures, polymer band structure and wave-function, Hartree-Fock calculations, integration procedures, 9–10
Allyl bond rotations, 1,4-polybutadiene force fields, 72–73
AMBER force field:
  PEO force fields, many-body polarization effects, 84
  polymer force field parametrization, repulsion parameter deviation, 61
Amorphous cells:
  isotactic polypropylene, mechanical properties, 233–235
  semicrystalline polymers, 225–226
Analytical models, crosslinking polymer simulation, rational design, 269
AN content, homogeneous polymers, mesoscopic simulation,

multicomponent phenomenological approach, 170–171
Angle-resolved ultraviolet photoelectron spectroscopy (ARUPS), polymer band structure, 20–23
Anisotropic crystalline solids, semicrystalline polymers:
  elastic constants, 227–228
  static deformation, 228–230
Approximate CCD (ACCD), polymer band structure and wave-function, Møller-Plesset second-order (MP2) perturbation theory (MPPT), 16–20
Atomic form factors, crystal cell modeling, semicrystalline polymers, 224–225
Avogadro's number, crosslinking polymer simulation, perfect networks, 253

Band structure, *ab initio* polymer quantum theory, 4–20
Bloch's theorem, 4–5
  density of states and photoelectron spsectroscopy, 20–23
  excitation energies, 23–28
  Hartree-Fock structure calculations, 8–12
    integration procedures, 9–10
    lattice summations, 8–9
    quasi-linear dependencies, 10–12
  helical symmetry, 12–13
  LCAO Hatree-Fock methodology, 5–7
  Møller-Plesset and wave-function-based electron correlation, 14–20
  unrestricted Hartree-Fock approach, 13
Basis set superposition error (BSSE):
  polymer force field parametrization, 53–54
  Hartree-Fock corrections:
    dispersion parameters, 61–63
    repulsion parameters, 60–61
  polymer force fields, parametrization algorithm, 69–70
Beaded polymer structure, inhomogeneous polymers, mesoscopic simulation, 172–173
  Kohn-Sham-like self-consistent field formalism, 180–182
  mean field approximation, 195
Bethe-Salpeter equation, polymer excitation energies, 27–28
Binary polymer liquids:
  homogeneous polymers, mesoscopic simulation, multicomponent phenomenological approach, 165–171
  Monte Carlo simulations, 96–98
    coarse-grained models, 99–101
    entropic contributions, χ parameter, 124–134
    monomer shape, non-additive packing, 124–127
    pressure and solvent density effects, 131–134
    stiffness asymmetry, 127–131
  films and two-dimensional systems, 134–136
  future research issues, 143–145
  interfaces, 136–142
  mean-field theory predictions, 101–106
  models and techniques, 106–111
  phase behavior, fluid structure and composition, 111–124
    Flory-Huggins parameter χ, symmetric blends, 111–120
    single chain conformations, 120–124
  strong segregation limit, 105–106
  weak segregation limit, 104–105
"Black box" technology, crosslinking polymer simulation:
  limits of, 271–273
  research background, 244–247
Bloch's theorem, *ab initio* polymer quantum theory, band structure and wave function, 4–5
Block copolymers, mesoscopic simulation, Kohn–Sham-like formalism, quasi-ideal ensembles, 191–194
B3LYP theory, polymer force field parametrization, partial charge determination, 55–59
Boltzmann density distribution, mesoscopic polymer simulation, dynamic density functional theory, 206–208
Bond fluctuation model, binary polymer liquids, Monte Carlo simulations, Flory-Huggins parameter χ, 113–120
  stiffness asymmetry, 128–131
Bond length alternation (BLA):
  1,4-polybutadiene force fields, 70–71
  polymer band structure and wave-functions, Hartree-Fock calculations, lattice summations, 8–9

INDEX

Bond polarizability model, polymer potential function, partial atomic charge and dipole polarizability, 56–58

Born-von Kàrmàn cyclic boundary conditions, polymer band structure and wave-functions:
  Bloch's theorem, 4–5
  electron polarizabilities/hyperpolarizabilities, 29–32

Bragg's theory, semicrystalline polymers, crystal cell modeling, 224–225

Branched polymers, crosslinking polymer simulation, scission of, 259–260

Brominated poly(isobutylene-co-p-methylstyrene) (BIMS), homogeneous polymers, mesoscopic simulation, multicomponent phenomenological approach, 170–171

Buckingham potential, polymer force fields, 50–51

Bulk properties, binary polymer liquids, Monte Carlo simulations, 108–111
  films and two-dimensional systems, 134–136

Canonical partition functions, binary polymer liquids, Monte Carlo simulations, Flory-Huggins parameter χ, 125–127

Car-Parinello type quantum mechanics, binary polymer liquids, Monte Carlo simulations, 97–98

Characteristic ratio, mesoscopic simulation, inhomogeneous polymers, Kohn-Sham-like formalism, quasi-ideal ensembles, 194

Charlesby-Pinner (C-P) equation, crosslinking polymer simulation:
  gelation, 248–250
  mathematical principles, 259
  rational design, 269
  research applications, 255–256
  sol-gel partitioning, 250–252
  spatial effects without molecular motion, 264

CH–CH and CH_2–CH nonbonded parameters, 1,4-polybutadiene force fields, 75–76

Chemical potential, homogeneous polymers, mesoscopic simulation, 160–163

C–H vector, 1,4-polybutadiene force fields, nuclear magnetic resonance spin-lattice relaxation, 79–80

Circular dichroism, polymer band structure and wave-function, 37–38

Clenshaw-Curtis (CC), polymer band structure and wave-function, Hartree-Fock calculations, integration procedures, 9–10

Closed-shell systems, polymer excitation energies, 24–25

Coarse-grained models, binary polymer liquids, Monte Carlo simulations:
  Flory-Huggins parameter χ, 114–120
  research background, 96–101

Comonomer distribution, crosslinking polymer simulation, 245

Compliance matrix, semicrystalline polymers, elastic constants, 227–228

Composition-dependent interaction parameter, homogeneous polymers, mesoscopic simulation, 163–165

Condensed-phase conformations, PEO structural and dynamic properties, 85

Configurational entropy, binary polymer liquids, Monte Carlo simulations, Flory-Huggins parameter χ, 120

Configuration interaction singles (CIS), polymer excitation energies, 25–28

Conjugated polymers, *ab initio* polymer quantum theory, 2

Constant-pressure, constant-temperature ensemble (NPT):
  isotactic polypropylene:
    amorphous phase, 233–235
    crystalline phase, 232–233
  semicrystalline polymers:
    amorphous cells, 225–226
    force field simulations, 223–224

Constant-volume, constant-temperature ensemble (NVT), semicrystalline polymers, force field simulations, 223–224

Constrained junction model, crosslinking polymer simulation, rubber elasticity, 253

Continuous field model, inhomogeneous polymers, mesoscopic simulation, 199–204

Convolution integrals, inhomogeneous polymers, mesoscopic simulation, Kohn-Sham-like formalism, quasi ideal ensembles, 189–194

Copolymer-solvent mixtures, mesoscopic simulation, dynamic density functional theory, 209–210

Correlational functional, inhomogeneous polymers, mesoscopic simulation, Kohn-Sham-like formalism, 182–183
Correlation-consistent polarizable basis sets, polymer force field parametrization, 53–54
Correlation hole effect, mesoscopic simulation, Kohn-Sham-like formalism, quasi-ideal ensembles, 192–194
Coulomb interaction:
  1,4-polybutadiene force fields, 70–71
  polymer band structure and wave-functions:
    electron polarizabilities/hyperpolarizabilities, 31–32
    Hartree-Fock calculations, lattice summations, 8–9
    LCAO Hartree-Fock calculations, 5–7
  polymer potential function, 49–51
  partial atomic charge and dipole polarizability, 54–59
Coupled cluster doubles (CCD), polymer band structure and wave-function, Møller-Plesset second-order (MP2) perturbation theory (MPPT), 16–20
Coupled Hartree-Fock approach, polymer properties, electron polarizabilities/hyperpolarizabilities, 29–32
Coupled-perturbed Hartree-Fock (CPHF), polymer band structure and wave-functions, quasi-linear dependencies, 12
Covalent bonds, polymer force fields, 63–64
Crank-Nicolson scheme, inhomogeneous polymers, mesoscopic simulation, Kohn-Sham-like formalism, quasi-ideal ensembles, 190–194
Critical micelle concentration (CMC), inhomogeneous polymers, mesoscopic simulation, continuous field model, 200–204
Critical temperature scaling, binary polymer liquids, Monte Carlo simulations, Flory-Huggins parameter χ, 119–120
  stiffness asymmetry, 129–131
Crosslinking polymer simulation:
  absence of spatial effects or molecular motion, 260–261
  applications, 253–256
  empirical models, 257–259

future research issues, 272–273
  mathematical theoretical models, 259
  network formation (gelation), 247–250
  network structure definitions, 250–253
  rubber elasticity, 253
  rational design:
    analytical refinement, 269
    structure and chemistry effects, 269–270
    theoretical refinement, 268–269
    validated simulation standards, 270–271
  research background, 244–247
  rubber elasticity, network formation/deformation, 267–268
  scission of linear and branched polymers, 259–260
  scission with spatial effects or molecular motion, 261–262
  spatial effects with molecular motion, 264–267
  spatial effects without molecular motion, 262–264
  taxonomy of categories, 256–268
Crystal cell modeling:
  isotactic polypropylene, 231–233
  molecular simulation-micromechanical models, 235–240
  semicrystalline polymers, 224–225
Crystal orbital strategies, polymer band structure and wave-functions, structural geometry, vibrational and mechanical properties, 32–36
Curemeter data, crosslinking polymer simulation, without spatial effects or molecular motion, 262
Cyclohexane-polystyrene (CH-PS) mixtures, homogeneous polymers, mesoscopic simulation, multicomponent phenomenological approach, 167–171
Cyclohexane-polystyrene-polyisobutylene (PIB), homogeneous polymers, mesoscopic simulation, interaction parameters, 171

Dangling chain end, crosslinking polymer simulation, 250–253
De Broglie wavelength, mesoscopic simulation, Kohn-Sham-like self-consistent field formalism, ideal systems, inhomogeneous polymer systems, 180–182

Debye function, binary polymer liquids, Monte Carlo simulations, mean-field theory, 104–105
Demixing temperature, binary polymer liquids, Monte Carlo simulations, 103–106
Density functional theory (DFT):
  mesoscopic simulation, inhomogeneous polymers:
    basic concepts, 178–180
    functional derivatives, 174–175
    Kohn-Sham-like formalism, 182–183
  polymer band structure and wave-function, 20
  structural geometry, vibrational and mechanical properties, 32–36
  polymer excitation energies, 27–28
Density of states (DOS), polymer band structure, 20–23
Diblock copolymers:
  mesoscopic simulation, Kohn-Sham-like formalism, quasi-ideal ensembles, 191–194
  Monte Carlo simulations, interfaces, 141–142
Different orbitals for different spins (DODS), polymer band structure and wave-functions, unrestricted Hartree-Fock approach, 13
Diffusion equations, mesoscopic polymer simulation, dynamic density functional theory, 205–208
Dihedral potentials:
  1,4-polybutadiene force fields, 75
  transferability, 76–77
  polymer force fields:
    parametrization data, 66–68
    valence bonds, 65–66
1,2-Dimethoxyethane, PEO structural and dynamic properties, 85
Dipole polarizability determination, polymer force fields, classical potential function, 54–59
  many-body polarizable model, 56–58
  transferability, 58–59
Dirac delta function, inhomogeneous polymers3, mesoscopic simulation, 173–174
Dispersion parameters:
  1,4-polybutadiene force fields, 70–71
  parametrization, 74
  polymer force fields, 59–63
Dynamical mechanical analysis (DMTA), isotactic polypropylene, molecular simulation-micromechanical models, 239–240
Dynamic density functional theory (DDFT), mesoscopic polymer simulation, 205–215
  copolymer-solvent mixtures, 209–210
  excluded volume effects, 208
  external potential dynamics, 210–215
  principles, 205–208

Eichinger simulation, crosslinking polymers, spatial effects and molecular motion, 265–267
Elastically active chain, crosslinking polymer simulation, 250–253
Elastically active junction, crosslinking polymer simulation, 250–253
Elastic constants. *See also* Young's modulus
  crosslinking polymer simulation, 253–256
  isotactic polypropylene, molecular simulation-micromechanical models, 235–240
  semicrystalline polymers:
    mechanical properties simulation, 226–228
    static deformation, 228–230
Electron-affinities (EAs), polymer band structure and wave-function, Møller-Plesset second-order (MP2) perturbation theory (MPPT), 17–20
Electron correlation, polymer band structure and wave-function, 14–20
Electron density distribution function, semicrystalline polymers, crystal cell modeling, 224–225
Electron-hole interactions, polymer excitation energies, 23–28
Electron polarizabilities, polymer properties, 28–32
Electron spectroscopy for chemical analysis (ESCA), *ab initio* polymer quantum theory, 2
Electrostatic potentials, polymer potential function, partial charge determination, 54–59
Empirical models, crosslinking polymer simulation, 257–259
Entanglement effects, crosslinking polymer simulation:
  network formation-deformation, 267–268
  spatial effects and molecular motion, 264–267

Enthalpy density of mixing, homogeneous polymers, mesoscopic simulation, 159–160
Entropic contributions, binary polymer liquids, Monte Carlo simulations:
 Flory-Huggins parameter χ, 124–134
 monomer shape and non-additive packing, 124–127
 pressure and solvent density effects, 127–134
 stiffness asymmetry, 127–131, 128–131
 mean-field theory, 102–106
Equilibrium bead density functions, inhomogeneous polymers, mesoscopic simulation, Kohn-Sham-like formalism, quasi-ideal ensembles, 187–194
Equilibrium bond lengths, 1,4-polybutadiene force fields, 74
Exchange integrals, polymer band structure and wave-functions, Hartree-Fock calculations, lattice summations, 8–9
Excitation energies, polymer band structure, gaps and excitons, 23–28
Excitons, *ab initio* polymer quantum theory, 23–28
Experimental calculations, force field parametrization, 53–54
Extended volume effects, mesoscopic simulation:
 dynamic density functional theory, 208
 inhomogeneous polymers, Kohn-Sham-like formalism, 184–187
External potential dynamics (EPD), mesoscopic simulation, dynamic density functional theory, 210–215

Feynmann decomposition, inhomogeneous polymers, mesoscopic simulation, Kohn-Sham-like formalism, quasi-ideal ensembles, 189–194
Fick's law, mesoscopic polymer simulation, dynamic density functional theory, 206–208
Field theoretic methods, inhomogeneous polymers, mesoscopic simulation, 182–183
Films, binary polymer liquids, Monte Carlo simulations, 134–136
Filon quadratures, polymer band structure and wave-function:
 Hartree-Fock calculations, integration procedures, 9–10
 structural geometry, vibrational and mechanical properties, 33–36
Finite-difference techniques, semicrystalline polymers, force field simulations, 222–224
Flory-Huggins parameter χ:
 binary polymer liquids, Monte Carlo simulations:
  coarse-grained models, 99–101
  future research, 143–145
  interfaces, 136–142
  mean-field theory, 101–106
  research background, 97–98
  symmetric blends, 111–120
 mesoscopic simulation:
  homogeneous polymers, 156–171
   chemical potential, 160–163
   interaction energy density parameter, 159–160
   multi-component phenomenological approach, 163–171
    composition-dependent interaction parameter, 163–165
    parametrization, 166–171
   notation for, 156–159
  inhomogeneous polymers, mean field approximation, 195
Flory-Stockmayer relation, crosslinking polymer simulation:
 gelation, 248–250
 without spatial effects/molecular motion, 260–261
Fluctuation technique, semicrystalline polymers, mechanical properties simulation, 230–231
Fluid function, inhomogeneous polymers, mesoscopic simulation, 173–175
Fock matrix elements:
 polymer band structure and wave-function, Møller-Plesset second-order (MP2) perturbation theory (MPPT), 15–20
 polymer band structure and wave-functions:
  Hartree-Fock calculations:
   lattice summations, 8–9
   quasi-linear dependencies, 11–12
  LCAO Hartree-Fock calculations, 5–7
Force field simulations:
 quantum-chemistry-based polymers:
  classical potential function, 48–70

covalent bond parameters, 63–64
dihedral potential, 66–68
dispersion and repulsion parameter determination, 59–63
existing potentials, 51
improper torsion, 68
parametrization algorithm, 68–70
  data set establishment, 69
  force field determination, 68–69
  validation, 69–70
parametrization data, 51–54
partial atomic charge and dipole polarizability determination, 54–59
  many-body polarizable model, 56–58
  transferability, 58–59
potential form, 49–51
valence bonds, 64–66
PEO and PEI-LiBF$_4$ polarizability, 83–91
  condensed-phase conformations, 85
  dynamics, 85–86
  force field development field, 87–88
    many-body polarizable potential, 87
    two-body polarizable potential, 87–88
  many-body force field development, 83–84
  molecular dynamics simulations, 88–91
    structural properties, 88–89
    transport properties, 89–90
  structural and dynamic properties, 84–86
  thermodynamics, 85
1,4-polybutadiene, 70–82
  dihedral potential transferability, 76–77
  model molecules, 71–73
  needs assessment, 70–71
  parametrization, 73–76
    CH–CH and CH$_2$–CH nonbonded parameters, 75–76
    dihedral parameters, 75
    dispersion-repulsion, 74
    equilibrium bond lengths and valence bond angles, 74
  potential function validation, 77–82
    molecular dynamics simulations, 77–78
    nuclear magnetic resonance spin-lattice relaxation times, 78–79
    potentials accuracy, 82
    single chain dynamic structure factor, 80–82
    static structure factor and gyration radius, 78
  semicrystalline polymers, 221–224
Fourier transform:
  binary polymer liquids, Monte Carlo simulations:
    interfaces, 138–142
    mean-field theory, 104–105
  inhomogeneous polymers, mesoscopic simulation, Kohn-Sham-like formalism, quasi-ideal ensembles, 190–194
  polymer band structure and wavefunction:
    Hartree-Fock calculations, lattice summations, 9
    Møller-Plesset second-order (MP2) perturbation theory (MPPT), 16–20
  semicrystalline polymers, crystal cell modeling, 224–225
Free energy functions:
  binary polymer liquids, Monte Carlo simulations, mean-field theory, 105–106
  inhomogeneous polymers, mesoscopic simulation:
    density functional theory, 179–180
    Kohn-Sham-like formalism, 182–183
  semicrystalline polymers, fluctuation technique, mechanical properties simulation, 230–231
Functional derivatives, mesoscopic simulation:
  dynamic density functional theory, 208
  external potential dynamics, 210–215
  inhomogeneous polymers, 173–175
    density functional theory, 178–180

Gauche molecular models, 1,4-polybutadiene force fields, 72–73
Gaussian bead-bead interaction, inhomogeneous polymers, mesoscopic simulation, Kohn-Sham-like formalism, 184–187
Gaussian chain structure:
  binary polymer liquids, Monte Carlo simulations, 102–106

Gaussian chain structure (*Continued*)
    Flory-Huggins parameter χ, 132–134
    single chain dynamic structure factor, 121–124
    mesoscopic simulation:
        dynamic density functional theory, 207–208, 209–210
        inhomogeneous polymers, 173
        Kohn-Sham-like formalism, quasi-ideal ensembles, 188–194
Gauss-Legendre (GL) quadrature, polymer band structure and wave-function, Hartree-Fock calculations, integration procedures, 9–10
Gelation, crosslinking polymer simulation:
    spatial effects without molecular motion, 263–264
    theoretical background, 247–250
Gel point, crosslinking polymer simulation, theoretical background, 247–250
Geometric optimization:
    1,4-polybutadiene force fields, 72–73
    polymer band structure and wave-function, structural geometry, vibrational and mechanical properties, 33–36
    polymer force fields, parametrization algorithm, 69–70
Gibbs-Duhem integration technique, binary polymer liquids, Monte Carlo simulations, 108–111
Gibbs free energy density, homogeneous polymers, mesoscopic simulation:
    chemical potential, 160–163
    interaction energy density parameter, 160
Ginzberg criterion, binary polymer liquids, Monte Carlo simulations, 119–120
    Flory-Huggins parameter χ, 126–127
Glide plane symmetry, polymer band structure and wave-functions, 13
GNU Scientific Library (GSL), inhomogeneous polymers, mesoscopic simulation, Kohn-Sham-like formalism, 185–187
Grand canonical ensemble, inhomogeneous polymers, mesoscopic simulation, 176–178
    Kohn-Sham-like self-consistent field formalism, 181–182
Grand potential, mesoscopic simulation, Kohn-Sham-like self-consistent field formalism, ideal systems, inhomogeneous polymer systems, 180–182
Green's function:
    polymer excitation energies, 26–28
    Scheutjens-Fleer model, inhomogeneous polymers, 197–199

Halpin-Kardos elastic moduli, isotactic polypropylene, molecular simulation-micromechanical models, 2237–240
Hartree-Fock calculations:
    ab initio polymer quantum theory:
        band structure and wave-functions, 8–12
        integration procedures, 9–10
        lattice summations, 8–9
        quasi-linear dependencies, 10–12
        LCAO methodology, 5–7
        research background, 2–4
    polymer excitation energies, band gaps and excitons, 25–28
    polymer force fields:
        quantum chemistry dispersion parameters, 61–63
        quantum chemistry repulsion parameters, 60–61
Helfand penalty functional, mesoscopic polymer simulation, dynamic density functional theory, 208
Helical symmetry:
    isotactic polypropylene, crystalline phase, 231–233
    polymer band structure and wave-functions, Hartree-Fock calculations, 12–13
Helmholtz free energy:
    mesoscopic simulation, inhomogeneous polymer mixtures:
        continuous field model, 201–204
        density functional theory, 179–180
    semicrystalline polymers, fluctuation technique, mechanical properties simulation, 230–231
Hermite integrals, polymer band structure and wave-function, structural geometry, vibrational and mechanical properties, 33–36
Hessian matrices, polymer band structure and wave-function, structural geometry, vibrational and mechanical properties, 34–36
Highest occupied crystalline orbital (HOCO), polymer band structure and wave-function, 17–20

High-impact polystyrene (HIPS), mesoscopic simulation, 154–155
Histogram-reweighting techniques, binary polymer liquids, Monte Carlo simulations, 110–111
Homogeneous polymer mixtures, mesoscopic simulation, 156–171
  chemical potential, 160–163
  interaction energy density parameter, 159–160
  multi-component phenomenological approach, 163–171
    composition-dependent interaction parameter, 163–165
    parametrization, 166–171
  notation, 156
Hooke's law, semicrystalline polymers, elastic constants, 226–228
Hückel approach:
  ab initio polymer quantum theory, 3
  polymer properties, electron polarizabilities/hyperpolarizabilities, 28–32
Hybrid exchange-correlation functions, polymer band structure and wave-function, structural geometry, vibrational and mechanical properties, 34–36
Hyperpolarizabilities, polymer properties, 28–32

Ideal systems, inhomogeneous polymer systems, mesoscopic simulation, Kohn-Sham-like self-consistent field formalism, 180–182
Improper torsions, polymer force fields, 68
Incremental chemical potentials, binary polymer liquids, Monte Carlo simulations, 108–111
  Flory-Huggins parameter χ, stiffness asymmetry, 130–131
Information entropy functional, inhomogeneous polymers, mesoscopic simulation, 182–183
Infrared (IR) spectra, polymer band structure and wave-function, structural geometry, vibrational and mechanical properties, 35–36
Inhomogeneous polymer mixtures, mesoscopic simulation, 172–204
  continuous field model, 199–204
  density functional theory, 178–180
  extended volume effectss, 184–187
  functional mapping, 173–175
  grand canonical ensemble, 176–178
  Kohn-Sham-like self-consistent field formalism, 180–184
    Gaussian chain model, 188–191
    ideal system, 180–182
    parametrization, 191–194
    quasi-ideal ensemble, 187–194
      definition, 187–188
    real system, 182–183
    schematics, 183–184
  Lagrange multiplier fields, 196
  mean-field approximation, 194–195
  notation, 172–173
  Scheutjens-Fleer model, 196–199
Integration procedures, polymer band structure and wave-function, Hartree-Fock calculations, 9–10
Interaction energy density parameter:
  homogeneous polymers, mesoscopic simulation, 159–160
    composition-dependent parameters, 163–165
  inhomogeneous polymers, mesoscopic simulation, 182–183
Interfaces, binary polymer liquids, Monte Carlo simulations, 136–142
Intermediate scattering function (ISF), PEO structural and dynamic properties, 85–86
Intermolecular forces:
  binary polymer liquids, Monte Carlo simulations:
    Flory-Huggins parameter c, 116–120
    mean-field theory, 103–106
  Scheutjens-Fleer model, inhomogeneous polymers, 197–199
Intramolecular chain distribution function, mesoscopic simulation, inhomogeneous polymer mixtures, continuous field model, 203–204
Ion mean-square displacements (MSD), PEO-LiBF$_4$, 89–90
Ion transport, PEO-LiBF4, 89–90
Isotactic polypropylene, mechanical properties, 231–240
  amorphous phase, 233–235
  crystalline phase, 231–233
  molecular simulation and micromechanical models, 235–240
Isotropic polariability model model, polymer potential function, partial atomic charge and dipole polarizability, 56–58
Isotropic semicrystalline polymers:
  static deformation, 228–230

Isotropic semicrystalline polymers (*Continued*)
  stress-strain behavior, elastic constants, 228

Jacquemin technique, polymer band structure and wave-functions, LCAO Hartree-Fock calculations, 5–7

Kinetic energy functional, inhomogeneous polymers, mesoscopic simulation, Kohn-Sham-like self-consistent field formalism, 182–183
Kohn-Sham equations, polymer excitation energies, 27–28
Kohn-Sham-like self-consistent field formalism, mesoscopic simulation, inhomogeneous polymer mixtures, 180–184
  Gaussian chain model, 188–191
  ideal system, 180–182
  Lagrange multiplier fields, 196
  mean-field approximation, 194–195
  parametrization, 191–194
  quasi-ideal ensemble, 187–194
  real system, 182–183
  schematics, 183–184
  Scheutjens-Fleer model, 196–199
Koopmans approximation, polymer band structure, density of states and photoelectron spectroscopy, 21–23
Kuhnian segments:
  binary polymer liquids, Monte Carlo simulations, Flory-Huggins parameter χ, 125–127
  homogeneous polymers, mesoscopic simulation, multicomponent phenomenological approach, 170–171

Lagrange multiplier fields, mesoscopic simulation, inhomogeneous polymers, 196
Lame constants, semicrystalline polymers, 228
Lamellar structures, isotactic polypropylene, molecular simulation-micromechanical models, 235–240
Langevin equation:
  mesoscopic polymer simulation, dynamic density functional theory, 206–208
  mesoscopic simulation, dynamic density functional theory, external potential dynamics, 212–215
Laplace transform, polymer band structure and wave-function, Møller-Plesset second-order (MP2) perturbation theory (MPPT), 16–20
Lattice spacing:
  binary polymer liquids, Monte Carlo simulations, 107–111
  Hartree-Fock structure calculations, polymer band structure and wave-functions, 8–9
Legendre polynomial:
  polymer band structure and wave-function, Hartree-Fock calculations, integration procedures, 9–10
  transform of Helmholtz free energy, inhomogeneous polymers, mesoscopic simulation, grand canonical ensemble, 177–178
Lennard-Jones potential:
  crosslinking polymers, spatial effects and molecular motion, 265–267
  mesoscopic simulation, inhomogeneous polymers, 186–187
  1,4-polybutadiene force fields, 71
  polymer potential function, 49–51
Lifshitz formula, binary polymer liquids, Monte Carlo simulations, mean-field theory, 105–106
Linear combination of atomic orbitals (LCAO):
  ab initio polymer quantum theory, 2
  Hartree-Fock calculations, 5–7
  polymer band structure and wave-functions, helical symmetry, 12–13
Linearized coupled cluster doubles (LCCD), polymer band structure and wave-function, Møller-Plesset second-order (MP2) perturbation theory (MPPT), 16–20
Linear-muffin-tin-orbital (lMTO), polymer band structure and wave-function, 20
Linear polymers, crosslinking polymer simulation, scission of, 259–260
Liquid-vapor phaes separation, binary polymer liquids, Monte Carlo simulations, 108–111
  Flory-Huggins parameter χ, 131–134
Local space approximation (LSA), polymer band structure and wave-function, 37–38

INDEX 287

Log-normal distributions, crosslinking polymer simulation, 250–253
London formula, dispersion-repulsion parameters, polymer force fields, 59–63
Loop, crosslinking polymer simulation, 250–253
Low density polyethylene (LDPE), crosslinking polymer simulation, 255–256
Lower critical solution temperature (LCST):
  binary polymer liquids, Monte Carlo simulations, Flory-Huggins parameter χ, 126–127
  homogeneous polymers, mesoscopic simulation:
    multicomponent phenomenological approach, 167–171
    multi-component phenomenological approach, composition-dependent interaction parameter, 163–165
Lowest unoccupied crystalline orbital (LUCO), polymer band structure and wave-function, 17–20

Macromolecular hypothesis, crosslinking polymer simulation, research background, 244–247
Many-body polarization effects:
  PEO force fields, 83–84
  PEO-LiBF$_4$, 87
  polymer potential function, 49–51
    partial atomic charge and dipole polarizability, 56–58
Mathematical theoretical models, crosslinking polymer simulation, 259
Maxwell-Boltzmann equation, semicrystalline polymers, force field simulations, 223–224
McMurchie-Davidson technique, polymer band structure and wave-functions, LCAO Hartree-Fock calculations, 6–7
Mean field approximation:
  binary polymer liquids, Monte Carlo simulations:
    coarse-grained models, 101
    predictions, 101–106
    strong segregation limit, 105–106
    weak segregation limit, 104–105
  mesoscopic simulation, inhomogeneous polymers, 194–195
Mechanical properties, polymer band structure and wave-functions, 32–36
Mesoscopic polymer simulation:

dynamic density functional theory, 205–215
  copolymer-solvent mixtures, 209–210
  excluded volume effects, 208
  external potential dynamics, 210–215
  principles, 205–208
homogeneous polymer mixtures, 156–171
  chemical potential, 160–163
  interaction energy density parameter, 159–160
  multi-component phenomenological approach, 163–171
    composition-dependent interaction parameter, 163–165
    parametrization, 166–171
  notation, 156
inhomogeneous polymer mixtures, 172–204
  continuous field model, 199–204
  density functional theory, 178–180
  extended volume effectss, 184–187
  functional mapping, 173–175
  grand canonical ensemble, 176–178
  Kohn-Sham-like self-consistent field formalism, 180–184
    Gaussian chain model, 188–191
    ideal system, 180–182
    parametrization, 191–194
    quasi-ideal ensemble, 187–194
      definition, 187–188
      real system, 182–183
      schematics, 183–184
  Lagrange multiplier fields, 196
  mean-field approximation, 194–195
  notation, 172–173
  Scheutjens-Fleer model, 196–199
research background, 154–156
Metallocene catalysts, crosslinking polymer simulation, research background, 244–247
Micromechanical models, isotactic polypropylene, 235–240
Miller-Macosko model, crosslinking polymer simulation, 259
Miscibility behaviors. *See also* Critical micelle concentration (CMC)
  binary polymer liquids, Monte Carlo simulations:
    films and two-dimensional systems, 135–136
    Flory-Huggins parameter χ, stiffness asymmetry, 130–131
    future research, 143–145

Molecular dynamics (MD) simulations:
 crosslinking polymers, spatial effects and molecular motion, 266–267
 PEO-LiBF$_4$, 87–91
  polarization, 90–91
  structural properties, 88–89
  transport properties, 89–90
 PEO structural and dynamic properties, 84–85
 1,4-polybutadiene force fields, 71–73
  potential function validation, 77–78
 semicrystalline polymers, force field simulations, 222–224
Molecular orbitals (MO), *ab initio* polymer quantum theory, 3–4
Molecular simulation-micromechanical models, isotactic polypropylene, 235–240
Molecular weight distribution (MWD):
 crosslinking polymers, without spatial effects/molecular motion, 260–261
 crosslinking polymer simulation:
  linear-branched polymer scission, 259–260
  rational design, 269–270
  research background, 244–247
Møller-Plesset second-order (MP2) perturbation theory (MPPT):
 polymer band structure and wave-function:
  electron correlation, 14–20
  Hartree-Fock calculations, lattice summations
  structural geometry, vibrational and mechanical properties, 33–36
 polymer force field parametrization, 53–54
  dispersion parameters, 62–63
  partial charge determination, 55–59
Monomers:
 binary polymer liquids, Monte Carlo simulations:
  coarse-grained models, 100–101
  Flory-Huggins parameter χ, 124–127
 mesoscopic simulation, Kohn-Sham-like formalism, quasi-ideal ensembles, 191–194
Monte Carlo simulations:
 binary polymer liquids, 96–98
  coarse-grained models, 99–101
  entropic contributions, χ parameter, 124–134
   monomer shape, non-additive packing, 124–127

  pressure and solvent density effects, 131–134
  stiffness asymmetry, 127–131
 films and two-dimensional systems, 134–136
 future research issues, 143–145
 interfaces, 136–142
 mean-field theory predictions, 101–106
 models and techniques, 106–111
 phase behavior, fluid structure and composition, 111–124
  Flory-Huggins parameter χ, symmetric blends, 111–120
  single chain conformations, 120–124
 strong segregation limit, 105–106
 weak segregation limit, 104–105
 crosslinking polymers, 256–257
  linear-branched polymer scission, 260
  spatial effects and molecular motion, 264–267
  spatial effects without molecular motion, 263–264
  without spatial effects/molecular motion, 260–261
 homogeneous polymers, multicomponent phenomenological approach, 171
 inhomogeneous polymers, mesoscopic simulation *vs.*, 186–187
Mooney-Rivlin model, crosslinking polymer simulation:
 empirical techniques, 257–259
 rubber elasticity, 253
Multicanonical recursion, binary polymer liquids, Monte Carlo simulations, 111
Multi-component phenomenological approach, homogeneous polymers, mesoscopic simulation, 163–171
 composition-dependent interaction parameter, 163–165
 parametrization, 166–171

Namur threshold scheme:
 polymer band structure and wave-functions:
  Hartree-Fock calculations, lattice summations, 8–9
  LCAO Hartree-Fock calculations, 6–7
  polymer excitation energies, 26–28

Network formation, crosslinking polymer simulation:
  deformation and (rubber elasticity), 267–268
  spatial effects and molecular motion, 264–267
  theoretical background, 247–250
Network structure, crosslinking polymer simulation, definitions, 250–253
Neutron spin echo (NSE), 1,4-polybutadiene force fields, 80–82
Non-additive packing, binary polymer liquids, Monte Carlo simulations, Flory-Huggins parameter χ, 124–127
Nonbonded energy, polymer potential function, 49–51
Nonlinear optics (NLO), *ab initio* polymer quantum theory, 2
Non-local functionalss, inhomogeneous polymers, mesoscopic simulation, 175
Nonoscillatory techniques, polymer band structure and wave-function, Hartree-Fock calculations, integration procedures, 9–10
Nonspatial models, crosslinking polymer simulation, 260–262
Nuclear magnetic resonance (NMR)-spin lattice relaxation times, 1,4-polybutadiene force fields, potential function validation, 79–80

Oligomer approach:
  ab initio polymer quantum theory, 3
  polymer force fields, parametrization algorithm, 69–70
One-particle theory, polymer band structure and wave-functions, Bloch's theorem, 4–5
Onsager coefficient:
  binary polymer liquids, Monte Carlo simulations, 144–145
  mesoscopic polymer simulation, dynamic density functional theory, 205–208

Parametrization data:
  homogeneous polymers, mesoscopic simulation, multicomponent phenomenological approach, 166–171
  inhomogeneous polymers, mesoscopic simulation, Kohn-Sham-like formalism, quasi-ideal ensembles, 191–194

1,4-polybutadiene force fields, quantum chemistry data set, 71–73
polymer force fields, 51–54
  algorithm, 68–70
  covalent bonds, 63–64
  dihedral potentials, 66–68
  dispersion-repulsion parameters, 59–63
  improper torsion, 68
  partial atomic charge and dipole polarizability, 54–59
  valence bonds, 64–66
Pariser-Parr-Pople (PPP) scheme, *ab initio* polymer quantum theory, 3
Partial atomic charge, polymer force fields, classical potential function, 54–59
  many-body polarizable model, 56–58
  transferability, 58–59
Pearson-Graessley relation, crosslinking polymer simulation, rational design, 268–269
PEO and PEI-LiBF$_4$ polarizability:
  homogeneous polymers, mesoscopic simulation, multicomponent phenomenological approach, 168–171
  quantum-chemistry-based force fields, 83–91
    condensed-phase conformations, 85
    dynamics, 85–86
    force field development field, 87–88
      many-body polarizable potential, 87
      two-body polarizable potential, 87–88
    many-body force field development, 83–84
    molecular dynamics simulations, 88–91
      structural properties, 88–89
      transport properties, 89–90
    structural and dynamic properties, 84–86
    thermodynamics, 85
Perfect network, crosslinking polymer simulation, 251–253
Phase behavior, binary polymer liquids, Monte Carlo simulations, 111–124
  Flory-Huggins parameter χ, symmetric blends, 111–120
  single chain conformations, 120–124
Photoconduction, polymer excitation energies, band gaps and cxcitons, 23–28
Photoelectron spectroscopy, polymer band structure, 20–23

Pluronic systems, mesoscopic simulation, dynamic density functional theory, 209–210
Poisson ratio, isotactic polypropylene, molecular simulation-micromechanical models, 238–240
Polarizability:
  PEO force fields, 84–85
  PEO-LiBF$_4$, 90–91
  1,4-polybutadiene force fields, PEO structural and dynamic properties, 85–86
Polybutadiene (PB), homogeneous polymers, mesoscopic simulation, multicomponent phenomenological approach, 170–171
1,4-Polybutadiene (PBD), quantum-chemistry-based polymer force fields, 70–82
  dihedral potential transferability, 76–77
  model molecules, 71–73
  needs assessment, 70–71
  parametrization, 73–76
    CH–CH and CH$_2$–CH nonbonded parameters, 75–76
    dihedral parameters, 75
    dispersion-repulsion, 74
    equilibrium bond lengths and valence bond angles, 74
  potential function validation, 77–82
    molecular dynamics simulations, 77–78
    nuclear magnetic resonance spin-lattice relaxation times, 78–79
    potentials accuracy, 82
    single chain dynamic structure factor, 80–82
    static structure factor and gyration radius, 78
Polymer properties and applications, *ab initio* polymer quantum theory, 20–36
  band structure, density of states, and photoelectron spectroscopy, 20–23
  electronic polarizabilities and hyperpolarization, 28–32
  excitation energies: band gaps and excitons, 23–28
  structural geometries, vibrational and mechanical properties, 32–36
Potential function:
  binary polymer liquids, Monte Carlo simulations, 107–111

1,4-polybutadiene force fields:
  accuracy, 82
  molecular dynamics simulations, 77–78
  nuclear magnetic resonance spin-lattice relaxation times, 79–80
  single chain dynamic structure factor, 80–82
  static structure factor and gyration radius, 78
  validation, 77–82
quantum-chemistry-based polymer force fields, 48–70
  classical potential form, 49–51
  covalent bond parameters, 63–64
  dihedral potential, 66–68
  dispersion and repulsion parameter determination, 59–63
  existing potentials, 51
  improper torsion, 68
  parametrization algorithm, 68–70
    data set establishment, 69
    force field determination, 68–69
    validation, 69–70
  parametrization data, 51–54
  partial atomic charge and dipole polarizability determination, 54–59
  many-body polarizable model, 56–58
  transferability, 58–59
  valence bonds, 64–66
Pressure effects, binary polymer liquids, Monte Carlo simulations, Flory-Huggins parameter χ, 131–134
P-RISM theory, binary polymer liquids, Monte Carlo simulations, 118–120
  future research, 143–145
  single chain dynamic structure factor, 121–124
Pseudokinetics, crosslinking polymers, spatial effects and molecular motion, 265–267
Pseudolinear dependencies. *See* Quasi-linear dependencies

Quadratic configuration-single and double substitution (QCSID) operators, polymer band structure and wave-function, Møller-Plesset second-order (MP2) perturbation theory (MPPT), 16–20

INDEX

Quantum-chemistry-based polymer force
  fields:
  classical potential function, 48–70
    covalent bond parameters, 63–64
    dihedral potential, 66–68
    dispersion and repulsion parameter
      determination, 59–63
    existing potentials, 51
    improper torsion, 68
    parametrization algorithm, 68–70
      data set establishment, 69
      force field determination, 68–69
      validation, 69–70
    parametrization data, 51–54
    partial atomic charge and dipole
      polarizability determination,
      54–59
      many-body polarizable model,
        56–58
      transferability, 58–59
    potential form, 49–51
    valence bonds, 64–66
  dispersion-repulsion parameters, 59–63
  parametrization data, 53–54
  PEO and PEI-LiBF$_4$ polarizability, 83–91
    condensed-phase conformations, 85
    dynamics, 85–86
    force field development field, 87–88
      many-body polarizable potential, 87
      two-body polarizable potential,
        87–88
    many-body force field development,
      83–84
    molecular dynamics simulations,
      88–91
      structural properties, 88–89
      transport properties, 89–90
    structural and dynamic properties,
      84–86
    thermodynamics, 85
  1,4-polybutadiene, 70–82
    dihedral potential transferability, 76–77
    model molecules, 71–73
    needs assessment, 70–71
    parametrization, 73–76
      CH–CH and CH2–CH nonbonded
        parameters, 75–76
      dihedral parameters, 75
      dispersion-repulsion, 74
      equilibrium bond lengths and
        valence bond angles, 74
    potential function validation, 77–82
      molecular dynamics simulations,
        77–78

    nuclear magnetic resonance spin-
      lattice relaxation times, 78–79
    potentials accuracy, 82
    single chain dynamic structure
      factor, 80–82
    static structure factor and gyration
      radius, 78
Quasi-ideal ensembles:
  dynamic density functional theory,
    207–208
  inhomogeneous polymers, mesoscopic
    simulation:
    Kohn-Sham-like formalism, 187–194
    Scheutjens-Fleer model, 197–199
Quasi-linear dependencies:
  polymer band structure and wave-
    function, Hartree-Fock calculations,
    10–12
  polymer band structure and wave-
    functions, electron polarizabilities/
    hyperpolarizabilities, 31–32
Quasi-particle levels, polymer excitation
  energies, 23–28

Radius of gyration, 1,4-polybutadiene force
  fields, potential function validation, 78
Raman intensities, polymer band structure
  and wave-function, structural geometry,
  vibrational and mechanical properties,
  35–36
Random phase approximation (RPA):
  binary polymer liquids, Monte Carlo
    simulations, mean-field theory,
    104–105
  mesoscopic simulation, inhomogeneous
    polymers, Kohn-Sham-like
    formalism, quasi-ideal ensembles,
    191–194
  polymer excitation energies, 25–28
  polymer properties, electron
    polarizabilities/hyperpolarizabilities,
    29–32
Rational design, crosslinking polymer
  simulation:
  analytical refinement, 269
  research background, 244–247
  structure and chemistry effects, 269–270
  theoretical refinement, 268–269
  validated simulation standards, 270–271
Reactive molecular dynamics simulations,
  crosslinking polymers, spatial effects
  and molecular motion, 266–267

Reduced weight average molecular weight, crosslinking polymer simulation, gelation, 248–250
Relaxation phenomena, 1,4-polybutadiene force fields, potential function validation, 77–78
Repulsion parameters:
1,4-polybutadiene force fields, 70–71
parametrization, 74
polymer force fields, 59–63
Ritz variational principle, polymer band structure and wave-functions, LCAO Hartree-Fock calculations, 5–7
Rotational isomeric state (RIS):
Metropolis Monte Carlo (RMMC) simulation, mesoscopic simulation, inhomogeneous polymers, Kohn-Sham-like formalism, quasi-ideal ensembles, 191–194
semicrystalline polymers, amorphous cell modeling, 225–226
Rouse dynamics equation, mesoscopic simulation, dynamic density functional theory, external potential dynamics, 210–215
Rubber elasticity, crosslinking polymer simulation, 253
network formation and deformation, 267–268

Saito-Inokuti model, crosslinking polymer simulation:
applications, 259
gelation, 249–250
log-normal distributions, 251–253
Scalar potential, polymer properties, electron polarizabilities/hyperpolarizabilities, 30–32
Scaling estimates, binary polymer liquids, Monte Carlo simulations:
films and two-dimensional systems, 134–136
single chain dynamic structure factor, 121–124
Scheutjens-Fleer model, mesoscopic simulation, inhomogeneous polymer mixtures, 196–199
continuous field model, 201–204
Schwartz functional approach, inhomogeneous polymers, mesoscopic simulation, 173–175
Schwartz inequality expression, polymer band structure and wave-functions, LCAO Hartree-Fock calculations, 7

Scission functions, crosslinking polymer simulation:
gelation, 249–250
linear and branched polymers, 259–260
without spatial effects or molecular motion, 261–262
Self-consistent field (SCF) techniques:
binary polymer liquids, Monte Carlo simulations:
future research, 143–145
interfaces, 136–142
mean-field theory, 102–106
single chain dynamic structure factor, 121–124
mesoscopic polymer simulation:
continuous field model, 200–204
dynamic density functional theory, external potential dynamics, 213–215
Kohn-Sham-like self-consistent field formalism, inhomogeneous polymer mixtures, 180–184
Gaussian chain model, 188–191
ideal system, 180–182
parametrization, 191–194
quasi-ideal ensemble, 187–188, 187–194
real system, 182–183
schematics, 183–184
research background, 155–156
Scheutjens-Fleer model, 196–199
polymer band structure and wave-functions:
electron polarizabilities/hyperpolarizabilities, 29–32
Hartree-Fock calculations, quasi-linear dependencies, 10–12
LCAO Hartree-Fock calculations, 7
Self-interaction energy (SIE), polymer excitation energies, 27–28
Self-polarization of molecules, PEO force fields, many-body polarization effects, 83–84
Semicrystalline polymers, mechanical properties:
amorphous cell preparation, 225–226
crystal cell preparation, 224–225
elastic constants, 226–228
force field simulations, 221–224
isotactic polypropylene, 231–240
amorphous phase, 233–235
crystalline phase, 231–233

molecular simulation and micromechanical models, 235–240
  research background, 220–221
  simulation approaches, 228–231
  fluctuation, 230–231
  static deformation, 228–230
Semigrandcanonical ensemble, binary polymer liquids:
  Flory-Huggins parameter χ, stiffness asymmetry, 129–131
  Monte Carlo simulations, 108–111
SimPolyMod simulation models, crosslinking polymers, 270–271
Single chain dynamic structure factor:
  binary polymer liquids, Monte Carlo simulations:
    Flory-Huggins parameter χ, 120–124
    mean-field theory, 104–106
  1,4-polybutadiene force fields, 80–82
Slater determinant, polymer band structure and wave-functions, LCAO Hartree-Fock calculations, 5–7
Sol-gel partition theories, crosslinking polymer simulation, 248–250
  applications, 255–256
  basic definitions, 250–253
  rational design, 270–271
  spatial effects without molecular motion, 262–264
Solubility parameters, crosslinking polymer simulation, 248–250
Solvent chemical potential, homogeneous polymers, mesoscopic simulation, multicomponent phenomenological approach, 166–171
Solvent density, binary polymer liquids, Monte Carlo simulations, Flory-Huggins parameter χ, 131–134
Spatial effects, crosslinking polymer simulation, 262–264
  molecular motion and, 264–267
Spherulite structures, isotactic polypropylene, molecular simulation-micromechanical models, 235–240
Spinodal decomposition, binary polymer liquids, Monte Carlo simulations, 144–145
Static deformation:
  isotactic polypropylene:
    amorphous phase, 234–235
    crystalline phase, 2322–233
  semicrystalline polymers, mechanical properties simulation, 228–230

Static structure factor, 1,4-polybutadiene force fields, potential function validation, 78
Statistical segment lengths, binary polymer liquids, Monte Carlo simulations, Flory-Huggins parameter χ, 128–131
Stern layers, inhomogeneous polymers, mesoscopic simulation, Kohn-Sham-like formalism, 184–187
Stiffness matrix:
  binary polymer liquid asymmetry, Monte Carlo simulations, Flory-Huggins parameter χ, 127–131
  isotactic polypropylene:
    amorphous phase, 234–235
    crystalline phase, 232–233
  semicrystalline polymers:
    elastic constants, 227–228
    static deformation, 228–230
Stress-strain behavior:
  crosslinking polymer simulation, 254–256
  network formation-deformation, 267–268
  semicrystalline polymers:
    elastic constants, 226–228
    static deformation, 228–230
Strong segregation limit, binary polymer liquids, Monte Carlo simulations, mean-field theory, 105–106
Structural geometry, polymer band structure and wave-functions, 32–36
Structure-property relationships:
  crosslinking polymer simulation:
    rational design, 269–270
    research background, 245–247
  semicrystalline polymers, research background, 220–221
Styrene-co-butadiene rubber (SBR), homogeneous polymers, mesoscopic simulation, multicomponent phenomenological approach, 170–171
Sum-over-states (SOS) expression, polymer properties, electron polarizabilities/hyperpolarizabilities, 29–32
Swelling properties, crosslinking polymer simulation, 254–256
Symmetric blends, binary polymer liquids, Monte Carlo simulations, Flory-Huggins parameter χ, 111–120

Tacticity, crosslinking polymer simulation, 245

Tamm-Dancoff approximation (TdA), polymer excitation energies, 25–28
Thermodynamic properties:
 binary polymer liquids, Monte Carlo simulations, Flory-Huggins parameter χ, 112–120
 inhomogeneous polymers, mesoscopic simulation, grand canonical ensemble, 176
 PEO structural and dynamic properties, 85
 polymer force fields, parametrization data, 53–54
Three-dimensional Ising universality class, binary polymer liquids, Monte Carlo simulations, 109–111
 Flory-Huggins parameter χ, 119–120
Time-dependent density functional theory (TDDFT), polymer excitation energies, 27–28
Time-dependent Hartree-Fock (TDHF), polymer band structure and wave-functions:
 electron polarizabilities/hyperpolarizabilities, 29–32
 quasi-linear dependencies, 12
Torsional potentials, 1,4-polybutadiene force fields, 75
Toyozawa's electronic polaron formalism:
 polymer band structure and wave-function, 17–20
 polymer excitation energies, 26–28
Transferability of potentials:
 1,4-polybutadiene force fields, dihedral potentials, 76–77
 polymer force fields:
  covalent bonds, 64
  dihedral potentials, 67–68
  dispersion-repulsion parameters, 63
  parametrization data, 51–54
  valence bonds, 65–66
 polymer potential function, partial atomic charges and dipole polarizabilities, 58–59
Transformation matrix, polymer band structure and wave-functions, Hartree-Fock calculations, quasi-linear dependencies, 10–12
Translational entropy, mesoscopic simulation, inhomogeneous polymer mixtures, continuous field model, 201–204
Trans molecular models, 1,4-polybutadiene force fields, 72–73

Triblock copolymers, mesoscopic simulation, Kohn-Sham-like formalism, quasi-ideal ensembles, 192–194
Two-body polarizable potential, PEO-LiBF$_4$, 87–88
Two-dimensional systems, binary polymer liquids, Monte Carlo simulations, 134–136

Uncoupled Hartree-Fock (UCHF) approach, polymer properties, electron polarizabilities/hyperpolarizabilities, 28–32
United-atom model, binary polymer liquids, Monte Carlo simulations, 98
Unrestricted Hartree-Fock (UHF) approach, polymer band structure and wave-functions, 13
Upper critical solution temperature (UCST):
 binary polymer liquids, Monte Carlo simulations, Flory-Huggins parameter χ, 126–127, 132–134
 homogeneous polymers, mesoscopic simulation, 163–165
 multicomponent phenomenological approach, 167–171

Valence bonds:
 1,4-polybutadiene force fields, 70–71
 equilbrium bond lengths, 74
 polymer force fields, 64–66
Valence effective Hamilton (VEH), *ab initio* polymer quantum theory, 3
Validated simulations, crosslinking polymers, 270–271
Van der Waals approximation, inhomogeneous polymers, mesoscopic simulation, Kohn-Sham-like formalism, 184–187
Vapor pressure measurements, homogeneous polymers, mesoscopic simulation, multicomponent phenomenological approach, 166–171
Vector notation, semicrystalline polymers, elastic constants, 226–228
Vibrational properties, polymer band structure and wave-functions, 32–36
Virial theorem, semicrystalline polymers, force field simulations, 223–224
Voigt notation, semicrystalline polymers, elastic constants, 226–228

Volume fraction differentials, homogeneous polymers, mesoscopic simulation, chemical potential, 161–163

Wannier functions (WF):
  polymer band structure and wave-function, Møller-Plesset second-order (MP2) perturbation theory (MPPT), 15–20
  polymer excitation energies, 26–28
Wave function, *ab initio* polymer quantum theory, 4–20
  Bloch's theorem, 4–5
  Hartree-Fock structure calculations, 8–12
    integration procedures, 9–10
    lattice summations, 8–9
    quasi-linear dependencies, 10–12
  helical symmetry, 12–13
  LCAO Hatree-Fock methodology, 5–7
  Møller-Plesset and wave-function-based electron correlation, 14–20
  unrestricted Hartree-Fock approach, 13
Weak segregation limit, binary polymer liquids, Monte Carlo simulations, mean-field theory, 104–106

Weight factors, binary polymer liquids, Monte Carlo simulations, 111
Wide-angle X-ray scattering (WAXS), isotactic polypropylene, molecular simulation-micromechanical models, 239–240
Wigner-Seitz unit cell, polymer band structure and wave-functions, Bloch's theorem, 4–5
Wolf's model, homogeneous polymers, mesoscopic simulation, multicomponent phenomenological approach, 166–171

Young's modulus. *See also* Elastic constants
  isotactic polypropylene:
    crystalline phase, 232–233
    molecular simulation-micromechanical models, 237–240
  polymer band structure and wave-function, structural geometry, vibrational and mechanical properties, 36